A CIVIL WAR SOLDIER
OF CHRIST AND COUNTRY

A Civil War Soldier of Christ and Country

The Selected Correspondence of John Rodgers Meigs, 1859–64

Edited by

MARY A. GIUNTA

Foreword by

CAROL BLESER

UNIVERSITY OF ILLINOIS PRESS
URBANA AND CHICAGO

Manufactured in the United States of America
C 5 4 3 2 1
∞ This book is printed on acid-free paper.

Library of Congress Cataloging-in-Publication Data
A Civil War soldier of Christ and country : the selected
correspondence of John Rodgers Meigs, 1859–1864 /
edited by Mary A. Giunta ; foreword by Carol Bleser.
p. cm.
Includes bibliographical references and index.
ISBN-13: 978-0-252-03076-5 (cloth : alk. paper)
ISBN-10: 0-252-03076-1 (cloth : alk. paper)
1. Meigs, John R.—Correspondence. 2. United States—
History—Civil War, 1861–1865—Personal narratives.
3. United States—History—Civil War, 1861–1865—
Campaigns. 4. Shenandoah Valley Campaign, 1864
(August–November) 5. Military engineers—
United States—Correspondence. 6. United States Military
Academy—Biography. 7. Meigs, Montgomery C.
(Montgomery Cunningham), 1816–1892—Correspondence.
8. Meigs family—Correspondence.
I. Giunta, Mary A. II. Title.
E467.1.M43A4 2006
973.7'81—dc22 2005031544

This volume is dedicated to my parents,
Anthony A. and Catherine A. Giunta,
who kindled my interest in history.

Be able to show yourself
the soldier of Christ & of your country.

—Montgomery C. Meigs to John Rodgers Meigs,
November 3, 1861

CONTENTS

FOREWORD

Carol Bleser

In early September 1859 Captain Montgomery C. Meigs, a military engineer, took his seventeen-year-old son, John Rodgers Meigs, to West Point to begin his military training. John's appointment had been long in coming due to a disagreement between Montgomery and Secretary of War John B. Floyd over the letting of government contracts. After a brief visit, Montgomery returned to Washington, D.C.. to continue his work as superintendent on the construction of the U.S. Capitol and of the Washington Aqueduct. Not long after, Floyd "exiled" him to the Dry Tortugas.

This volume, edited by Mary A. Giunta, contains 238 letters and documents, most of which were written by or to John Rodgers Meigs while he was at West Point or serving in the Civil War. They provide information about the lives of the family of his father and to a lesser extent the family of his mother. The letters disclose the interpersonal relationships among John and his immediate family—his father, Montgomery; his mother, Louisa Rodgers Meigs; his sisters, Mary and Louisa; and his brother, Monty—as well as his extended family. They give insight and provide a vivid picture of the domestic and military concerns of a prominent military family during the Civil War.

Montgomery Meigs has been much written about as a military man. The family correspondence reveals his personal side. He was a doting father when it came to his children, especially John, the first-born. Even following his appointment as quartermaster general of the Union Army, when his time was especially taken up with his military responsibilities, he was a stern but patient and consoling parent. The picture that emerges of John's mother, the daughter of a naval commodore, is that of a woman determined to see that her son did not disgrace the family, especially his father. Her letters are filled with admonitions and reminders for John to be responsible and reliable at West Point and in society.

Following John's graduation in June 1863, he served in Maryland, in West Virginia, and in the Shenandoah Valley campaign of 1864. In addition to the personal correspondence, previously unpublished, the volume contains examples of his official correspondence depicting his various responsibilities as a military engineer and aide-de-camp.

When John Rodgers Meigs entered West Point he had many of his father's characteristics. He was intelligent, self-reliant, independent, strong willed, headstrong, and high tempered. He did not hesitate to stand up to authority when he believed that he was in the right. He was constantly in difficulty for his failure to obey the rules, thereby receiving more and more demerits. He came close to dismissal from the academy for fighting. It was only through the intervention of Edwin M. Stanton, Abraham Lincoln's secretary of war, that his court-martial sentence was reduced to confinement to camp for the whole summer. Not daunted but chafing at his detention, John took advantage of a presidential visit to West Point to advise Abraham Lincoln that "you can do me . . . a great favor . . . if you will only sign your name on this card," which called for the sentence to be remitted. His natural brashness was perhaps enhanced by the important position his father held.

With perseverance and continued pressure and support from his parents, Meigs graduated first in his class in June 1863. He now had the training that his parents wanted for him, and he was prepared to fulfill the admonition of his father to "be able to show yourself the soldier of Christ & of your country." Following graduation, he entered active military service. He saw much action and was promoted rapidly. The letters bring out the ferociousness of the war and the frustrations of army life. In one of his last letters home, John wrote to his mother: "At the beginning of the war when such things were new soldiers wrote letters to their friends and made little arrangements about their watches etc before going into action. Now they march upon an expected battlefield with little more emotion than an ordinary parade ground." A month later, John was killed on a rainy night early in October 1864. In letters written after his death, his mother described her deep grief. For Montgomery there was no solace. His was a silent grieving. When Montgomery moved John's coffin to Arlington Cemetery in 1880, he placed atop a sarcophagus a bronze image of his son's dead body in the position in which it was found, the lasting tribute of a father's pride.

This is a compelling story of a young man coming of age during the most devastating war in the history of the United States. These beautifully crafted letters provide a lasting memory of a vibrant, living John Rodgers Meigs. This skillfully edited and gripping volume displays the brashness and heroism of youth, bringing home to the reader the wanton destruction and horror of war.

PREFACE

More than any other national event, the United States Civil War continues to receive the attention of the general public, study groups, academic researchers, writers, and others. Biographies for most if not all of the major participants, histories of battles and military units, studies of the impact of the war, and edited document collections are available in many formats, including presentation on the Internet. Still there are more lives to write about, additional stories to uncover, and more documents to explore. Each account helps to complete the mosaic that is the war's history.

This documentary edition adds another piece to that mosaic by providing information on the military career of John Rodgers Meigs and on his experiences as a cadet at the United States Military Academy, through the publication of his selected correspondence and related documents for the period 1859 to 1864. The documents covering his West Point years, from September 1859 to June 1863, describe his personal accomplishments and distresses there, including being court-martialed. They also provide an intimate picture of his family's full involvement in his education and life. Manuscripts pertaining to Meigs's wartime service as commander of the armored railroad cars used to keep open communications and protect the Baltimore and Ohio Railroad during July 1863, his experiences during Brigadier General William W. Averell's raids later that year, and his participation in the Shenandoah Valley campaign of 1864 flesh out the picture of his involvement in the Civil War. Many of the documents are charged with human interest and poignant emotion. They leave the reader with a sense of sadness, almost like that experienced by family members in the face of profound personal loss. When read as a whole, they provide a vivid account of the life and times of a brave young man, his family, and a country at war.

ACKNOWLEDGMENTS

Without the assistance of many individuals, this volume would not have been prepared for publication. Three persons in particular deserve special words of appreciation. Fay Helsley Kidd gave unselfishly of her time in transcribing documents, inserting footnotes, retyping headnotes, and reading the materials at various stages. When not working on the actual documents, she searched for Meigs materials at the Library of Congress and the Harrisonburg Rockingham County Historical Society. William D. Mohr transcribed the shorthand documents. He worked with meticulous care and patience, ever ready to rework shorthand outlines, investigate historical sources, or build a word or phrase from what at times appeared to be random lines and curves hurriedly written across a page. He assisted others with shorthand transcriptions and worked with the U.S. Senate Historical Office in transcribing Montgomery C. Meigs's journal entries pertaining to the U.S. Capitol. Without his efforts the shorthand documents would not be available for research and writing. Samuel A. Tucker gave much retirement time to this project. He stayed the course through his search for historical materials, preparation of annotations, and trips to local libraries, providing information and reports on visits to sites associated with Meigs. He graciously shared what he learned about Meigs and was persistent in his search for facts. His ability to track elusive documents in the extensive records in the National Archives is unexcelled. His encouragement and "finish the job" approach helped me to do so.

Members of the staff of the National Archives and Records Administration helped in many ways. Michael P. Musick, Howard Wehmann, Lida Churchville, Jeffery T. Hartley, Randall Fortson, Maryellen Trautman, Carah M. Smith, and Rick Peuser provided information, found volumes and documents, and secured library loans. Mike and Howard were especially helpful in finding Meigs items, and providing citation information. Dale Floyd, of the National Archives and Records Administration, and the United States Army Corps of Engineers shared information on fortifications and pontoon bridges. Patricia Eames, formerly of the National Archives and Records Administration, assisted with early document transcriptions and Meigs's genealogy. Her encouragement in the early stages of the project is greatly appreciated.

Members of the staff of the Manuscript Division of the Library of Congress were always available. These include Fred Bauman, Ernie Emrich, Jeff Flannery,

Joseph Jackson, Patrick Kerwin, and Bruce Kirby. John Sellers and Mary Wolfskill provided encouragement and were always ready to help. Other Library of Congress employees provided information and assistance. These include Virginia Woods and Rebecca Jones, who researched difficult-to-locate publications.

Staff members of the Special Collections Division and the Archives of the U.S. Military Academy at West Point provided valuable assistance that enhanced my research. Alan C. Aimone, Suzanne M. Chrisoff, James McEnery, Dorothy Rapp, and Judith Sibley identified series and answered questions in helping to locate Meigs materials at West Point.

Mary E. Herbert of the Maryland Historical Society located and transcribed documents pertaining to the construction of ironclad railroad cars.

Jean Martin, archivist at the Shenandoah County Library in Edinburg, Virginia, assisted in locating reference sources and even helped to title this volume. She provides a "scholarly home" for many in the library's Shenandoah Room.

Claire M. Bettag of Washington, D.C., checked references and prepared the genealogical chart. Her assistance and support, especially in the latter stages of the project, were critical.

Three individuals read early versions of the manuscript. Frank L. Byrne provided valuable suggestions and further leads to relevant materials. Richard Duncan directed changes in the presentation of materials and provided editorial continuity to the placement of source notes. John McDonough edited headnotes and made suggestions that helped to balance the overall work.

My colleagues at the National Historical Publications and Records Commission were most supportive in reading headnotes, making suggestions, and providing professional encouragement.

Dane Hartgrove, Timothy Connelly, and Michael Meier provided information on Meigs items. Dane read the manuscript in its early stages and provided research assistance in the latter stages. His help was invaluable.

I thank Carol Bleser, who wrote the foreword to this volume. She is a true believer in documentary publications and has been a major supporter of various projects, especially in her role as a commission member of the National Historical Publications and Records Commission. Her own work with scholarly editions deserves special recognition.

John L. Heatwole shared generously his extensive knowledge of Shenandoah Valley history. His knowledge of the Burnt District, the area around the spot where Meigs was killed, is especially valuable. The map of the area reproduced in this volume is the first of its kind that shows the extent of the damage done in retaliation for Meigs's death. I thank Katherine Tennery Bennett for her permission to reprint the map. Betty Martin of Harrisonburg, Virginia, granted permission to use the picture of Frank Shaver, who believed himself responsible for Meigs's death. I am grateful to Ms. Martin and also to Russell Hicks of Altavista, Virginia, who provided the picture of George W. "Josh" Martin and gave permission for its use in this volume. As a result of my research, I have had the opportunity to share John Rodgers Meigs's life with members of the Alger family, descendants of the Meigs family. The Honorable John Rodgers Meigs Alger and his wife, Judy, were most generous in sharing the picture of John Rodgers Meigs that appears on the cover of this book, and in providing the copy of the last letter he wrote to his mother, dated September 18, 1864. The Algers' daughter, Louisa Alger Watrous, extended her support and encouragement in the final stages of the project. She served graciously in a liaison role between her parents and the editor. I thank her for all her efforts.

Joan Catapano and Carol Betts of the University of Illinois Press played significant roles in the publication of this volume, especially in the presentation of the documents and the editing of the manuscript. Joan's role is particularly critical to the publication of documentary editions. I thank them both.

I owe much to the above mentioned individuals. The merits of this volume are shared with all. However, I am solely responsible for the headnotes, the annotation, and the presentation of documents.

DOCUMENTARY SOURCES AND EDITORIAL METHOD

The documents presented herein, with few exceptions, are located in the Montgomery C. Meigs Papers and in the Rodgers Family Papers at the Manuscript Division of the Library of Congress; in the records of the National Archives and Records Administration; and in the Archives and Special Collections Division of the library at the United States Military Academy at West Point. The majority come from the Montgomery C. Meigs Papers and the National Archives.

Montgomery C. Meigs was a prodigious record keeper. He not only saved the "family archives"—including personal correspondence—but recorded his own activities and those of his son, John Rodgers Meigs, and other family members in daily diaries, journals, and letterpress copy books. In addition, after the Civil War, the elder Meigs retrieved copies of his son's documents from the various military offices. His efforts to collect and save the letters and documents resulted in an important source for information on John Rodgers Meigs, other members of his family, several members of the Rodgers family, his own role in the planning and construction of the U.S. Capitol, other public building projects, and his involvement in the Civil War.

As a military officer, John Rodgers Meigs reported for duty, received orders, wrote for permission to act in his military capacity, requested funds and materials to construct bridges and fortifications, and reported his actions as he carried out his official duties. As a military engineer, John Rodgers Meigs's responsibilities included drawing maps and preparing surveys of areas in which he served. Information concerning these documents fills his official correspondence and, to a limited extent, his private correspondence. Several of the maps and surveys are preserved in the records of the National Archives and Records Administration and in the collections of the Library of Congress. Even today, following his surveys and maps, it is quite an adventure to walk or ride over sometimes now forgotten roads and battlefields. In one field notebook, he recorded his surveying methods and results as he rode horseback between different points. It was part of his practice to measure the stride of the horse and then count the horse's steps from one landmark to another. A turn in the road, a freestanding building, or a town center served to mark the distances. By today's standards this method is crude and inaccurate; however, his maps and surveys are considered some of the best drawn during the Civil War. Some of his sketches and drawings are presented in the letters in this volume. They have been redrawn for clarity of presentation.

Among the items in the Montgomery C. Meigs Papers, there are many private letters mostly written to John Rodgers Meigs by his primary correspondents—his father, mother, sister, aunts, grandfather, and friends. These and his own letters cover his experiences at West Point and as a young officer in the Civil War. Some of this private correspondence—approximately forty items—was written in Pitman shorthand. These have been identified as written by John Rodgers Meigs or to him by his father. In addition to the Pitman shorthand letters, Montgomery C. Meigs kept journals and diaries, many of the entries of which are in Pitman shorthand. Some of these relate to John Rodgers Meigs. Once transcribed, the letters and entries provide an intimate picture of the young Meigs, uncover the concerns of family, and report war actions.

This volume includes what the editor believes are John Rodgers Meigs's most important private correspondence and a selected group of his official correspondences and documents. The writings of his father, Montgomery C. Meigs, and others are included to provide additional information, especially during times when the younger Meigs was at home, and at the time of his death and afterwards. These writings help to present a fuller picture of his life and the circumstances surrounding his death.

The dictum "let the documents speak for themselves" has been the guiding principle in the preparation of this volume. With very few exceptions, the letters and documents are from original sources and are presented here for the first time in print. Each edited document follows as closely as reasonably possible the original written document. Modifications to the form of the document were made to assure ease of reading. Excessive punctuation was eliminated. Periods were used to complete sentences, and paragraphs were used to segregate subjects. Misspellings are as written. Superscript letters in dates and other abbreviations were dropped to the line. Words positioned by the use of carets or written between lines were dropped to the line. In instances where a word or series of words were not legible, a conjectured word or words were placed in brackets. Some of the documents were written entirely in shorthand; others contained shorthand passages or notes. When shorthand forms could not be identified, the plain script transcription was treated as if written in English.

Headnotes are included to provide general contextual background or to provide transitional information helpful to the reading of the documents. To further assist the reader, a glossary of names, places, and phrases has been provided in the back of the book for a fuller understanding of the context of a letter or document, or to further identify individuals. The names of individuals not identified are omitted from the glossary.

ABBREVIATIONS

CDMC	Charles D. Meigs Correspondence
JRM	John Rodgers Meigs
JRMC	John Rodgers Meigs Correspondence
JRMD	John Rodgers Meigs Diary
LC	Library of Congress
LRMC	Louisa Rodgers Meigs Correspondence
MCM	Montgomery C. Meigs
MCMC	Montgomery C. Meigs Correspondence
MCMD	Montgomery C. Meigs Diary
MCMJ	Montgomery C. Meigs Journal
MCMP	Montgomery C. Meigs Papers
Ms. Div.	Manuscript Division
NA	National Archives and Records Administration
OR	*The War of the Rebellion: A Compilation of the Official Records of the Union and Confederate Armies*
RG	Record Group
USMA	United States Military Academy

JOHN RODGERS MEIGS'S RECORD OF SERVICE

Cadet at the United States Military Academy from September 7, 1859, to June 11, 1863. While on leave of absence, he served as volunteer aide-de-camp to Colonel Israel B. Richardson, commanding the Second Michigan Volunteers at the First Battle of Bull Run, July 21, 1861.

After graduation he was promoted in the United States Army to first lieutenant in the U.S. Corps of Engineers.

He served as assistant engineer responsible for the construction of the defenses of Baltimore, Maryland, and for opening communications along the Baltimore and Ohio Railroad Line to Harpers Ferry from July 5 to July 23, 1863.

He was an engineer from July 23 to November 23, 1863, on the staff of Brigadier General Benjamin Franklin Kelley, commander of the Department of West Virginia. In this position he was engaged in making surveys, reconnaissances, and works of defense.

As engineer and acting aide-de-camp to Brigadier General William W. Averell, he participated in the combat of Rocky Gap, near White Sulphur Springs, West Virginia, on August 26, 1863; in the battle of Droop Mountain, on November 6, 1863; and in pursuing Confederate forces, from November 6 to November 18, 1863.

He served as chief engineer of the Department of West Virginia, from November 3, 1863, to August 17, 1864. He was active in the Salem Raid from December 8 to December 25, 1863.

He served under command of General Franz Sigel in the raid on Staunton, Virginia, in May 1864 and participated in the battle of New Market, May 15, 1864.

Under the command of General David Hunter, he engaged in the Lynchburg raid, in June 1864; in the combat at Piedmont, June 5, 1864; and in the combat of Diamond Hill, near Lynchburg, Virginia, June 17–18, 1864.

During the remainder of June and in July and August 1864, he participated in the movement of Hunter's army up the Shenandoah Valley and down the Kanawha River.

He served as chief engineer of the Middle Military Division, and aide-de-camp to Major General Philip H. Sheridan in the Shenandoah campaign, from August 17 to October 3, 1864, the date of his death.

He was active in the battle of Opequon, Virginia, September 19, 1864, for which he was brevetted captain for gallant and meritorious service, and in the battle of Fisher's Hill, Virginia, September 22, 1864, for which he was brevetted major for gallant and meritorious service.

A CIVIL WAR SOLDIER

OF CHRIST AND COUNTRY

INTRODUCTION

John Rodgers Meigs was born February 9, 1842, in Washington, D.C., to Louisa Rodgers and Montgomery Cunningham Meigs. His direct ancestors include members of the Rodgers and Meigs families prominent in American history. His mother's father was Commodore John Rodgers, patriarch of the Rodgers naval family. His paternal grandfather was Charles D. Meigs, a prominent Philadelphia physician and son of Josiah Meigs, the brother of Return Jonathan Meigs of American Revolution fame. In addition to John, three other Meigs children—Mary Montgomery Meigs, Montgomery Meigs Jr., and Louisa Rodgers Meigs—reached adulthood. Two other children—Charles D. Meigs and Vincent T. Meigs, aged eight years and two years, respectively—died in 1853. A seventh child, an unnamed girl, was stillborn in 1856.

As members of an upper-class family with strong moral values, Montgomery and Louisa felt the weight of ancestral heritage and parental responsibility both to raise John as a Christian devoted to country and to ensure him a career that would guarantee his success in life. While John's siblings received the love and care of their parents, and while they were expected to succeed in the roles family and society had carved out for them, the Meigs family placed highest expectations on their first born.

Soon after John's birth, the family began their travels to military posts where Montgomery worked as a military engineer. Young John Rodgers Meigs spent part of his earliest years in Detroit, Michigan, where his father supervised the construction of Fort Wayne on the Detroit River. As a "prattling babe," the youngster exhibited a temper as well as a "fondness for books" and the alphabet. By his fourth Christmas, he was reading books and writing to Santa Claus. After a brief return to Washington, D.C., in 1849, where Montgomery served as assistant to the chief engineer, Joseph G. Totten, the family moved again. In October 1850, the elder Meigs was in charge of the construction of Fort Montgomery, near Rouses Point, Clinton County, New York, just south of the Canadian border. For young John, childhood play often turned to bullying his siblings and other childhood misbehavior. His mother wrote to her own mother, Minerva Denison Rodgers, that she had to use the "whalebone and closet . . . to reduce him to obedience and to punish misdameanors." Though his parents were concerned about his conduct, they were pleased that he exhibited intelligence and the ability to write at an early age.[1]

On November 2, 1852, Montgomery C. Meigs was ordered back to the capital to superintend the planning and construction of the Washington Aqueduct. The Meigs family now had a permanent place of residence and the opportunity to begin John Rodgers Meigs's formal education. When the youngster entered a school in Washington, it must have been of some relief to Louisa, who had earlier written to her younger sister, Ann Minerva "Nannie" Rodgers Macomb, that John had bad study habits. Louisa wrote, "I think he will find it very hard work to be put into the regular routine and discipline of a school. I shall embrace the first favorable opportunity to place him where he will be compelled to exert himself more than he now does."[2]

Ever aware of their parental responsibility, John's parents monitored his behavior. When words failed to stop his fighting with siblings, his father disciplined him in a very strict manner. On one occasion, Montgomery C. Meigs locked his son in a bedroom after he first bound his arms and then tied him "to the leg of the wardrobe and left him there, without dinner, all the afternoon." When he returned to the locked room and found that John had untied himself, his father whipped him with the rope.[3]

Louisa and Montgomery continued to oversee his education. On John's thirteenth birthday, February 9, 1855, his father described him as having "intelligence" and being "quick in his studies. Too apt to be running from one to another [subject], but knows more than boys of his age generally know. Has a decided taste for scientific study . . . [and] . . . has much more than average ability." With his parents' encouragement, in the fall of 1856, John entered Columbian College (later renamed George Washington University) as a freshman. However, his time there was to be relatively short. At the end of 1857, John Rodgers Meigs, supported by his mother and somewhat less so by his father, decided to seek a West Point appointment. As an army officer's son, he was to send the application through the secretary of war to the president of the United States, a factor that would prove significant.[4]

Several prominent political leaders, including Jefferson Davis, Stephen A. Douglas, and R. M. T. Hunter, found John qualified and supported his application. Davis wrote that young Meigs "possesses high mental and physical qualifications. . . . [He has] a fondness for scientific pursuits and gives promise of a career worthy of his illustrious Grand Father Commodore Rodgers." However, even with this backing, the appointment was denied. John applied again in November 1858. In his youthful hand, he called attention to his ancestral roots in the history of the country and to the many public officials who supported his application. His father went directly to President James Buchanan, who referred the matter to the War Department; once again the appointment was denied. It became clear that the appointment was not given because of the opposition of Secretary of

War John B. Floyd, with whom the elder Meigs was at odds over finances and contract-letting for the Aqueduct and for the United States Capitol.[5]

With determined effort, Montgomery pursued the appointment. Armed with the knowledge that two or three new cadets selected for the academy had failed their requirements, the elder Meigs visited President Buchanan on June 23, 1859. The president was reluctant to make the appointment over the opposition of his secretary of war but felt that the two men could work out their differences if they had the opportunity to meet. At the president's request, Montgomery met with Floyd. Though the two remained at odds, after the failures of other cadets the appointment was made in early September 1859.

This was the start of John Rodgers Meigs's military training and career. It was also the beginning of his formal education as an engineer. Had the Civil War not occurred, this part of his training alone would have prepared him for a respected profession and a comfortable private life, for West Point graduates commonly became civil engineers and thereby gained secure financial positions for themselves and their families. But the war did occur. John Rodgers Meigs joined Union ranks as a military engineer shortly after his West Point graduation in June 1863. He was soon engaged in wartime activities in Maryland and Virginia.

The Civil War was well under way when John Rodgers Meigs entered it as a newly graduated officer. The restoration of the Union was the primary war aim of the North, while the Southern states fought for recognition of the independence and sovereignty of the Confederacy. Other issues such as the abolition of slavery and foreign recognition awaited resolution. On the battlefield, the overall Northern military strategy was to blockade the seceding states' coastlines; to seize Richmond and control the area east of the Appalachian Mountains; to control the trans-Mississippi region with ground troops; and to use gunboats to control the Mississippi and Tennessee Rivers. For the Confederacy the plan was to seize Washington, D.C., drive into central Pennsylvania, and divide the northeast from the northwest, thus forcing the North to seek peace terms. However, most of the Southern troops were used to defend Southern territory while its navy was used to defend Southern waterways. But these plans were the grand designs of government officials and generals and lay mainly outside of the immediate concerns of many soldiers and officers.

What lay before the officers was heavy fighting in sometimes desperate battles, including Shiloh, Antietam, Chancellorsville, Vicksburg, and Gettysburg. For John Rodgers Meigs, who in June 1863 was engaged in the fortification of the western and southwestern defenses of Baltimore, the immediate significance of the Battle of Gettysburg was that Rebel forces were en route south after being defeated and that he was part of the effort to keep lines of communication open along the Baltimore and Ohio Railroad, and to help stop the Confederate retreat. Thereafter his military actions were confined to protecting the railroad along its route; fighting Confederate troops in Maryland, Virginia, and West Virginia;

and participating in the Shenandoah Valley campaign of 1864. While this latter area was not the main theater of military action, it was important nonetheless, made more so by the need to keep the railroad open and because of the valley's importance in the war.

The Shenandoah Valley was of significance to both the Union and the Confederacy. Union troops marching up the valley—that is, in a southern direction, "up the Shenandoah River"—could move through the Blue Ridge Mountains, cut the ties between the eastern and western Confederate states, and threaten Richmond from the west. When Confederate troops moved down the Valley—that is, in a northern direction, "down the Shenandoah River"—the valley became the main part of the "highway" into central Pennsylvania and the route to the national capital. (Confederate troops followed this route in September 1862, July 1863, and July 1864.) For the Confederacy, the valley was also a major source of grains, beef, pork, poultry, and other vital supplies. It was not without reason that Union troops were ordered to burn crops and storehouses filled with grain and to kill livestock in the Shenandoah Valley campaign of 1864. By the end of that campaign, the loss of grain and other provisions and the loss of Confederate control of the valley contributed to Union victory.

The other area of John Rodgers Meigs's military exploits was present-day West Virginia. In 1863, he served with the cavalry commander William W. Averell during his raids through the mountainous terrain that at the beginning of the Civil War had been part of western Virginia. In 1864, Meigs was in the same area, and though the new state of West Virginia had been admitted to the Union in June 1863, its political status did not guarantee full support for Union troops. In addition to Confederate soldiers, bushwhackers fought Federal troops throughout the war. (It is interesting to see on area surveys drawn by Meigs the "R" and "U" designations for the individual Rebel or Union households.)

Although the life of John Rodgers Meigs was prescribed by family expectations and custom and by ancestral heritage based in early U.S. history, he was "his own person," to use the modern vernacular. His boyhood fighting temperament carried over into adulthood; his childhood curiosity about many subjects extended into manhood; and his love of fun and adventure continued into battle. He was also an intelligent, serious young man endeavoring to serve his country and fulfill the plans for his success made by himself, his parents, and other family members.

The letters and documents in this volume describe John Rodgers Meigs's life as a young military cadet and a soldier. They cover his student days at the academy, reveal his anguish when court-martialed and disciplined for his conduct there, report his actions while still a cadet at the First Battle of Bull Run, and record his wartime exploits as he built bridges and fortifications, prepared surveys and maps, and fought the enemy. They supply information about the daily routine of a military engineer and aide-de-camp. They also provide much insight into his relationship with his family, especially with his father. It was the elder Meigs who advised young John to prepare himself well to "be able to show yourself the soldier of Christ & of your country."[6]

Montgomery C. Meigs was an army officer, an architect, an engineer, a scientist, and a West Point graduate. He reported at various times to the secretary of war and had direct contact with members of the United States Congress and Presidents Buchanan, Lincoln, Johnson, and Grant. The senior Meigs's life and professional training were greatly influenced by his own father, Charles D. Meigs, and the rigors of the United States Military Academy, where "uncompromising devotion to orthodoxy" prevailed in every aspect of cadet life. According to his biographer Russell F. Weigley, Montgomery was "self-willed like his grandfather Josiah." Weigley writes that Montgomery's mother found her son "'high tempered, unyielding, tyrannical toward his brothers.'" He had a "keen mind" and was an observer of all things around him. He possessed the Meigs family's "tradition of a stubborn self-reliance and implicit devotion to duty, a loyalty to country . . . [and] a fascination for things military and a deep interest in the mechanical and scientific." The same characteristics were passed to his son.[7]

At times, the sharing of traits by a father and son can produce rancor and hostility or smothering filial loyalty. Not so in the case of Montgomery C. and John Rodgers Meigs, who enjoyed a strong bond of love and devotion. The father was a willing teacher once John had matured beyond his earliest childhood years, and John was an eager student. They had common interests, especially photography and engineering. Both had a strong sense of duty and loyalty to country and family. While Montgomery expected his son to soar to heights he himself had not attained at West Point, he was patient and consoling; he tempered his expectations of above-average achievement as he encouraged and led John Rodgers Meigs to a successful stint at the academy, a promising military career, and, afterwards, a successful engineering career in the civil sector. It was clear that what Montgomery failed to achieve for himself—a first-place position in his graduating class at West Point, and a secure career by which to support his family without parental assistance—he wanted for his son.

John Rodgers Meigs was devoted to his father. He relied on him in times of crisis, confided in him, and, although he was somewhat shorter than his father, was fast becoming his father's carbon copy. Yet there were times when he felt his father ignored him, especially when he did not receive letters or expected visits. But these feelings were short-lived as both father and son found strength in their devotion to each other and in common interests as they carried out their responsibilities either at West Point, in public works, or in the war effort.

John's mother, Louisa Rodgers Meigs, was the daughter of Commodore John Rodgers, America's early naval hero. As wife and mother, she was the moral and spiritual guide of her family. Early in her marriage to Montgomery, she sought self-improvement for herself and her husband. Self-indulgence and time wasting were to be abhorred; there had to be time for her husband's social company and the development of her mind. If Montgomery's role was to secure for his son the basic foundation of a man adept in the pursuit of a successful career, Louisa would be the one to make her child a gentleman. She would provide him with the training to stand well in society. Yet when Montgomery was not at hand to supervise and instruct John, she drafted the stern words and constant reminders of responsibility and reliability. She supported John's decision to seek an appointment at West Point and, when the appointment was initially denied, sought a private interview with President James Buchanan to argue her son's case. Maternal devotion and responsibility contrasted with her natural desire to soften her words in promoting the promise of her firstborn. Though she herself punished John, she sometimes felt Montgomery disciplined her son too harshly, and she disagreed with his stringent methods. The watchful keeper of her son's spiritual compass and a champion of his social growth and experiences, she was second only to his father as a great influence on his life.

While John fervently acknowledged his love and devotion to his mother, he often chafed under her instructions and admonitions. In mild retaliation, he resisted her proddings by subtle, indirect means—especially by not responding to questions in her letters. More directly, he complained of being punished when she failed to write to him. In each they met their match. But in the end, it is filial love and maternal love that help to smooth out the process of a boy becoming a man. John wrote in gratitude of her sacrifices for him, and Louisa was ever present to assure the transition through the most difficult times. With this special relationship in mind, it is easy to understand the overwhelming effect his death had on her.

Two other family members—his sister, Mary Montgomery Meigs, and his maternal aunt, Ann Minerva "Nannie" Rodgers Macomb—played prominent roles in the life of John Rodgers Meigs. Mary, a year and a half younger, was a confidant and friend. She admired her brother immensely but chided him to obey the rules and regulations at the academy, to write to their mother, and to remember her with letters and tokens of affection in the form of uniform buttons. John, the loving and loyal brother, described for her his many activities, accepted her coquettish ways, tolerated girlhood foibles, welcomed her letters, and anticipated with pleasure her visit to West Point. Ann Minerva "Nannie" Rodgers Macomb was what might be described as a second mother to John Rodgers Meigs. Her closeness to his parents provided the foundation for this role. Her relationship with young Meigs was established early in his childhood and continued into adulthood. She was concerned about his welfare and success. She too chided and cautioned him regarding his responsibilities and reminded him of his father's pride in him. She in turn was his favorite aunt and a mainstay in his life. Other family members, including his grandfather, Charles D. Meigs, and uncle, Colonel Robert Smith Rodgers, had various degrees of influence on his life. In these and other family members, he found support mixed with high expectations for his success. Regardless of his failings or successes, in all he found unconditional love and pride of family.

John Rodgers Meigs was killed in October 1864. His death was but one episode in the history of the war, one death among tens of thousands. To his family and to his commander, however, he was not a war casualty but a murder victim. Both Montgomery C. Meigs and John's commander, Philip H. Sheridan, believed young Meigs had been bushwhacked outside Dayton, Virginia. In retribution, Sheridan ordered the burning of the town of Dayton and all buildings and houses within five miles of the scene of the killing.[8] Montgomery C. Meigs sought the murderers through private and public means. He hired a private detective and utilized the resources of the Adjutant General's Office of the United States Army. But pertinent documents reveal that a fair encounter had taken place between soldiers.

The final resting place of John Rodgers Meigs, along with his parents and many of his relatives, is Arlington National Cemetery in northern Virginia. A sarcophagus containing his remains is topped by a bronze sculpture of his body, posed as it was found at the site of his death. His reclining figure is clothed in his Union uniform and cloak. His face is turned upward as if seeking heaven. His service revolver, showing an empty chamber to indicate that he had fired at his opponents, lies close at hand. This carefully detailed scene of his death was intended as a loving tribute to a heroic and beloved son, a soldier of Christ and country.

Notes

1. Russell F. Weigley, *Quartermaster General of the United States Army: A Biography of M. C. Meigs* (New York: Columbia University Press, 1959), 43–44, 46, 49; Louisa Rodgers Meigs to Minerva Denison Rodgers, July 9, 1842, and October 10, 1843, Correspondence 1834–76, Rodgers Family Papers, Ms. Div., LC.

2. Louisa Rodgers Meigs to Ann Minerva "Nannie" Rodgers, September 22, [1847], and n.d., in Correspondence 1834–76, Rodgers Family Papers, Ms. Div., LC.

3. MCMJ, March 23, 1856 (quote), MCMP, Ms. Div., LC. See entries for July 7, 1854, September 25, 26, 1854, November 30, 1854, December 22, 1854, and January 1, 1855, for additional information on John Rodgers Meigs's early youth.

4. MCMJ, February 9, 1855, January 1, 1858, MCMP, Ms. Div., LC.

5. Jefferson Davis to John B. Floyd, December 16, 1857, in *Jefferson Davis Constitutionalist: His Letters, Papers, and Speeches,* ed. Dunbar Rowland, 10 vols. (Jackson, Miss.: Mississippi Department of Archives and History, 1923), 3:122; Lewis A. Edwards to Jefferson Davis, January 8, 1859, Papers of Jefferson Davis, reel 1, Ms. Div., LC; and MCMJ, March 10, 1858.

6. MCM to JRM, November 3, 1861, MCMC, MCMP, Ms. Div., LC.

7. Weigley, *Quartermaster General,* 24, 29, 32–34.

8. Before Sheridan cancelled the order, an area of about three square miles was burnt. See map of "Burnt District" in photo section.

CHAPTER 1

Early Days at West Point

Following a lengthy application process made even longer by a feud between Montgomery C. Meigs and Secretary of War John B. Floyd, President James Buchanan presented John Rodgers Meigs with the much sought after appointment to West Point. The education standards at the institution fitted handsomely into the plans Louisa and Montgomery had for their firstborn. The financial security that an engineering career could give the young man, the pride in following in his father's footsteps, and the discipline that Louisa had long wanted for him awaited his arrival at the academy. Although the school was probably little different from other institutions of higher learning of the time, its attraction was enhanced for this Christian military family—they were Episcopalians—by the fact that the academy's faculty was attempting to educate not just the Christian gentleman but the Christian soldier. Duty, honor, loyalty, and courage were virtues more important at West Point than elsewhere.

On September 5, 1859, John and his father left Washington, D.C., for New York City. The next day, after breakfast, they visited Central Park, took in Barnum's museum, and enjoyed the sights of Broadway. At 5:00 P.M. they boarded the Hudson River train for West Point, arriving there at 7:30 P.M.

On September 7, 1859, under the direction of Superintendent Richard Delafield, John reported to the adjutant's office. He was assigned a room on the lower floor of the barracks and met his roommates. At one o'clock that afternoon he reported for his medical examination at the Cadet's Hospital, where he was examined by two surgeons, Richard S. Satterlee and Samuel P. Moore, and an assistant surgeon, John Campbell. The Medical Board recorded John Rodgers Meigs's physical characteristics. He had light eyes, light-colored hair, and was 5 feet 7½ inches tall, with a good vaccination mark. A medical report was prepared, one with unexpected results of the examination. The board found a varicocele on the left side of his scrotum. Because of the nature of the varicocele, members of the board questioned whether young Meigs would be able to perform his cadet duties. Accordingly, they recommended that he be admitted on probation for one year. If he was unable to carry out his duties as a cadet or became ill during that time, he would be discharged. Following this decision and his successful academic examination, he was sworn in as a cadet. Montgomery C. Meigs recorded the events along with a brief description of the academy in one of his journals.

Montgomery C. Meigs Journal Entries of September 7 and 8, 1859 (Shorthand Transcription)

September 7, 1859

The next morning I took John round. Major Delafield directed us, sent his adjutant (Holabird) with us to show us the way.

John made his report at the adjutant's office, that of the commandant, Colonel Hardee, and was assigned a room in the barracks upon the lower floor. He has for his roommates a young Frenchman, Olivier, of New Orleans, and Gibson of Richmond, who must, I think, be a son of my old friend Ellen Eyre of Philadelphia and of Dr. Gibson's son. I saw Olivier, who looks like a gentlemanly youth.

Leaving John here to get inducted into his new position, I went with Delafield all over the place. . . .

. . . . The apparatus for instruction, the library, the museum, the observatory—all have been much improved since my cadetship. But I fear that the timber of the cadets has not improved in preparation, from all that I heard while there. I concluded that there is a bad feeling between officers and cadets or, rather, on the part of cadets towards the officers. This, I think, is in great measure due to the system of demerit, which breaks down to abuse, interest and ambition. It is run into the ground.

John alarmed me by telling me he had been to the hospital to undergo his medical examination, that the surgeon asked him some questions about his health which showed an opinion that he was subject to some diseases of the spermatic cord. This took me as much by surprise as it did him, for I had never supposed that he had anything the matter with him, having never known him to complain or to show signs of suffering from fatigue. I spoke to Delafield about it, who said that such an affliction was very common among boys who they generally thought were admitted on probation, being likely to outgrow it, that I should therefore not feel uneasy.

It proved, when I saw the report, that they found him to have varicocele of such an aggravated character that it was doubtful whether he would be able to discharge the duties of a cadet. They therefore recommended that he be admitted on probation until September 1860, to be discharged from the academy if in the meantime he should become, by reason of this ailment, disqualified from the performance of any of his duties.

I have since inquired from the army surgeons and read the books upon this subject, and most think it is a very common disease, an enlargement of the veins. Indeed, Henderson quotes someone who says that if all recruits with this disease—or, rather, informed that it is a varicose state of the veins of the spermatic cord or of the scrotum—are to be rejected, we might as well close every recruiting rendezvous. The use of suspensory bandage is prescribed for it, and this, with cold bathing, will, I am told, effect a cure. If not, an operation by tying the vein is effective and easy. Dr. Pancoast tells my father that he has performed it in hundreds of cases without ever a single bad result. It takes some two weeks to prepare for it and undergo it.

It was pleasant to me to find so many of the old professors still at the academy greeting me so cordially and telling me, after the examination of my son, which took place at 5 P.M., was over, that he had passed a very good examination—indeed; several of them said the best they had seen for two or three years, that he reminded them of my appearance and mode of answering in the classroom for Bartlett, and many seemed to recall me perfectly, though it is 23 years since I was under their instruction. All the officers and professors professed good will to my son and told me to send him to them, if he got into any trouble, for advice, and promised to do their best to lead him in the right way.

I spent a very pleasant day. John fell into the hands of the drillers soon as he left the examination hall, and they had him trotting round the plain in the double-quick drill at once. He found the usual fate of the newcomers to such a place, a crowd of cadets teasing and tormenting him, and this he said he did not mind too much. He has, I think, sense and courage to bear it sufficiently well. He is well prepared for his course, having studied the whole course of pure mathematics already. . . .

September 8, 1859

. . . . I dined with Delafield, then went to see John. We had a talk. He is much mortified to find that he has any disability which he had not supposed before. I encouraged him as much as I could, though I felt much aggravated at it at the time, myself, but I have since learned that it is probably not so serious a matter as I thought.

After attending the evening drill parade, I bade him goodby and went across the river, taking the train upon the Hudson River Railway; and at a speed of 30 miles per hour, I left my son to pursue his career.

God preserve and guide him, thus for the first time left alone away from the family home and the family ties and restraints. He is a good boy, moral and free from vice, and I hope good things of him, provided that his unfortunate disease is cured.

MCMJ, MCMP, Ms. Div., LC.

John Rodgers Meigs was assigned to the third section of the fifth class in mathematics and in English studies, and to the second section in the use of small arms. While at West Point, young Meigs received many letters from members of his family encouraging him to do well, keep up his spirits, and perform his duty. Among his most frequent correspondents were his father; mother; sister, Mary Montgomery Meigs; grandfather, Charles D. Meigs; and maternal aunt, Ann Minerva "Nannie" Rodgers Macomb.

A boyhood friend, Henry B. Ledyard, wrote him a letter shortly after his application to attend West Point was approved.

Henry B. Ledyard to John Rodgers Meigs

My dear John Washington, September 7, 1859

I went over to your house the other day about two hours after you left, was'nt that too bad? I found Mrs Macomb at the door & she told me the news. A moment or two after your Mother came down wearing your shoes as she had no others up here. I have thought all along that you would get the appointment & when you used to talk about it down at Baileys. I was ~~at~~ on the point of telling you several times. I am sorry you have gone as you were the only person in Washington I could write to. I cannot write to Mary & Lizzie will be so busy with her parties, & I do not know anybody else so you see I will come back to Washington an entire stranger. Mary is spending a few days here. Sue is very glad to have her here as she was lonely without her, as a matter of course I am also very glad to have her here. I go a week from today. I suppose you will be so busy with your studies & that you will forget that there is such a person in the world as myself but I will never forget you or Mary as you have done me a great deal more good than you think for. If you find time to write please direct to E. Charlier. 48. East 24 street N.Y.

Good Bye old fellow, take care of yourself

yrs truly H.B.L.

Autograph letter initialed, JRMC, MCMP, Ms. Div., LC.

Charles D. Meigs to John Rodgers Meigs

My dear John Hamanassett, Saturday Evening, September 10, 1859

Being in town to day I was much surprised at 1. o'clock to learn that your father was in the city & when I afterwards, at 2 ½ P.M.[, met] with him in the Media Car, I was as much delighted as I was surprised to learn the cause of his absence from Washington City. The intelligence of your appointment to and your success in the West Point examination elicited from all your friends here loud expressions of joy, for besides and beyond the pleasure they felt in knowing that your anxious wish to partake of the advantages of a W.P. education they, or at least the oldest of them were moved to gladness by the consideration that this, so far, good fortune opens up to you a clear career in [a] tumultuous struggle of this modern world for the power & distinction that wait upon success in such a pursuit as that to which you are now fairly addressed.

I hope the acquaintance that you & I made last year at this place has left upon your mind a conviction of the complacency I had in your company—conversation and conduct—& indeed, I doubt not you are fully aware of the strong attachment I have for your father, which ~~above~~ alone might suffice to show you that nothing can interest any one of his children without in an almost equal degree interesting me—for it is quite true, & if you live long you will not *believe* but *know* it, that a man's life & happiness are wrapped up in & ruled by the life & happiness of his children. So, you may see dear John that what you are doing is most anxiously considered by many friends who truly & dearly love you. We are, all of us, so dependent for happiness on the esteem as well as the love of those with whom we are connected by ties of kindred or blood that I suppose you will not receive with indifference these assurances from me of the warm interest we all do now & shall continue to take in your work at W. Point. I hope at least that you will not only believe it, but gladly believe it and that, should any troubles or griefs, or any vexations arise in your course, to obstruct or hinder either your way of life at the Institution you will turn with a feeling of support and consolation to those who have a right & are moreover obliged to love you. I may well say obliged, since your visit here won all the hearts at Hamanassett from your grandma's and Auntie Mollie's down to Alexander & Ann's.

I am not unmindful that the difference of age between mine & yours does in fact serve as a great gulf to divide and separate us, and that one so young and redundant of hope shd. naturally distrust & avoid a person so old as I am—& yet this distrust is rather an act than a law of nature—for there is in fact no difference between an old and a young man, as regards their sentiments & affection, other than that which renders the old one less bashful of the imaginations & hopes with which the young are bewildered & united, while on the other hand the young are too apt to suspect that the old all became intolerant of youthful exuberances and soured as to pleasures their age no longer enables them to enjoy—or if enjoyed, then without either much happiness in them or any true dignity or propriety. However true this is, still it seems to me that even a very young person might well consent to reflect that age gives a privilege to tender counsel & point out the way to true happiness—& at the same time not be chargable with acerbity of temper in any unreasonable intolerance of youthful fervor or any disposition to preach or cant.

I presume you have not separated yourself from my dear Mont & Lou & Nookie & all the rest without first forming good resolutions. I take it for granted you have done so with the sincerest & most earnest intentions to make yourself a distinguished member of the clases at W Point.

Pray dear John do'nt think me so unreasonable when I ask you to sometimes remember us here—and our love for yours and you—& our ardent wish hope & even expectation of seeing your name among the stars at the Graduation. Neither I pray you my dear John think me an obdurate old man or old fogie for the conviction I now feel that if you do not come out No. 1 it will be no person's fault but yours—for I have that confidence in the double blood that is in your veins that leads me to the conclusion that you *can* do whatsoever you may mind to do at the Point in the way of excellence of attainments—in character etc—probity—firmness of purpose—resoluteness of will and intelligence equal to the fulfillment of all demands to be made upon you by the government that educates the family that live in the Country that may a far distant day rely upon your patriotism & wisdom.

There my dear boy! There is what I think of *you* & what I hope for from you. Dont interpret this as scolding & as *exigeant*—but as a true representation of my heart & mind which look to you for still greater happiness than that of which I can well boast of out of the conduct of my children.

May I not here add that I consider the position of a new cadet as a most critical one & his whole career there as [illegible] and accompanied with a hint of danger. And that the first steps he takes are those that confirm & settle him in his path whether that path is to lead to good or to evil. Hence I beg you to be at first circumspect but not suspicious—mixing together both carefulness & trustfulness—ever looking forward to see where you will go & above all things Respecting yourself. Pythagoras' golden verses contain the injunction to his followers—"Keep thy truth & especially & above all things respect thyself"

σὲβον όρκον - μὰλιστα δε πάντων αισχύνου σεαυόν

Wont you write to me sometimes to let me know how you *get on.* I wish I had your heart open in my hand—I believe I would hold it as holy & sacredly and tenderly as the vainest girl you may hereafter have as your sweetheart.

Aby Sharples was in the Car today—& much pleased to hear of you.

He is to go on Monday next to Brandywine apprentice to the flouring miller. They hope greatly to enlarge their mill at Astin so as to [make] it the foundation of his Estate. He is in many regards a good youth & if he can but be earnest in life he will become a man of wealth & power & will always be inclined to take sides with you.

Mollie says lots of love from us all. She will write you a letter.[1] Aunt Emily &

all the rest send love & congratulations. Pray ever look to me as a true friend & most interested relative.

Ch D Meigs

Autograph letter signed, CDMC, MCMP, Ms. Div., LC.

1. Mary Craythorne Meigs wrote on September 10, 1859, encouraging John to "Be a man . . . one of those true, honorable, open, nobleman, of whom I cannot offer you a better example than my dear brother Mont." JRMC, MCMP, Ms. Div., LC.

Louisa Rodgers Meigs to John Rodgers Meigs

My dear John, Washington, Tuesday, September 13, 1859

Your father arrived last night from Phila and drove down immediately to the "Chateau" where we were all delighted to see him. I had received but one letter from him and none from you and I was impatient for further news of you. I need not tell you my dear boy how gratified we were to hear of the successful result of your examination and of your being installed in all the dignities of your new position. I received the letter one morning which announced the fact. Thomas brought it down himself finding it was a letter from your father. I suppose he felt some anxiety himself to know how you were getting on. I called to Mr. Pyne after I had read the letter to tell him of the result of your examination. He expressed his gratification to hear such good accounts of you etc. I forgot Thomas, but heard him below the porch very much pleased apparently at something, telling something to the servants. He went up to town and conveyed the news to Mary at Genl Cass with a great air of triumph and satisfaction as if he had been charged with the intelligence. Mary says they all seemed very much delighted at the Ledyards; all very kind and sympathy in your success. I came up to town this morning with your father and found a letter from you which I was very glad to receive. I judge from it and from what your father says of the fiery persecutions that you are undergoing among your new comrades, that you think a soldier's life has its hardships *even* from the beginning and perhaps you even doubt, whether after all your "*appointment*" was a thing to take so much trouble for. Mr. Trowbridge told me you would find it *very disagreeable* for a few weeks until you became accustomed to the life and then that you would begin to take great pleasure in it. I can easily fancy that any one leading the life you have been doing lately will find very great difficulty in conforming to the discipline of *hours, implicit obedience* etc but remember it is a necessary part of the life you have now embraced that you learn to *Obey.* To command others it is necessary first to have learned obedience. It is the *one thing needful* for you and will I dare say give you more trouble than any thing else. It has cost me many sighs I might say tears, to teach you that lesson of "prompt *unhesitating* obedience." I was never successful, but I trust my dear boy that you will remember that in the profession you have adopted, you are not permitted to question authority. It is your duty to "*obey orders*" without question and without comment. I am sorry you have to get up so early. No doubt it seems very hard to you who are so fond of late hours, but I had to do it for four years, winter & summer when I went to school at the Convent. I soon got accustomed to it. I feel quite *outraged* though I cannot help feeling amused to think how those young *demons* exercised their inginuity in tormenting you. Your father says he saw them at work. I suppose I should have been fired with indignation to witness such conduct to *my son*, but do not let them make you lose temper. It will all pass off in time. Those initiatory rites seem to be practiced every where among boys and are a sort of test of courage and endurance, two very necessary qualities of a man but especially for a soldier. We all think of you and talk of you a great deal, and shall anxiously watch your progress. Always look on the bright side of every thing. Cultivate a cheerful spirit and determine to do your duty manfully remembering that there are loving hearts at home watching over your career and praying for your welfare and happiness. Your father tells me of some physical ailment you have had to contend with which I was distressed to hear as it took me most completely by surprise. I had not the least idea that you ever had any disease of any kind. You have always seemed so well and have been accustomed to take so much exercise. Do attend *most carefully* to the Doctors directions and endeavor to get rid of any malady which may interfere with your success at the Academy. If you are *careful* you may outgrow it. Your Grandmother sends her love to you. She is in town with me. Harry and Fred and Fannie Bailey are down stairs. They came to see Mary. Harry and Sue will go to New York on Saturday. I think I shall come up from Baileys on Monday to remain. I hope to hear from you very often. Do not let any thing interfere with your *devotions*. Let nothing prevent you from finding time for *prayer*. It is so necessary to begin and end the day with. I must conclude this for I ought go down to the Chateau tonight if possible.

Ever your afft Mother

Autograph letter signed, LRMC, MCMP, Ms. Div., LC.

Henry B. Ledyard to John Rodgers Meigs

My dear John — Washington, September 13, 1859

I was very glad to hear that you had passed your examination so well. Mary was at our house when Thomas brought the news & she & I danced around the room like crazy people. I went down last night to Baileys. I found Mary lying down & your Mother, Mrs. Macomb & Mrs. Pyne out visiting. Mary was foolish enough to get up & come down to see me. After supper we took a long ride. I leave next Saturday for New York. Mother & Sue go on with me. Your Father got home last evening. I met him on the road going down to Baileys. As I suppose you will hardly have time to decipher this scrawl I will come to an end.

yours truly H. B. Ledyard

Autograph letter signed, JRMC, MCMP, Ms. Div., LC.

The elder Meigs's concerns for his son were reflected in his letters to John Rodgers Meigs. In addition to coping with medical problems, the younger Meigs had difficulty adjusting to his new environment, including military discipline and the ever-present possibility of demerits.

Montgomery C. Meigs to John Rodgers Meigs (Shorthand Transcription)

My dear son: Washington, September 16, 1859

I have, since I last wrote to you, been reading and inquiring in relation to the disease under which you are laboring. The idea that you have any possible disability is, of course, very distressing to me. I find in Henderson (Henderson on Examination of Recruits), which I suppose is in the library of the Academy, that at page 113, Marshall, a British writer upon the diseases of recruits, had the occasion to admit to the hospital a patient suffering under [sirsocele?], enlarged veins of the spermatic cord, which Dr. Coolidge thinks is your disease. They speak of it as a very light thing generally, though there are cases when the enlargement is such as to produce distention and unfit a man for the military service.

A Dr. Tripler has published a book upon the duties of a medical officer in which he quotes Henderson and makes nearly the same statement.

Varicocele, which the surgeons give as your disease, is an enlargement into all the veins of the cord to the scrotum, or bag. This they speak of as a still lighter thing. Which have you, in fact—sircocele or varicocele? You can distinguish for yourself.

Dr. Coolidge showed me these bags and advises that you be very careful to *change* your bandage regularly and to bathe as though taking a sitz bath, in cold water, morning and night. The effect of the application of cold water is to produce contraction. Debility or anxiety such as the idea of being mortified by an examination, which you suffer, aggravate the appearance of the disease for the time. In hot water it is worse than in cold.

Dr. Coolidge tells me that he had it at your age but that the use of cold baths and a suspensory bandage has cured him almost entirely. He says that every 5th man of my acquaintance probably has a little of it.

Should it become troublesome to you, Dr. Coolidge thinks that [illegible] of the service, there would be no difficulty in procuring, after the January examination is over and you have taken a good stand in your class, a leave of absence for two or three weeks, during which time you could visit Philadelphia to enter the care of your grandfather and uncle and submit to the operation by Dr. Pancoast, who says that he has performed it in hundreds of cases, without any one of them having any uncomfortable complication.

Your mother and sisters and brother will come to town on Monday, as now arranged. It rains today. The weather is not yet cold here. Your mother is altering for you some flannel shirts which will be sent by express when ready.

I have sent to Philadelphia directions with the money to send to you by mail a copy of the Greek testament you asked for, as it was not to be found here.

With this you will receive some stereographs. Those taken upon our journey to West Point I think were [illegible] too long exposed. All exposed over one minute seem to be less distinct and sharp than those which had only one-minute's exposure.

We are all well. Write to me constantly. Do not let the feeling that you understand the subject you are now reciting upon induce you to neglect a careful preparation of each lesson. You ought to be able so to recite as to get the best mark in your section, which can scarcely have in it anyone who has studied as much as you have of mathematics and of French. It is a good training of the mind to have the ability to do something that is not interesting. How much labor in which I take no interest have I gone through in the last 7 years.

Your Uncle John and aunt are much pleased to hear of the good examination which you passed. I hope that you will not disappoint our expectations. I am glad that you have joined Mr. French's bible class. He is a good man and will do you good by his teaching. Sunday at West Point may be properly spent, and as a day of rest after the constrained work of the week is delightful.

From your Father, M. C. Meigs

Autograph letter signed, MCMC, MCMP, Ms. Div., LC.

Henry B. Ledyard to John Rodgers Meigs

My dear John — Washington, Sepember 16, 1859

Is there such a thing as a circular of West Point? If there is I wish you would send me one to the care of E. Charlier 48. E. 24. Street New York. What I want by the circular is to find out what you are examined in when you enter &

what studies you go through with during your five years. I leave tomorrow for school. Mother & [illegible] & Susie go on with me. I went down last night to Baileys to say GoodBye. Fred went down with me. I found them all very well. ~~W~~ Tuesday evening Mary, Sue, Fanny Baily, Fred & I drank ~~to~~ standing to the health of John Meigs Cadet. I suppose you are so busy that you cannot find time to write to a fellow so I will say GoodBye.

ever yours truly H B Ledyard

Autograph letter signed, JRMC, MCMP, Ms. Div., LC.

Montgomery C. Meigs to John Rodgers Meigs (Shorthand Transcription)

My dear son: Washington City, September 19, 1859

I reached home last evening and went at once down to Baileys to see your mother and the rest of the family. I find them all well.

In Philadelphia, I saw your grandfather and mother and told them of your being at the Point. They are all much pleased with your success in entering.

I consulted with your grandfather and with your Uncle John as to your case. They both say that it will not trouble you if you are careful and wear the suspensory bandage constantly; that you ought to have at least three of them, so that if one gets soiled or torn or worn out, you may at once replace it, as a week without the bandage may undo all the good which a year's wearing may have done.

My father says that the disease can be cured in a week's treatment by a simple and safe operation which I believe consists in putting a ligature around the vein which is swollen; that Dr. Pancoast, a celebrated surgeon of Philadelphia, has often talked with him about this and told him that he has cured hundreds of cases, without a single failure or injury to the patient. He promised to see Dr. Pancoast and make some inquiry of him about it. He thinks that if you could withhold until the Christmas holidays, it may be possible to get a leave of absence of a week or ten days, during which you might visit Philadelphia and be cured entirely.

In the meantime, it is of the greatest importance to you to use the bandage all the time. Never be without it, except when lying down at night. John says that a man in Philadelphia makes these bandages and [illegible] if you have any difficulty in getting several of them or in getting them to fit you, you must write to him, and he will send some to you by express.

He says, moreover, that he has seen a great number of cases, some of them very aggravated, and has never had a young man be disabled from work by this disease. How bad yours is, I do not know, but my brother says that sometimes the enlargement is so great as to be very inconvenient and troublesome but it does not disable. I do not think my brother has as much confidence in the cure as my father, though he said he had not known of Pancoast's practice. Both think the bandage of great importance.

Now, to relieve you of a great trouble about this, I must tell you that one of the officers at the post told me, in showing me the report of the surgeons, that he had himself had this trouble, and yet he is now one of the stoutest men upon the post. My brother John says that it may be caused by a blow, any little accident; that when he was a young man traveling in Europe, he made a pedestrian excursion through the mountains of Switzerland. After two days' walking, he found that he had been wearing a pair of thick pants which hung very loose and low and had not supported him, and he had a varicocele. This gave him much pain; it ached. He drew up his pantaloons so as to give him some support, and he was cured, and he walked two days more without any inconvenience. He was cured.

By the way, I [illegible] have forgotten to tell you that the use of cold water bathing every morning and evening, a topical bath, is of importance. The effect of the application of cold water then is to contract the parts which are affected by the disease. It may be at first disagreeable to use this bath in the room with your fellows, and after a time you will find that whatever is done with modest feeling is not immodest and that nothing is immodest which is necessary to health. It is of great importance to your health that this be effected. Your whole career in life depends upon it. I wish I had known of your ailment at the time it could have been arrested with ease. But I had no idea that you were not in perfectly good health, nor did I know that such a disease as this existed. It seems from what I have now learned that it is common.

Major Lee, to whom they show at the department [illegible] everything that concerns me, from his known interest in me and mine, told me upon my return that he had seen the report of the board of surgeons and that he had himself something of this affliction. Do not, of course, speak of these cases to others. I tell them to you for your own comfort, and for that only.

I suppose that by the time this reaches you, you will have got through with the testing and have begun to make your position known to others as well as to know it yourself. You ask whether to take the advice of the cadets to have nothing to do with the officers. I can honestly say that it becomes a gentleman and a Christian to accept kind offers and kind service when offered frankly and be thankful for them. These officers are my friends. They are gentlemen of high standing in the community and the service which you are now entering. What they say is to be taken as true and is to be acted upon.

I know there is a strong feeling of antagonism between the cadets and officers, and I think it is a great pity. The advice of your cadet companions is to keep this up. It injures the Academy. It injures the cadets. It makes them misunderstand the intentions of the acts of the officers and professors, as well as for the professors and officers, the opinions of the cadets. If from my connection to the gentlemen who there sit over you, you are admitted to such an intercourse as they have invited you, you ought, in my opinion, to avail yourself of that offer.

Do not, of course, ever torment yourself to speak of the opinion of this or that cadet, which would be telling tales out of school, but speak very carefully and do it with proper respect of the feelings which you may know to be general upon any subject among the corps. This public opinion the officers ought to know. It is no harm to tell it. It may prevent mischief at some times. Always believe that the officers have a true regard for the welfare of the cadets.

You heard Colonel Delafield and Colonel Hardee and Mr. Williams speak with me upon these matters. You know what an interest they showed for you and your welfare. For all this, they have a right to your credit and confidence. Your intimate companions must always be of your own class. But a cordial and kindly association with the officers is also right.

Go over to the house of Colonel Delafield, go to him when you meet him in your walks, as you ought with Colonel Crozet or his [older?] man. Go and see the Frenchs. When you have liberty to do so, make acquaintance with Mr. Mahan and Mr. Bartlett. Keep up your knowledge of Mr. Saxton. Behave like a man and not like a sheep.

Do not complain of the regulations. They have stood the test of many years' trial and have been the means of educating some good officers who have done honor to our country. They turn out every year a set of young men whom the general voice of the public declares to be superior to those who cadet at any other institution in the country. Live up to them as closely as you possibly can. Keep the hours of the Academy. Many do go to bed after reveille. Do not do so yourself. If you are sleepy, you can find other hours in the course of the day in which you can take a short nap. The drill will not be so unpleasant when you have learned a little more of it, and close attention to it will soon get you into the battalion of those of distinction between yourself and the others of your class. Try your best to become a good fencer. It is a delightful exercise. I wish I had the time and opportunity to take lessons in it.

Write often to us. We love you.

From your affectionate Father

When you write next, tell me if you have difficulty reading this.

Autograph letter signed, MCMC, MCMP, Ms. Div., LC.

Ann Minerva "Nannie" Rodgers Macomb to John Rodgers Meigs

My dear John, Bailey's, Saturday–Wednesday, September 17–21, 1859

I am very glad to learn by your Past letter to your mother that you are beginning to get over your *homesickness—that feeling*—the heights & depths of which I know so well, but which, you were wont to consider, beneath the notice of such an independent fellow as John R Meigs of the *Royals!*

There is one thing very certain—we never know how to appreciate home & friends until we are parted from them—then when miles of land & water lie between us & our kindred & we are far from those who met us daily at the table, and at the fireside how many days of pleasure we remember, how many acts of kindness & self denial crowd upon the memory. We think of the virtues & the charms only of our friends at home, & wonder how we ever *could* have been undutiful & ungrateful.

We were all very much delighted at your *first* success—though indeed we *hoped* you could spell.

I must here do you the justice to inform you that your brother Monty & not yourself "tore *his pance*" some years ago. The fact is reported in a letter to me, which ought to be preserved among the family archives.

We are still in the country, your mother expects however, to leave us on Monday with her brood.

Your grandmother, & myself will follow in about two weeks.

Your father is in a great hurry to have the big house repeopled. He is very lonely & misses you constantly. He speaks of you so tenderly & thinks of you so much it is evident he has much hope & pride garnered up in you. I trust you will not disappoint him. I know my dear boy that you will never willingly do so.

The children at the present time, under the patronage of Monty Meigs & Frank Bailey are playing horse up & down stairs & halls. We are in the midst of a storm of wind & rain, which prevents the children from indulging in their usual sport of gathering chinquapins which abound now in the woods & on the roadside.

Mary has taken leave of her "true lorzen[?]," Mr. Harry Ledyard, who was to depart for New York this morning. He drove down here in his usual dashing style two days ago, accompanied by Fred Baily who very generously withdrew from the tender scene & entertained Mrs. Pyne all the evening. I hope Harry will not take "*cold pison.*"

Now Sue will next week be alone at the "*Spingler*" won't she be homesick? Poor thing, she has never been away from her mother before, except on her late summer trip.

The foregoing sheet was written on Saturday past & this is Wednesday afternoon.

The heavy rain prevented your mother from going home on Monday as she expected for the Potomac was so swollen for two days that the cause-way communicating with the Long bridge was covered with water, the water indeed reached on Monday to the "*Pavilion*" at the corner where we leave the [Alexandria and Washington] railroad in coming here. The "wagon" heavily laden with your mothers & Mrs. Pynes goods & chattels got as far as the "*Pavilion*" on Monday under the guidance of Ray & Montie & they came slowly back here much to the chagrin of Mr. Bailey who had to try it again today. We have just heard from your uncle Gus that he is ordered to California. He is coming on to make his mother a visit. He does not intend to take his family with him. Your dear grandmother feels sadly at the prospect of parting again with her Benjamin but the fact is we must take a practical & disinterested view of the matter then we shall see that he has a better opening in California than any where on this side of Continent. He will be more likely to gain fortune & distinction in California than in Washington or on the North River, so I say God speed him there & home again!

I have tonight received your very welcome letter. If I had known you considered me so severe a *critic* I should have edited these lines less hastily & with a greater attention to the "Polite letter writer" style. *Don't* make a bugbear of me. Write to me whenever you feel inclined to have a little home chat & never fear that the "schoolmistress is at hand" & *don't* "commend the poison'd chalice to my *own* lips!"

No I do not think you are conceited. We all hope to see you at the head of your class *some* day. Perhaps all are now equal to the appointed tasks & I suppose as you are the last comer, you must not hope immediately, to take precedence. Do not despise small things, aim at excellence in military matters as well as in studies, buckle all your armour on & bide your time, be, in the academic race what Agathos was in the Heavenly, armed *cap a pie* [head to foot] even the sandals, fast bound—& I think you will come in the *Conqueror* at last, then we shall strike up at home, "See the *Conquering* Hero comes" scene closes, etc.

I shall endeavor to execute satisfactorily your commission as to the *Silk*—but meantime send my letter.

I cut off a little sliver of my daughter's golden curls, the other day, & will enclose you one when I get the photographs. I send you your ring I hope it will fit you & "din'na [do not] forget" the *little things* or your old Aunt Nannie.

[Unsigned]

P.S. About the daily lessons—you know this is Wednesday Sep. 21st [illegible] look in the table of lessons you will find opposite 21—the letter *A.* so *the* proper lessons for the day—or of any other day in this year if you know the day of the month you can find its lessons. Mark in your prayer book any special day, as "*to* the Sunday letter" then you can always find the table as I do for a Calendar.

Autograph letter unsigned, JRMC, MCMP, Ms. Div., LC.

Montgomery C. Meigs to John Rodgers Meigs (Shorthand Transcription)

My dear son: Washington, September 22, 1859

I am sorry to hear that you do not yet get the highest mark in your classroom. I remember, for myself, no difficulty in getting to understand the military mode of reciting. I do not think that anything more is needed than to show that you understand and know the lesson and to speak plainly and explain clearly, so as to show this knowledge.

How you can fail to show that you understand and know a lesson in addition and subtraction is a mystery to me. I fear that you do not study enough the lesson you have to recite upon. I caution you against thinking that a general understanding of a subject would be taken instead of a knowledge of the lesson set you for the day's work. No knowledge which I have or which any man, even as good a mathemetician as Colonel Crozet, would enable him to recite upon a lesson in a book which he had not studied. Attend to this.

Your mother and myself were amused with the description of the interview with Lieutenant Saxton. We could understand precisely what it meant. How often have I had to reprove you for coming so near as to tread upon my toes. You never seemed to be willing to take the reproval as deserved and thought me perhaps very unreasonable.

You said you went to Lieutenant Saxton, who is your superior and commanding officer, for proper military and personal respect and he rebuked you. Let it be a lesson to you for your future conduct and do not feel unkindly towards him. I know, from the manner in which you speak, of his views in regard to the discharge of his duty to the cadets as only kind feeling and regard for their welfare. Do not let the same bad manners—for it is more bad manners resulting from want of proper feeling of the respect due to age and station—make you lose the regard of Major Kendrick, of Mr. French, Mr. Delafield, Mr. Mahan, or Lieutenant Saxton. Believe that if any of these gentlemen speak reprovingly to you, they do it because you violated some propriety, perhaps ignorantly, but it is to the advantage to use and be taught, however roughly, to know what is the proper conduct of a gentleman towards others. And if you get some hard rubs, remember that it is better to be taught while young by reproves from those who have a right to give them than to be avoided in society as a man of bad manners or perhaps be taught by knock-down arguments that things are offensive.

The shirts will be sent to you by the proper channel. Pillows will be asked and obtained from the department. Do not allow yourself to be involved with any smuggling, which, like all things done clandestinely, is of the nature of meanness.

You say you have no time to wash every day. I never found myself unable to wash. At the first beat of the drum, you can certainly wash in a basin. This takes a moment only. At night, you have from 9 ½ till 10 P.M. to put down your bed and make ready for sleep. In this ½ hour you can certainly find time to sit down over a basin or bucket and wash yourself with cold water. All the doctors with whom I spoke upon the subject say this is very important to you. Your health is of the greatest importance. Whatever is necessary to be done should be done. As for the presence of your roommates, do you suppose the whole world gets along without washing? And yet, how few people are able to have separate rooms for dressing in. I take it your roommates are gentlemen and will not offensively interfere with your ablutions.

Your mother and sister came up on Tuesday. They could not get up on Monday. The causeway at the end of the Long Bridge [illegible] from flood by a great freshet. The water at the Little Falls [of the Potomac] Bridge was higher by 3 or 4 feet than when I went up to look at last spring. At Great Falls, it was high enough to obliterate the falls, which became a simple rapid. It has done damage to the canal, which will perhaps interrupt navigation for some months. The aqueduct did not suffer.

The flood was so great that I rode up to the falls to examine its effects. The embanked part of the Long Bridge was flooded between the brick walls. It is probably the greatest flood for the last 6 or 7 years.

I ordered the Greek testament sent to you from Philadelphia, as I could not find one here. As I paid the money, I suppose you have it before this time.

I send you, as I do not think I sent them before, some proofs of the photographs taken on our last journey. They are, I think, all over-exposed. The best are those exposed only one minute. The others vary from 1 ¼ to 1 ¾ and seem to be fogged, as though too long exposed to light.

I shall direct Thomas to get the weekly *Star* sent to you, in order that you may know what is going on in the town. I think you had better subscribe to *Littel's Living Age,* which is now published weekly. It is a paper containing the best articles of all the foreign reviews and treats thus of the subjects of the day, giving the thoughts of the best writers of the age. This is instructive. For daily papers, I think that it will be as well to pay but little attention to them. The politics of the day are not for you and could only distract your mind from your duties and studies.

Write anything. It is practice for me to read it. Write plainly and carefully. Your hand is not good.

Mother and Mary send their love. Monty pleaded so to stay a little longer in the country that I left him until next Saturday.

[Unsigned]

Autograph letter unsigned, MCMC, MCMP, Ms. Div., LC.

Montgomery C. Meigs to John Rodgers Meigs (Shorthand Transcription)

My dear son: Washington, September 23, 1859

I have received today from your grandfather a letter of which I copy a part.

"My dear Montgomery:

"Being at a hotel today to meet Dr. Pancoast in a consultation, he told me that John's disorder is always, and certainly easily, curable by a slight operation, unattended with risk or any mutilation, that he cured two such last week, and, to confirm, took me to a room in a hotel where he showed me a young gentleman, aged 19, on whom he operated day before yesterday. He certainly has no pain, no fever, and without any sign of the varicocele left.

"If John will come here, he will cure him in about two weeks, at least so he said. By means of a surgeon's needle, armed with a ligature, he places the ligature around the varicose vein and strangles it, whereupon the swelling immediately vanishes away. When the ligature comes away, the patient is cured. I believe he makes no cut, not the slightest. It is all a mere dexterous needle work. I can and do believe what he says. Therefore, our concern about our dear boy may henceforth be laid aside. I am looking for a letter from him."

Dr. Satterlee writes that there can be no occasion, at least for some time, for you to submit to an operation. He says that after some months' observation, Dr. Moore, the surgeon at the Academy, will be better able to judge of the propriety and necessity for such an operation.

So be careful, and if you feel any inconvenience, you had better at once let me know what it is. If so bad as to be very uncomfortable to you, you can, of course, speak to the doctor about it.

We are all well. I am not able to take time to write Father this morning.

From your affectionate father, Meigs

Autograph letter signed, MCMC, MCMP, Ms. Div., LC.

Montgomery C. Meigs to John Rodgers Meigs (Shorthand Transcription)

My dear son: September 28, 1859

I send you a receipt for a package of shirts sent to the care of the Quartermaster at West Point. I also send you a couple of photographs which I took the other day. They are of the Little Falls Bridge. They are exposed 1 ½ minutes and upon one of the plates which I prepared before going to West Point. It has been to West Point and back and was about 20 days old when used.

The Georgetown Aqueduct [illegible] is upon one of the 12 plates which I prepared with Anthony's clear, colorless collodion. I exposed 4, of which 2 were good and 2 were spoiled by use of iron developer. I am quite proud of the success of my attempts to coat plates. They are Fothergill.

All well at home.

From your father, Meigs

Autograph letter signed, MCMC, MCMP, Ms. Div., LC.

Montgomery C. Meigs to John Rodgers Meigs (Shorthand Transcription)

My dear son: Washington, September 30, 1859

I send you a report from Dr. Wood, who, at my request, wrote to Dr. Satterlee, the senior surgeon of the board which examined you. Read and return it to me, as it may be of importance hereafter and for a record of the opinions of the senior surgeon.

We are all well. Your mother is very busy putting down carpets, et cetera. The hall is being paneled with oak paper and looks very well.

Write to me regularly.

What books are you now studying? What are your recitations? How do you get on? Do not elect a moment's idleness, losing you the chance for taking the highest stand at the Academy. Remember that every day tells. In the course of human life, what is gone cannot be recalled. If you take the first place, you will be an engineer, and if the second, you will probably not be one.

When I was a cadet, I looked upon the Engineer Corps as unattainable. I did not work as hard as I should have done had I known it was possible to enter it and what it was. You know more of these matters than I did, and you should do better. You are older, too.

Ever yours, Meigs

Autograph letter signed, MCMC, MCMP, Ms. Div., LC.

Thomas C. L. Hatcher to John Rodgers Meigs

Dear Meigs Columbian College, October 5, 1859

It is needless to say that I was extremely shocked and grieved when I arrived at the Columbian to find that you had deserted us. And although I was glad to hear of your success in getting the appointment to West Point; I could not help shuddering when I thought of the weary hours I must spend in the old Columbian without any one to beat me at a game of chess. I played two games on yesterday with Riordan but I can beat him every time and there is no fun in that.

We have two new seniors from Georgia but yet our class is small. I wish all were in it now who have been in it since I entered College & I think then we could compare very favorably, in respect to talent as well as numbers with any other class in college.

We have more students now than I have ever seen in college at one time & I hope the old College will come out.

I like Dr. Sampson very much. He is *nearly* as nice a man as *your old friend* Dr. Binney. Shute is a little green & I expect (for I have never recited to him) is not half as well qualified for his situation as Huntington, but he is a very good sort of a man.

We are taking up a course of law under Pete & religious history under Sampson in addition to the regular Senior studies, which keeps us quite busy.

We expect to have a valedictory and salutatory this year at our commencement and as there is considerable honor attached to it. I expect to study tolerably hard this year, but as Pollard always beats me on composition and declamation I do not expect to succeed in getting the valedictory.

How do you stand in your class at West Point? I suppose that you have plenty of competition there for the first rank in mathematics, but my good opinion of myself makes me certain that you will excell all rivals.

I am very sorry that I am not acquainted with Miss Hattie so as to speak a good word for you in your absence, but I suppose the mail affords you an opportunity of attending to your own affairs in that quarter. I sincerely hope that she will remain faithful. But I do not wish to occupy too much of your time with my nonsense & will therefore conclude.

Hoping soon to recieve a long letter from you I remain as ever, your true friend

T. C. L. Hatcher

P.S. I heard from Chambers yesterday he is at Marietta College in Ohio & sends his respects to all our class.

Autograph letter signed, JRMC, MCMP, Ms. Div., LC.

Montgomery C. Meigs to John Rodgers Meigs (Shorthand Transcription)

My dear son: Washington, October 6, 1859

I have just received your letter of the 4th instant.

I have only time to say that I am sorry that you had not sufficient force of the character to study a book which did not interest you. By not doing it, you have failed to get into the first section at a time when it is most easy to enter it.

Remember that you are not the judge of the propriety of the course of studies or of the books which you are commanded to study. The government has arranged this by its proper officers and they know more than you do about it. Even if they have made mistakes, you are paid for obeying their directions, if they are mistaken.

I send you a view from the library window. It was taken upon a plate which I prepared, and, having put yellow glass into the window of the bathroom, 4 panes, I made a trial of its qualities by exposing one-half of this plate upon the seat under the window, the outside blinds being closed, with no curtain being used. I left it there for about 2 hours, then put it away in my drawer, and the next day I placed it in the camera. You see the result.

Remember that now is the time to work. They are now looking out for those who are fit to be promoted to the first section. If you do well now—that is, do best—you will go up. If you are left down now, you will thus get a bad character, and it will be impossible to get up without doing far better than it would now be necessary to do in order to get the highest place. A bad character is a heavy weight upon a boy or man, a student or a businessman.

From your affectionate father, M. C. Meigs

Autograph letter signed, MCMC, MCMP, Ms. Div., LC.

Ann Minerva "Nannie" Rodgers Macomb to John Rodgers Meigs

My dear John, Friday, October 7, 1859

You are not forgotten in the household although so far away from us. I do not suppose there is a cadet in the corps more thought of, talked of, or more surrounded by good wishes, than you are.

The remembrance of this, must be the best, incentive you can possibly have to honorable exertion.

Major Sibley told me the other evening that I must write to you, & urge you to study carefully, for the next few weeks that you may get into the first section in all your studies before the January examination when, if you attain a high position you will probably be able to keep it during your stay at West Point—&

that you would certainly find it much more easy to get into first sections, *now* than hereafter. He says, you perhaps think yourself so well "up" in some of your studies that you do not give them the same diligent attention that other boys not so confident do thus enabling them to excel you in prompt & *exact* recitations. On the other hand, your father tells me, that one book you voted "a bore," & he feared you were tempted to neglect it. Don't let the fact of your not relishing a book prevent your being perfect, in reciting it & don't think I am a bore my dear boy. You must know that nothing but tender interest in your success could induce me to write all this. I asked your father if I should give you the Major's advice. He said "*do* by all means. I agree with the Major in thinking that John may be too confident."

Mary has just gone back to Miss Brookes'. She is to come out of school at 1 oclock in order to give her time to practice. Fanny Bailey continues her devotion to Mary who as usual calmly receives it. She went down however with Fanny the other evening, to take tea & returned rather late under the escort of Mr Fred Bailey. Mary came in, with her quiet manner & took a seat at the table working at her embroidery, so calmly—her *Pa,* who had just been enquiring for her, "did not know she was out."

Sue Ledyard has begun to feel the pangs of homesickness. They miss her very much at home. Harry sent Mary the "Spirit of the Times" which your father told her not to read. A picture of Flora Temple, & a sporting newspaper are delicate love tokens to a fair maiden are they not?

Morgan Ogden I am sorry to tell you is turned away from the Naval School. Mary says for *drunkenness.* Your mother wishes to know why you have not written to Harry Ledyard.

Your Uncle Gus sailed from New York on the 5th leaving his dear little wife & baby behind him. This was a hard parting but I trust that he may return safely to his family next year.

Monty has gone to school to Dr Steele Mr Harrolds successor. He studies as much as ever! Kate is still with us. Mr. & Mrs. Trowbridge, & Fred were here a few [days] ago. Fred T looks quite manly with a fine pair of whiskers but he is still at the Michigan University.

Do you read the lessons appointed in the prayer book? If you do you are with us every day as Uncle Gus, myself, & Mary read them daily. Good bye God bless you dear boy Your affec

Aunt Nannie

Your mother sends her love & I send you a letter from my other John. I think it will amuse you. Did you get the mouth [illegible] sent with the shirts?

Autograph letter signed, JRMC, MCMP, Ms. Div., LC.

Montgomery C. Meigs to John Rodgers Meigs (Shorthand Transcription)

My dear son: Washington, October 9, 1859

We have a cold, rainy day. I have been to church this morning, but it rained so hard that your mother did not go.

I send you Fothergills taken by Mr. Wood, which you will recognize. He told me the other day that he found that he could not print the moon upon his plates. He had to take a Fothergill as being more sensitive. So that I believe he has at length come around to the side of Fothergill.

I took out some plates which he prepared and tried them in comparison with the last of the dozen which I prepared with Anthony's white collodion on the 17th September, which were, therefore, on the 7th instant, 20 days old. Mine give the best results, being free from blisters, while those prepared by Mr. Wood blistered so much that I had difficulty in fixing them, and some of them are quite spoiled.

The views were of Bridge 6 [of the Washington Aqueduct], in which we have put the key pipes. The arch, in the course of the day, in which the key pipes were put lifted itself off its center by the expansion of the iron and stood alone. The edges have never been removed, and the arches above stand free.

Yesterday, we filled the soft earth with water, to see if it was tight under pressure. There was not leakage enough in all its joints to get a drink from. It held 143,000 pounds of water, and under this load sank down at the ground $\frac{9}{64}$ of an inch only.

This is a compression in the whole length of the arch, 207.64, of $\frac{9}{32}$" or of $\frac{1}{9000}$. The pressure due to the weight of the water is at a mean between the pressure at the key and at the spring, 1580 per square inch of section of the pipe. And the compression of the iron, as deduced from this experiment, upon a grade scale, is, for every pound of pressure per square inch, $\frac{1}{14,281,143}$ of the length.

I ordered the *Star* sent to you weekly many days since. Do you not receive it?

I send you with this a *Spirit of the Times,* containing a good article upon the value of training of missiles. Endeavor to be skillful in the use of the weapons of your trade. If you get a chance, learn to box; and in all the games—of football, et cetera—try, without making such exertions as might be dangerous to you, to take a high stand. It is good to have a sound mind in a sound body. I should never have been able to do the work of mind and body which I have gone through in the last 7 years at Washington had I not been fond of field sports, of ball play, of fencing, boxing, and all other athletic games within my reach. I wish I had been able to learn more of boxing. I should introduce it at West Point as part of the drill, if I were Superintendent.

You have a good gymnasium, got up for the cadets by the kindness of Colonel Delafield. For this, they ought to know that they are under great obligations to him for proper use of the opportunity it affords to make all the boys athletes. I hope you will use it always—however, with discretion, so as not, by over-exertion, to injure yourself.

Your mother and all are well and send a great deal of love. Try to send me a better report of your next month's work. Get into the first section in everything. It is much easier now than hereafter.

From your father, Meigs

Autograph letter signed, MCMC, MCMP, Ms. Div., LC.

Montgomery C. Meigs to John Rodgers Meigs (Shorthand Transcription)

My dear son: Washington, October 12, 1859

I send you the photographs which I believe are what you asked for in your memorandum, which I sent to Mr. Wood.

I am too busy today to write to you more than to say that I have your last letter and I am glad to find that you intend to study hard whether you like your books or not.

I am very anxious that you should do well in the Academy, and I have sufficient parental partiality to believe that if you study hard, you can do best. It is only by aiming at the best that we can ever hope for much success. And it is only by coming out at the head of the class that you can be certain of getting a position in the Corps of Engineers, which I take it is your ambition. I do not suppose that you care to go to the western country to spend your days on the limits of civilization, among trappers and hunters, do little or nothing except at times exposing your life in long marches after the Indian robbers. This is a duty and creditable and necessary to be done; but it is not one requiring talents or knowledge of a high order, and it is one which may, therefore, be as well or better left to men who are capable of nothing higher, nothing more useful to their race.

We are all well here. I am very busy and much bothered, as usual.

Ever your affectionate father, Meigs

Autograph letter signed, MCMC, MCMP, Ms. Div., LC.

John Rodgers Meigs to Louisa Rodgers Meigs

My Dear Mother, West Point, Sunday, October 15, 1859

I have been debating whether to write to you this evening, or go to church, and I think that it is more pleasant and better to do the latter.

It seems to me that you do not write many letters to me, though I think that after all I am pretty well off in that respect. My room mate who comes from Washington Territory has had no letter for a month and a half.

I was talking about home yesterday evening with Bruton, a nice little fellow in our class, he is only 5 foot 1 inch, and he said that he never had any home, his father and mother died when he was very young.

His uncle took care of him for some time but he was very sickly and he knew he was a great trouble so he left as soon as he could.

He was working in a store before he came here, and one day a congressman came to the place he lived in to make a speech and having heard that he had an appointment to give he applied for it but found that it had just been filled. He was promised the vacancy however if the fellow was found, and so after a while he got a letter telling him that he might have the appointment. He had kept all this to himself but now he went and told his guardian and he said he thought it was a very lucky thing for him and gave him a hundred dollars and told him to go.

He had never seen a railroad in his life but he got on the cars and went to the sea where he went aboard a vessel and got to New York. He was very seasick of course.

He got lost in New York and thought he could find his way by following the omnibuses with Broadway on them. He was afraid to ask any one lest they should get him into some bad quarter and steal his money.

He finally got here safe. He is a little bit of a fellow about Monties size and looks about his age. It seems very strange to me that he should have got his appointment himself and found his way here all alone.

I am in the first section in everything now as I was transferred last evening.

I was invited to tea at the French's last night and stayed there about two hours. Miss Clara is quite pretty and very pleasant. I saw a first classman there who behaved in such a manner that I thought surely he was a cousin. Upon asking to day however whether he was a relation I was told that he was not but probably would be soon.

The tea was a great deal better than that we get at the Mess Hall. The meals there are a pet topic for the cadets to grumble about.[1]

They are not as good as those they get at home.

We take our dessert out in the woods on Saturday evening. It consists of hickory nuts and butter nuts and chestnuts; I suppose that the latter won't last longer than next week.

I am to enter the battalion next Monday when we will fire powder at our drills. I believe that we will have only skirmish drills ~~hereafter~~ for the rest of this year.

The last battalion drill the cadets fired. The noise of the guns was quite exciting to me there were over 200 fired at once. But they say that they sound like pop guns compared with the artillery.

Good bye ever your affct Son
John R. Meigs

Autograph letter signed, LRMC, MCMP, Ms. Div., LC.

---

1. The food at West Point was considered very bad in the nineteenth century. It included boiled potatoes, meat, bread pudding and other food stuffs served, at times, with bugs. A local hotel owner by the last name of Cozzens provided food for the cadets. Another source was food prepared by the "Misses Thompson," destitute daughters of a Revolutionary Army soldier.

Henry B. Ledyard to John Rodgers Meigs

My dear John New York, October 16, 1859

I received your letter last night & was very glad to see you answer your letter so quickly. I met Captain Saxton in the street yesterday & he gave very good accounts of "young Meigs." I was very much surprised to hear that Mary had gone to Miss Brooke's, as she had not the slightest idea of it when I left W. I do not think that West Point in comparison to our school can be a worse place for swearing here every body almost without exception swears horribly although if any of the fellows heard me they would say in the words of Scripture "Physician heal thyself." I heard from Fred Bailey the other day he was very well & said that all your family were well. There is one thing that I have no fear of that is of your ever falling into the habit of swearing. Do you Know a fellow at West Point by the name of Toutant he has a brother here do you Know besides, a fellow from Detroit by the name of J. Benson Williams I used to know him quite well. But the bell is ringing & I must close. Remember me to all at W. when you write. Good Bye old fellow.

Ever your true friend HBLedyard

Autograph letter signed, JRMC, MCMP, Ms. Div., LC.

Montgomery C. Meigs to John Rodgers Meigs (Shorthand Transcription)

My dear son: Washington, October 18, 1859

I am glad to learn by your letter of Sunday to your mother that you have at last got into the first section in English. I hope you will stay there and make it your business to take the first place in your class if your abilities are sufficient. They determine that if you do not get the first place and the engineer graduation, it will be because you have not the talent and because you are careless and indolent.

We are all well here. I am busy writing my annual reports. Yesterday, I went down to Annapolis for the first time for over a year, and I was much disappointed at the state of things there.[1] My assistant has not done his duty and has not got from the expenditure so useful a result as he should have done. For this I am in some degree responsible, though had the government left me the proper authority to visit the place once a month, as I used to, I have no doubt that I should have saved the expenditure of over a thousand dollars and have got a better result. But some regulations are made to prevent abuse by those who are not to be trusted, and the public service will sometimes suffer. It is my duty and that of every officer to submit to the regulations and do the best I can for the public interest consistent with them.

I send you a photograph of the bridge over Rock Creek, which has for a week or two been standing free from the centers, and one of the operation screwing up the last pipe, the key pipe. You will see a pretty good likeness of the back of the engineer.

Mary is practicing downstairs. Her voice is so much like her mother's that I constantly take it for her mother's.

Write regularly to us. We are always disappointed when many days pass without a letter.

I saw Fred Rodgers at the Naval School yesterday. They say he is smart. I do not know how he stands.

Ever your affectionate father, M. C. Meigs.

Autograph letter signed, MCMC, MCMP, Ms. Div., LC.

1. On September 14, 1854, Montgomery C. Meigs had been assigned the responsibility of supervising the construction of the battery at Fort Madison in the Annapolis harbor, Maryland.

Thomas C. L. Hatcher to John Rodgers Meigs

Dear Meigs Washington, October 20, 1859

I received your very welcome letter a few days since. I can assure you I have been exceedingly lonely this year with no one to beat me at a game of chess. Have you found anyone to play with at West Point? Riordan is a young law student of 20 & a great friend of Dr. Jones. J.O.K. & McVeigh room to-gether now, & Betts & I room in Betts' & Dunaways old room; you see as I board in the city I don't care what room I have.

We do not recite in Greek or Latin now, as we have 3 recitations a day without, but we have to study them privately & will be examined on them. I beat Prof. Shute playing chess last week.

I have been to the theatre twice this year, but don't expect to go again, I think it a bad habit. There are about 70 students in College & 40 in the prep. Griffin manages the prep. firmly.

I heard from Chambers last week. He says that he takes about as good marks at Marietta College as he used to do at the Columbian.

We all like Sampson very much. I think him decidedly a nicer man than Prof. Huntington used to be, but yet I don't like him as much as Fristoe.

By the way old Fris sends his respects to you.

I am afraid that you will have to give up all idea of Miss Hattie now as I consider it quite probable that a lady as intelligent & attractive as you think her to be, will be married before five long years can roll round. But I hope if such should be the case you will not take it as hard as Gus Owen did last year, for he cried when Miss Cattie (his lady love) got married & I don't think it would look pretty for you to cry.

The college is extremely quiet now. Phil Strother did not return & Betts Moore & myself are boarding in the city.

I stayed all night at college last week & tried to pull some of the fellow's toes but the strings which we had were not strong enough & we could not pull much before the strings would break.

Freeman was in Washington last Sunday. He said that he remembered how you looked at him the day he chucked you out of his window. He is now at Cambridge Mass. I also heard from John McDaniel last week, he is getting along very well. You must excuse the brevity of this letter & also the mistakes as I am too unwell to attend college & consequently don't feel much like writing. But I will try & do a little better next time "deo volente." Have you gotten acquainted with any ladies yet about West Point? I am thinking of going to the University of Virginia next year to take a course of Mathematics History & Moral Philosophy. Hoping to receive a long letter from you very soon.

I remain as ever your true friend Hatcher

P.S. Mrs. Granbury (Miss Jennie Massy) died last week.

Autograph letter signed, JRMC, MCMP, Ms. Div., LC.

Montgomery C. Meigs to John Rodgers Meigs (Shorthand Transcription)

My dear John: Washington, October 22, 1859

This is Saturday evening. I have been hard at work all day in the little office, in which for the first time, as it snowed and is cold, I have had a study set up. Snow and rain mix today in the air and remind me of the first day I spent at West Point, a nasty day, of which you will remember the picture.

I send you a picture taken on the 20th by Mr. Wood. It is a pretty good likeness, I think. You will be better able to judge of it at a distance than we are here. A likeness for characteristic is recognized better at a little distance.

We are all well and are all pleased to find that you have at length entered the first section in English as well as in mathematics. I, for my own part, am not displeased that you have found it necessary to study a little harder than you expected. I hope that you will set your mark high. This is no mere strive of ambition for success in beating your fellows, but it is a struggle for a definite object—rank, pay, and position in the Army for life.

It makes to you the difference between, if you go into the Army, spending your days in the wilderness and in the modes of civilization and refinement of arts and sciences. It makes the difference between being a captain, in case of active service in time of war, of a company of infantry, with the opportunity only of honoring the reputation of a brave man, and nothing more, and of being in such a position, in the attack or defense of a position as an engineer, that the whole country will know of your success—a success requiring as much courage and much more brains than the other. Instead of commanding a few men doing little or nothing, you will, in the engineers, have occasion to employ all the faculties and all the knowledge you can acquire.

Mr. Wood thinks now that the wet collodion and the Fothergill are the only process. He told me that in printing the [gran?] upon his collodion albumin, he had to use Fothergill, as more sensitive.

I send you a [gran?].

Ever your affectionate father, M. C. Meigs

Autograph letter signed, MCMC, MCMP, Ms. Div., LC.

Montgomery C. Meigs to John Rodgers Meigs (Shorthand Transcription)

My dear son: Washington City, November 8, 1859

You have seen by the newspapers that I have at last come to the end of my connection with the public building. The Secretary has relieved me from the charge of the Capitol extension, Post Office extension, and new dome.[1]

Do not let this affair weaken your resolve in whatever condition of life it may please God to call you to, to do your duty, regardless of all considerations which might seem to make it expedient to swerve from the path which your conscience and honor point out to you.

You know that you have seen how much labor and zeal I have bestowed upon the works upon which I have spent the last 7 years and know that you have heard me say how much I prize the honor of completing them in a manner which should leave my name so connected with them as to cause my children to point to them hereafter and say that "these were constructed by our father."

In defending the right, as I firmly believe, I have given offense to the head of the War Department, and he having insisted upon my being relieved, the President reluctantly consented. I have no reproaches to make to the Secretary or to the President. I hope they did what they thought right. I am sure that I did what I thought right, and I expect the consequences which I have for a long time foreseen. I am not discouraged in duty. I hope that I shall still, in my duty committed to me by the authorities of the country to whom I am amenable, answer faithfully to do my best, looking for the reward of a good conscience, which now enables me to take this injustice, as I must think it, cheerfully and calmly.

I am thankful that the matter was not decided before I had finished all the great problems of engineering and science involved in the construction of these great works. The lighting, heating, ventilating, and acoustics of the Capitol are not theories and skill less but entail constructions they have in operas and are successful. These belong to me. I do not think that of the credit, whatever it may be, which is attached to the successful solution of these problems, anyone will be able to rob me.

Much remains at the Capitol, to put up the columns and entablature of the porticoes. I should have liked to crown my work by putting the statues, which but for me would not have been created by the chisel of the sculptor Crawford, into a place in the pediment and upon the pinnacle of the dome. But this seems now to be given to another. Yet, the people will know that I gave Crawford the opportunity to create these great works of art, that but for me they would not have been.

At the office, I had completed the design for the heating and ventilating apparatus and commenced its construction. It will be carried out precisely as I had designed it. I believe that Captain Franklin, who succeeds me, does so with a strong sense of the injustice that has been done to me and with a desire to act honorably towards me as well as towards others. He will carry out my designs so far as he is able to do it.

Let this be an example for you of the propriety of doing your duty. Let the feelings of indignation which will naturally arise in your bosom when you see injustice can be done by anyone, however high in power or place, without outraging someone who will carry the remembrance to his dying day. Avoid all injustice and tyranny, and think that no present gain can compensate a man for the injury to his own name and reputation and unhappiness certain to follow an act of injustice.

Write to me often. Work hard, but take necessary exercise. You can yourself skilfully perfect all the mental and bodily training for which you have been sent to the military academy, and set before you as an object to do justice, love mercy, fear God, and keep a good conscience, which will enable you to meet all the injuries of life with a calm and firm mind.

Ever your affectionate father, M. C. Meigs

The picture of Old John Brown in this paper of Harper's, I sent to you today, is said by Lieutenant Stuart, who was there, to be just like him, just as he looked when lying in bed wounded.

Autograph letter signed, MCMC, MCMP, Ms. Div., LC.

---

1. Secretary of War John B. Floyd relieved Montgomery C. Meigs of his responsibilities for the construction of the Capitol and for the Post Office on November 1, 1959.

Ann Minerva "Nannie" Meigs Macomb to John Rodgers Meigs

My dear John, Wednesday, November 30, 1859

Your mother received your letter this morning & I have possession of my ring. It was not necessary for me to say anything more on the subject as when we parted I told you I would send you another ring in the place of the one, I put on your finger, which I could only *lend* because it was a gift to me.

We all think of you a great deal & regret that we cannot expect to see you at Christmas. You will be obliged to let your imagination, partake of the Christmas turkey, and hear the joyous shouts of the children, when the glories of the *Tree* & Table are, revealed to their gaze.

We hope to have Capt. Macomb with us & I have written to beg John N. Macomb *jun* to join us, so I shall especially miss you if he comes. He is a nice boy; I think you would like him. He is full of mirth & amiability like his father though when he was Monty's age he had a furious temper excelling even Monty's.

Gus has been suffering for three days with a toothache which has left him with a face swollen to gigantic proportions, & hideous to behold.

He does not allow any laughing on the subject so I am obliged to avoid looking at him.

Your father looked so grim this morning at the breakfast table after reading your letter, that your mother asked him "why he did not eat his breakfast, & stop thinking." He said "I feel concerned about John. I did not tell you that I heard yesterday at the office that his professors think he is not doing himself justice. He understands his lessons they think but his recitations are not perfect. I hoped to see him doing better, he ought to be at the head of his class."

Your father says that "stopping short of perfection" has been always your fault in studying but that at West Point a general *understanding* of a lesson, is *especially not enough.* "When John was a boy at Rouses' Point I used to punish him (his mother thought, unduly) for imperfect recitations, but I failed in my object, that of giving him a habit of *reciting,* with precision & accuracy. I punished him because I *knew* he *could* master his lessons & *he* did *almost.*"

I asked your father "why he did not pass at the head of his class." He gave me the history of his course at West Point which will I know be interesting to you. He knew nothing of West Point, nothing of the peculiar studies there. His father one day asked him what he wished to do, he answered "I should like to be a ship carpenter." His father said "You are going to West Point" so forthwith he went. He had never heard anything about demerit marks, knew nothing of the manner in which his studies & conduct would affect his position in the army & had actually been three years at West Point before he knew there was an "engineer corps!" He saw Capt Fred. Smith of the Engrs. in church in uniform & noticed the device on his collar. He asked what it meant & learned of the existence of the corps for the first time. *Still* he says he was 2d in his studies, except one year when he was *1st* & that his *demerit* put him down. He was head of the drawing class & marching a squad of 50 or 60 boys, to class. As their Captain he often saw some of them twist their shoulder, or lift a leg in merriment & he could not bear to report them the consequence was that *he* was always reported for such failure in duty. He never knew how his standing was affected by these *delinquencies,* so as he could not get his demerit taken off, he was put down to 5 in his class, "was put into the artillery to march about the country! but by a fortunate increasing the Eng. corps was recommended to it & got in with the man next below him." Your father was 16. years & two months old, when he entered. All this he told me this morning also that *he* was very *modest,* & never hoped to pass high, he was always astonished to find his standing in his classes so good, "whereas John went to West Point, expecting to stand *very high* without any trouble."

Now my dear boy, I tell you all this that you may see why your father, now so indisputably *preeminent* among the men of his date was not at the head of his class. With your experience, he would undoubtedly have been so & he wants you, to give him the satisfaction & honor, which he failed to obtain himself.

I saw this morning even at this late day, how much that position in his class, annoys, & mortifies him. He feels that he did not do himself justice yet I think his success, was wonderful when he considers his entire ignorance & inexperience regarding West Point & the Army. But his mortification will be added to yours, if you fail to distinguish yourself. He thinks he knows your mind thoroughly & he says there is nothing you want but *attention.* My dear John you should read *Agathos* again as he ran his spiritual course, so must you the academic, that is with the *whole* armor, on, for the Dragon fearfully punished those soldiers, whom he found with even a sandal untied. May, you dear boy, be *Agathos* when you terminate your course at West Point, & may you continue *Christs* faithful soldier & servant unto your life's end, so that you may receive like Agathos, your reward.

Major Sibley was married on Saturday last to Miss Churchill whom you have probably seen at Mr. Trowbridge's.

The Major has entered into the matrimonial state now for the fourth time which as your Uncle John says "speaks well for the institution."

Mary is very busy, she seems to be studying hard this winter. She looks remarkably well. Fanny Bailey is very constant in her attentions to her, so is Fred but we do not encourage her going there. Susan Gillandel is spending the winter at your cousin Min's.

Anne Wadsworth is at Mrs Bryans' with Helen her sister in law. Annie looks quite pretty. She does not speak of her engagement, as none of us mention it to her. It would be wrong for us to speak with approbation of her marriage to a Unitarian minister. All other sects I can tolerate, but how *can* we, sympathyze with those who leave out the vital part, the tenderest, sweetest, part of our faith, the *sacrifice* of the Son of God for us?

I forgot to sign my name it is that of

your affectionate Aunt Nannie

Autograph letter signed, JRMC, MCMP, Ms. Div., LC.

Hattie L. Coyle to John Rodgers Meigs

Dear Johnnie, Washington, [Early] December 1859

This is the fourth day of the continuance of a most tiresome rainstorm, and as I sit beside the blazing wood fire in the parlour, where everything seems cheerful, notwithstanding the gloom without, I should like, instead of answering your letter, to transport you hither to spend the evening with us. Would *you* like to come I wonder? I was so much surprised to hear of your departure for W.P. I obstinately refused for a time to believe the report, and when finally I discovered that the statement was true, that you were really gone, and without ever coming to tell us goodbye, you may be sure, I scolded you very emphatically in my heart. But your letter cleared *your character* by assuring me it was impossible under the circumstances.

I think W.P. must be a perfect heathendom. I never heard of such atrocities as are there practiced. I wonder you are alive at all, that you are capable of surviving such persecutions. My sympathies were yours, as I read your account of the numerous trials you were subjected to without redress. I am glad you did escape the triteness of encampment. Truly "all is not gold that glitters." While at W.P. myself, *uninitiated I,* imagined the cadets must be the happiest people in the world with such magnificent scenery surrounding the place itself such a perfect fairy land of beauty, and that enchanting music, what could be wanting to increase their contentment? But you have disclosed the scene to me, under a new light, and showed me a darker side, than could have imagined, existed. Leo is seated beside me at the piano, practicing "Don Juan" for the next soiree, which meets at the house of Mr. Branham across the street, a musical gentleman, and a neighbor in every way desirable. Did you ever meet him one evening here when he came with his flute? He is a great acquisition to our soirees, but one we are destined to lose very soon unfortunately as he returns with his family next week, to his old home in Georgia. He presented me with a beautiful piece, the other day, a "Fantasie from Charles 5th," very long and difficult, but which I believe I have succeeded in acquiring to his satisfaction. Mr. Louis Carusi has ceased to favour us with his valuable society at the soirees, but the other evening at Mrs. Sherman's Mr. Carusi introduced his daughter Geaninne (don't know how to spell it) who is a very beautiful girl. His troublesome son is at home again, and ~~takes~~ collects his father's bills for him. We have ceased taking lessons, until Brother's next vacation. Brother laments your absence from College continually, Charlie Bradley has gone too, to the Indiana ~~College,~~ and Mason Noble to William's College, Andrew B. fills Charlie's place at the Col, but Leo finds very few congenial friends.

Did you hear of Charlie Hassler's sad accident. You know he started off so full of hope for his career at Princeton. About two months ago, while practicing in the gymnasium there, he fell and injured his head and back so badly, that the physicians say he must give up all study for the winter certainly, and perhaps longer. His mother is with him at the north, and as soon as he is able to bear the journey, I suppose he will return home. His sister was telling me the other day, that the loss of so much time was a terrible disappointment to him.

Among some of my researches a few days ago, *you* were vividly recalled to my remembrance, by the discovery of the *enigma,* you so ingeniously contrived upon my name, and sent in so mysterious a manner. Do you remember it?

When will your muse awake again? You must not allow the rigors of W.P. discipline, to put it to silence.

Washington's [illegible] have scarcely begun as yet, though strangers are beginning to assemble very fast. The Smithsonian lectures are about to commence again, and are to be opened by a series of lectures on the subject of agriculture.[1] I shall miss you, when I walk over there this winter. Lissie de [illegible] is in town, but *your* friend and *Harry's,* Miss Mary, is in N.Y. for the winter. Miss Ellen Hall is engaged to be married to a Mr. Curtis from Ct. Harry Sherman continues to have charge of the organ at Trinity, and is enjoying the prospect of having a new one, by Christmas. Our church is not finished as you supposed, and I'm afraid there is little prospect of its being finished, for a long time to come yet. In the mean time, the organ is silently packed away in a garret room, and Leo is obliged to content himself with playing for Harry occasionally.

I have no very interesting news for you I'm afraid. There is only one topic of conversation, and that is Charlestown, and Charlestown tragedies, and of this I am so tired. Even while I write, I hear the music of some of the troops, returning from the execution of Brown. These companies have become a sight quite familiar to us of late, and everywhere I go, I am regaled by a vision of the murderous weapons known as Brown's pikes.

Congress has reassembled, but I was prevented from attending the opening, by *the rain.* We are anticipating a stormy session this winter. I must close this *long letter.* You see I appreciate *yours,* by answering it so quickly.

Whenever you have time, and are not *too tired,* remember I shall always be glad to hear from you. Leo, Mama, and *Aunt Sarah,* all join in remembrance to you.

Ever your friend Hattie

You did not tell me how to direct your letters. I suppose this will reach you, but I should like to know if it is correct.

Autograph letter signed, JRMC, MCMP, Ms. Div., LC.

---

1. The Smithsonian Institution of Washington, D.C., sponsored public lecture series during the 1850s and 1860s. Joseph Henry, secretary of the Smithsonian, commented about the 1859–60 lecture series in his annual report: "While a large number of persons regularly attend the lectures for the sake of the advantage to be gained from them, others, and particularly young persons, attend as a mere pastime, or assemble in the lecture room as a convenient place of resort, and by their whispering annoy those who sit near them." Annual Report of the Board of Regents of the Smithsonian Institution, 1860, U.S. House of Representatives, 36th Cong. 1st sess., H. Misc. Doc. 90 (Washington, D.C.: Government Printing Office, 1860), serial set 1066, vol. 7, pp. 53 (quote), 119–94.

Thomas C. L. Hatcher to John Rodgers Meigs

Dear Meigs Washington, December 3, 1859

I received your letter last week & have until now been prevented by my numerous studies, from answering it. We have to study very hard this year at least I do. I have studied nearly twice as hard as I ever did before in my life. They commenced last Monday to mark us with reference to the Valedictory & Salutory honors at the commencement & as composition & declamation are to be taken into account & as we have no mathematics of any importance I expect Pollard will get the Valedictory. I intend next Wednesday (which is the day for us to speak) to put on my best coat, black my boots & holler as loud as I can & see if Jillson will not give me a better mark, the old fellow gave Pollard 10 last time: it is the first 10 that has ever been given for declamation since I have been in college.

Dr. Sampson is a splendid man I like him very much, he is not in the least like Dr. Binney.

My Lord Jillson is talking about leaving us, he is trying to get the [illegible] auditor's office I believe. I hope he will & that we will get some one better. He & I never were very fond of each other. Shute does not know one tenth as much about Latin & Greek as Prof. Huntington used to. I beat him several games of chess not long ago. I have played very little this year, for want of some one to play with me. Murphy is now in the city but I have not yet seen him. I want to go down & meet him before he leaves.

I have dropped half of my acquaintances this year. I found out that they took too much time from my studies.

I expect that next summer, when you get to flirting with the ladies & have nothing else to do, which tends more than anything else in the world to make a young man fall in love, you will forget all about Miss Hattie if you have not done so before this time. I am certain that I should, I never was in a place for a month in my life that they did not think me dreadfully taken with some young lady there. I suppose the reason that the Cadets dance ~~the round dances~~ with so much pleasure is that it is a recreation after study & probably nearly their only recreation. I never thought that most persons who danced the round dances had wrong thoughts while dancing them. I believe that thousands are too pure minded, but then on the other hand, there are thousands who have & I don't approve of too great familiarity any way between gentlemen & ladies & on *that* account, I entirely disapprove of a lady waltzing with any one except her brother.

J.O.K. did not show much feeling at Mrs. Granbury's death & don't know the reason. The fellows in our class are: Bond, Forrer, Gorman (of Georgia, my chum), Hatcher, Kirk, McClenny, Neal (also from Georgia), Pollard, Willson, Clampitt & Johnson from Western Va. & direct from Marietta college in Ohio. Forrer, Clampitt & Johnson are [B.Bs?]. We have (as you see) 8 regular seniors, & 3 scientifics quite a good class.

It is so dark that I can hardly see the paper & will therefore have to conclude. Hoping soon to receive a long letter from you I remain as ever your true friend

Hatcher

Autograph letter signed, JRMC, MCMP, Ms. Div., LC.

Montgomery C. Meigs to John Rodgers Meigs (Shorthand Transcription)

My dear son: Washington, December 4, 1859

I have not had a letter from you for some time. We all look for letters from you daily and are disappointed when much time elapses without hearing from you.

I rode for the first time in a month up to the reservoir on Friday. The water is clear and beautiful and still runs out at the west. The use of water is increasing in the two cities. I have laid in Georgetown the last pipe of the water distribution and in a few days expect to have the pump in the abutment of Bridge 6 at work supplying the upper parts of the town which are still without water. The water flows through the pipes of the bridge which approaches the completion of the iron work. But the masonry can't wait for another appropriation; and as the weather yesterday changed to cold, sleet and storm, I fear that the iron work and hod work of this bridge will also be left until next spring, before completion. However, I shall be able to take down the center and thus to show the beauty of the bridge to all who pass. It will be very beautiful.

The city is full of excitement about the approaching meeting of Congress. The members have been holding caucuses for the last few days to determine upon their plan of operation and names of those whom they will support for the offices of the House, et cetera. In all this I take no interest. I look for a stormy session, one in which much slander will be uttered by members of both parties, in which, however, I hope to see truth prevail in the end.

As Christmas approaches, we all worry if you would like anything sent that has a Christmas box. I suppose that I could get permission from the Department for you to receive a box of Christmas gifts, if of a proper kind. Do the boys ever receive such things? I forget whether I ever did or not, but I think not. However, I do seem to have some recollection of a box of cookies coming to me from Philadelphia while I was a cadet.

I hope that you have lately given greater attention to the important subject about which I wrote to you last—perfection in your recitations. No less than perfection should satisfy you if that is attainable with study, and in your case I think with very hard study. Nothing less will insure your taking a commission in the engineers.

I find myself still pretty busy, for I have much to bring up from the records of the past. Little, however, to do out of doors. I have almost finished the waterworks of Georgetown, and I have got in Georgetown, perhaps, the best water supply in the United States. They tried one of the new fire plugs the other day and the water flow over the highest mill in the street, Water Street.

Love from all whom my dear son ever affects.

Your affectionate father, Meigs

Autograph letter signed, MCMC, MCMP, Ms. Div., LC.

Montgomery C. Meigs to John Rodgers Meigs (Shorthand Transcription)

My dear son: Washington, December 6, 1859

I have your letter of the 2nd. I hope you will not give way to any feeling of shame or mortification at not taking at once the highest stand in your class. This indulgence would lead not to success but discouragement and failure. No one will have any right to reproach you if you take a stand among the five who are noted as distinguished. Our partiality may make us think you hope to do better than your relative power, when compared with your neighbors and rivals, really would enable you to do. And our love for you and my strong desire to see you take a position in the Corps, of which I am a member, make me very desirous that you should take such a stand in your class as will insure this beyond doubt.

I do not wish to urge you beyond your strength and health. I wish only to caution you against carelessness, which I fear has always been a besetting sin with you. If in your conscience you are studying as much as your health will permit with safety, you, yourself, do not, by shame or mortification, be led to go further. Shame, mortification and envy are not incentives to exertion. Do what you can because it is right, and your best interests will be subserved by it. Do not work merely to avoid the shame of having another exceed you.

Write to your grandfather and tell him how you stand and how you are getting on. Fear not that even if you do not take the stand we hope for, you will be less loved by us. No boy who goes through West Point above the middle of his class has cause to be ashamed before the world. But one who has the ability, properly exerted, to graduate in the Engineers and who does not do it will regret for life his failure. Setting this before you as the object of your honorable ambition, not as the subject of mean emulation, envy, and jealousy, strive for the end. Do not be ashamed or discouraged by present ill success. Read the Arabian Nights story of Murad the Unlucky and learn a lesson from him.

Be on good terms with your competitors. Do not let emulation lead to envy or strife. I lived with the best men in my class. Welcker Mason and myself lived in the same rooms for two years. We were good friends then and ever since. We helped each other in our studies with a free and liberal feeling which was far better than the mean jealousy which has to put a wall into the intercourse and rivalry of such a contest. Avoid this. It is not honorable. It is not Christian.

Write to us often and let us know your difficulties, your disappointments, your success. The crown and reward is for him that persists unto the end. Many start fair and fail for want of perseverance. Five years is a long race.[1] He who is steady and never becomes careless will succeed. Look only to the things that are better. Those that are past cannot be undone. But better things may be accomplished, and those overbalance, in the end, and present failures.

Ever your father, Meigs

Autograph letter signed, MCMC, MCMP, Ms. Div., LC.

---

1. The course of study at the academy was extended to five years in 1854 but reverted to four with the onset of the Civil War.

Montgomery C. Meigs Jr. to John Rodgers Meigs

Dear John — Washington, December 15, 1859

As I have nothing to do just now I thought that I would write to you, and so I got a pice of paper from Mr. Steel, which you may see is not the kind of paper that I generally write on. Yesterday or wednesday It snowed very hard and the ground is covered with snow.

I saw a horse and sleigh coming down 14th street just before I came down to school, and it seemed to run very well though there is but about an inch of it. The pond out at the brick-kiln has been frozen over for several days and I have been skating there right often. I had a good deal of fun yesterday evening, and one of the boys of our school has just gone out of the house with his sledge to slide down the hill at Public-lot. My sledge is getting rather week and I broke one of the runners when I was slidining but I am not any worse off than some other boys as there were 7 of them broken yesterday. We~~re~~ are all very well here excepting Louly and Mary. M~~has~~ary has had a head-ache and Louly a sthomac-ache or something of that kind, but she seems to be very well to-day. I will have go home now. So good bye. Your affectionate brother

Monty

This letter of Monty's he gave me to send last week. You ought to write him a few lines of acknowledgement. Mother

Autograph letter signed, JRMC, MCMP, Ms. Div., LC.

Montgomery C. Meigs to John Rodgers Meigs (Shorthand Transcription)

My dear son: Washington, December 20, 1859

I received today two papers which have given me much anxiety. One is your letter to your mother in which you tell her that you are writing by candlelight after taps; that you have a blanket up to the window, and that there is no great harm or risk in it, as there is no danger of being caught.

You have not forgotten already that the officer in charge entered your room and caught your companion [Robert Catlin] absent. Why may he not do the same when you have a light burning?

But the matter is worse than the danger of being caught. You have signed a promise to obey the regulations. One of them is that you must not have a light after a first signal. Now, it is true that the engagement to obey the regulations does not make it a lie to fail through inadvertence to do all that the regulations require; but when you deliberately and with intent set yourself to do what is prohibited by this promise, do you not violate your solemn engagement and promise? How do you reconcile this to truth?

I hope that you will think carefully of this and never again thus violate a solemn promise, one which you made not under compulsion but of your own free will, one which is a part of the compact and contract and agreement between yourself and the United States Government, which is giving you an education which you can get nowhere else and for which it asks no recompense, no pay, but your obedience to the regulations and your fulfillment of the conditions of your appointment. One of them, you write to your mother, you are deliberately violating.

The other paper was the report for November, by which I see that in this month you have got 13 demerits. This is at the rate of 156 a year, and, added to the 3 which you got in October, makes your demerit for the session thus far 16. 13 is a large number for one month and I fear shows that you are growing much more careless than before.

You say, too, that your room is always filled in recesses by company, and I think, from the style of your letter, with company that you do not find very improving or agreeable. Avoid this. It leads to idleness and to failure in your studies and carelessness and demerit.

If your papers attract more company than is good for you, I think the best thing I can do is to stop sending them to you. While they are instructive, there is no instruction in them that will make amends for a falling off in your conduct or studies.

I fear that your roommate is not a very good companion for you. He goes out after dark and gets caught. Do not let his influence injure your conduct and character. I have always believed you to be able to resist the temptations of such society, having found you inclined to study science and nature more than to frolicking and idleness.

I have seen at West Point very talented young men throw away themselves and all their parents' hopes from pure carelessness and [illegible]. I do not fear that you will do this, but I ask you to think and not to let any evil example induce you to relax for a single moment the resolution with which I believe you entered the Academy, to do your best, to obey its regulations and to endeavor to take its highest rewards.

I have just written a note to Colonel de Russy, asking him to have the proper permission to receive your mother's box forwarded to the Academy. I have just carted up the box, which contains some [illegible], et cetera, for Christmas. I hope that you will enjoy it. I have put in my skates, believing that you will have more pleasure in using them than I shall this winter.[1]

Your mother has spent some time and money and your sister some time in purchasing and making up a set of curtains for you. When you [illegible], it will be better to wait until after January. They are put away. They were to have been sent at this time but will be kept for you, should you need them hereafter.

I do not think this putting up of curtains is a good thing, myself, and when I was a cadet, no such luxury was permitted. It leads to envy and jealousy between those who are lucky enough to be able to buy them and those who are not.

You say your pants are worn out. Have you not enough money to buy another pair upon your account? If not, let me know how much they cost, and I will send you the means if you need it. That is, I will send it to the treasury of the Academy.

Ever your affectionate father, Meigs

Autograph letter signed, MCMC, MCMP, Ms. Div., LC.

---

1. Regulations required the superintendent of the U.S. Military Academy to give permission for a cadet to receive packages from home. In addition to the skates mentioned by Montgomery C. Meigs, the box contained drawing materials, gloves, and a Christmas cake, among other items.

Louisa Rodgers Meigs to John Rodgers Meigs

My dear John, Tuesday, December 20, 1859

I sent you a box this morning and wish you a Merry Christmas and a very happy New Year.

We all think of and wish for you at this Season, when the family circle draws closer together and the loved and absent ones are most tenderly remembered.

I think this is the first Christmas you have spent away from me since you were a little boy and it makes me very sad to think that the time is now come when we are to expect to spend few such anniversaries together. It is pleasant to you I know to look back on those which are past. Those at Rouses Point when we *were all together* and all of you old enough to enjoy the mystery of watching and waiting for the visits of "Santa Claus" recur very vividly to my recollection. But we are all changed since that time.

You have come to man's estate and a Christmas to you spent as the last, without any expectations or prospect for the future, would not afford you enjoyment. If you could come home as Henley does to spend a few weeks only, it would of course be very delightful to you. We shall all think of you on Christmas day and I am sure it is pleasant to you to be so remembered. We shall have the presents arranged as usual. The children are on the *qui vive* and anxiously expecting the day.

I should never have desired to see you enter the Service if I thought you were to belong to the Line and were to lead the idle garrison life which ruins so many young men. I have never thought of anything else than your taking the head of your class. And graduating in the Corps where your father has distinguished himself. This morning at breakfast your father received your report for last month, and I only wish you could have seen his countenance when he read 13 demerits. He handed it to me without a word of comment, but his *silence* was *expressive.* It was passed silently around the table. It is the first time you have had so many demerit. What can be the cause? Are you beginning to relax in your effort? I am afraid the influence of your room mate is bad upon you since you tell me that you had been writing "after taps." Your father and Capt Macomb seem to think that a grave offense, and your father always seems *disappointed* and *mortified* when he finds you guilty of any breach of discipline, or want of success in your studies. I believe that ~~Rowland~~ gets his advantage of you by being more attentive & more persevering than you. I feel unwilling to think for one moment that he is your superior in anything. The son of such a father as his ought not to take precedent of you whose descent and whose associations are such, as particularly favor you. I am sorry to hear you have so much company in your room. Your father thinks it very unfavorable for you. I wish that your companion was a different sort of person. It will certainly injure your standing if you continue to disobey orders and accustom yourself to think it of no consequence. I hope you will never again write to me after hours. Your father says he does not recollect having done it whilst at the Point. You might do it many times without being found out, but if you *were* it would at the least be exceedingly mortifying to you. Perhaps it is something of the kind which gives you so many demerit this month. We had a letter from your Uncle Gus this morning. He inquires about you and wants to know how you are getting on. He says he knows you are doing "*well of course*" but he expects to hear of a brilliant success from you. You see how much the whole family expect of you. I say nothing of your father and myself but I beg you at this time to think of him more than of any one else and do nothing to mortify him, but rather do all that you can to justify his fatherly pride. Major Lee a friend of your father's had two sons who graduated at the head of their class at West Point. We dined at the President's a short time since. I had not seen him before since he gave you the appointment. I thanked him for it and told him that I hoped you would *do credit* to his appointure as I ~~hoped~~ thought you were aiming for the head of your class. He said he was glad to hear you were doing so well, hoped you would be successful etc.

My eyes are getting weary so I must stop. Heaven bless you my dear boy and make you all that our heart can desire is the prayer of your Mother. I hope you will be pleased with the box. I did not put so many eatables as there were some other things to go in the box. Your father sent you his skates and hopes you will have some enjoyment with them. The cake I think is a very nice one. It cost me $2. The candy too is excellent. The apples are the last of a barrel which was sent your father and are unusually fine. I should have sent a larger box but it costs so much to send by express. That cost me $1.50 besides hack hire. I had gone to the expense of buying your curtains before I received your last letter, but I shall not send them until you know whether you want them. I hope you will change your roommate in Jany. Your Aunt Kate left us yesterday for Detroit.

Your Mother

Do not open your box until Christmas Eve or Christmas Morning so that we may all enjoy our Christmas at the same time. Ever Yrs.

Autograph letter signed, LRMC, MCMP, Ms. Div., LC.

Montgomery C. Meigs to John Rodgers Meigs (Shorthand Transcription)

My dear son: Washington, December 26, 1859

The children have had a good time over Christmas. The presents were placed upon the table on Saturday night, and your mother, as usual, was busy in making ready for their pleasure until 2 ½ A.M.

Loulie waked me up early in the morning with her stocking of goodies, which she insisted that I should taste. After breakfast, they were admitted to the treasures of the table in the front parlor.

We went to church and spent all day quietly, except for the noise of the children at home.

Captain Macomb's son John has arrived to spend his Christmas holidays with his father. He seems to be a good boy, very manly, of good address.

I have this morning read a letter from my mother in which she asks whether your mother can make room for a visit from Harry Meigs, your cousin. He seems not to be well, and they want to send him away from home for a few days. We are, of course, inviting him and will make room for him, either in the spare room with John Macomb or in the little room with Monty.

I do not know whether I have written to you that I have had the water pressures engine at the Rock Creek Bridge at work for some weeks past. It pumps the water up to the top of the hill of Georgetown and gives a very satisfactory supply of water to the people who live above the aqueduct level. It is a curious thing to see this machine so quietly push forward and send water into the pipes which terminate in the chambers and bedrooms of the houses on the top of the hill, 200 feet above the machine. I am much pleased with it.

I hope that you got your box safely and enjoyed it, and that you do not, in fact, have any doubt that you do not allow yourself to be led away by bad example to take any part in the revelry which you describe as common on Christmas Day at the Academy. Keep before you always the fact that you are receiving the bounty of the government, that with an interest of others, ready to pledge themselves, as you have done, to obey the regulations, made for the better preparing of boys and young men for the army; and they having taken this pledge, it is in the nature of a contract, in which it is dishonest to avoid doing your part and, moreover, it is a violation of your pledged word to wilfully neglect to obey any or all conditions of your promise.

With love from all at home, your father, M. C. Meigs

Autograph letter signed, MCMC, MCMP, Ms. Div., LC.

John Rodgers Meigs to Louisa Rodgers Meigs

My dear Mother, West Point, Tuesday after Christmas, December 27, 1859

I am very much obliged to you for the care and the kindness and trouble you took for me. Though I did not spend as merry a Christmas as the last it was a very pleasant one in some things. I had a good opportunity to think about home. On such days there are always little things happening that remind you of what you did the year before. I thought much about the little center table in the parlor where I supposed you were all gathered in the morning, of Loulie embracing her new dolls and Monty Macomb capering about and squealing out his delight.

I got my box on Saturday. That cake is delicious, it came uninjured. I am going to take part of it over into Rowland's room this evening. Thank Grandmother for that book it is just what I wanted, it is very nicely written and then the matter and the engravings are very interesting to me. The candy, almonds and raisins, were the first I had tasted since leaving home.

The match boxes from Mary & Aunt Nannie were very convenient, and much needed. We only had the paper ones before. I am much obliged to Aunt Jerusha for her paper Knife, she always remembered me at Christmas when I was at home.

Father's present's I shall often use and always think of him; he knows how I value them. I set to work with them immediately. I have made several likenesses of Cat and some others of my friends, and some of them really do look like them. I was unlucky in being on guard both Sunday and yesterday.

On Sunday the chapel was beautifully dressed with evergreens, hanging in graceful festoons over the pulpit and winding up the pillars. There was a large cross and star over the pulpit but no motto as there used to be at St. John's.

The choir was very much improved by the addition of some female voices, some of Col Delafield's daughters Miss Hardee and some others sang. They sang a beautiful hymn commencing

> "Shout the glad tidings exultingly sing
> Jerusalem triumps Messiah is King"[1]

I heard it two years ago at Mr Pyne's and have always wanted to hear it since.

Yesterday we had *turkey* and *mashed potatoes* for dinner which wonderfully good fare put all the cadets into good humour, though I confess it did not set me crazy with joy especially as I was told that the plebes always bore the expense of the Christmas and New Year dinners which cost ~~them individually~~ every one of them $4.00 extra. At this rate one turkey and a dish of mashed potatoes costs over 5 dollars, which I thought a high price, don't you? The French's kindly invited me to dine with them, but I did not find it convenient. Rowland went but I have [not] heard his report yet. I expect he had a nice time.

The reason I have not got my pantaloons yet is because it is against the rules to allow the plebes two pair of pants before January, not because I have not enough money.

I am very much ashamed of my stupidity about those curtains. I wrote to you not to make them because I thought that perhaps if Catlin should be found I would room with Chapman who has curtains, and who was then ~~destitute of~~ without a roommate. He has one now however, besides this Catlin is almost certain of getting through so that I am certain of wanting those curtains.

My stupidity has prevented me from having them *now* however and has made you think that your trouble was taken for nothing.

I asked you not to make them, because I thought that you certainly had not commenced them so soon after I wrote and that by rooming with Chapman, which would have been disagreeable, I could save you the trouble and expense entirely.

I am very sorry that I will not be able to send you a report in return for your presents placing me at the head of the class. I lost 1/10 last week in Math and 2/10 in English. The latter counts less in the general standing than the former. Rowland lost nothing in either and Michie nothing in Math.

Ever your affct Son J R Meigs

Autograph letter signed, LRMC, MCMP, Ms. Div., LC.

---

1. Verses and refrain written by the American Protestant Episcopal clergyman William Augustus Muhlenberg (1796–1877) in 1826 and set to a tune by the English composer and organist Charles Avison (ca. 1710–70). The hymn was published in William Augustus Muhlenberg and Henry U. Onderdonck, *Hymns of the Protestant Episcopal Church in the United States* (1827).

CHAPTER 2

The First Two Years at the Academy

Following the Christmas holidays, John Rodgers Meigs continued his studies at the academy. The new year 1860 brought frequent letters from his father, his mother, and his sister Mary. Each wrote with increasing concern over his class standing and his number of demerits. Of further worry to the family was Montgomery's departure for the Dry Tortugas in October 1860 at the order of Secretary of War John B. Floyd, who wanted the elder Meigs out of Washington.

Lighter moments are revealed by the letters. Mary wrote about Hattie Coyle—"Miss Hot Cakes"—an admirer of John. And Louisa instructed John on personal hygiene and the necessity of having a clean handkerchief, especially in the company of young ladies. She wanted to hear that he had stopped biting his nails and spooning tea. But most important, he was to make the family proud of his efforts at West Point.

National events became personal for the family. At West Point, John's fellow cadets spoke of politics and disunion. Louisa reported that all conversations at home focused on secession. Montgomery wrote that he would not be surprised if a southern confederacy or the state of Florida would ask the United States for the Dry Tortugas and Fort Jefferson, a fort he had no means to defend.

By April and the beginning of the war, practically everything changed for the family. Montgomery came back to Washington from the Dry Tortugas. Charged by Lincoln with the relief of Fort Pickens, upon his return he became an important figure in the war effort. John Rodgers Meigs came home on furlough in June 1861. He fought in the First Battle of Bull Run and became something of a hero to his fellow cadets.

Montgomery C. Meigs Jr. and Montgomery C. Meigs to John Rodgers Meigs

Dear John Washington, January 5, 1860

I got your letter on Monday and was very much pleased with it, especially with the pictures. I could not make out what you ment about that boy whether he had the staff shot out of his hands or not. It has been very cold here and the ice is about 4 inches thick. I have just come up from the canal where the ice is very good, only that it has snow on it an inch thick that fell yesterday. I suppose, that you know that Harry Meigs is staying at our house, and has been staying here a week yesterday.

We went down to the Capitol the other day, I forget what day it was, and he seemed very much pleased particularly with the staircases. We went to the Smithsonian too and he said that he thought it was the most beautiful that he had seen, on the outside. We play chess every day and he has taught me how to play back-gammon too. I got 15 presents. 10 of them were books, 1 was a gold pencil that Aunt Annie gave me. Montie Macomb gave me an ivory pen-holder but I broke it by letting it fall. We are all very well now. Nothing more now.

Your affectionate brother Monty

At school, 1/4 of 1 o clock

My dear John:

I send you Monty's letter and add a postscript to say that I hear from the Engineer Department that you are 3 in English as well as in mathematics and that I am pleased with this, as I believe you would not have got so good a standing in English without much attention and study.

From your affectionate father, Meigs[1]

Autograph letter signed, JRMC, MCMP, Ms. Div., LC.

1. MCM postscript is a shorthand transcription.

Montgomery C. Meigs to John Rodgers Meigs (Shorthand Transcription)

My dear son: Washington City, January 7, 1860

I suppose you are through with your examination by this time, and I hope that you have passed satisfactory to yourself. We are all much bound up in your success at the Academy, and we do not permit ourselves of a doubt that you will do all that is in your power not to disappoint our hopes.

The Almighty gives us all talents for which we are responsible. We are not responsible for those which we have not been endowed with. Industry and care we are bound to give, to improve to a good use those which He has given us, and we are bound by our own hopes of success to endeavor, by great industry and toil, to make up any deficiencies in our mental and moral organization. It is within the power of every man to be true, industrious, honest, faithful, to fulfill the obligation of his position. To fail in this is to own fault.

Of course, but one can be at the head of his class. To come next or third or fourth or fifth is a success. Two things are to be sought at West Point. In the first place, knowledge that you will acquire. In the second place, the place of an officer of engineers. This cannot be obtained of certainty unless you get a position as head of the class. Sometimes several members of a class graduate into the Corps.

It is not from mere gratification of vanity, of pride in my child, that I desire you, therefore, to take the highest place possible. In the peculiar studies for which West Point is famous, as they study your tests and talents, I have no doubt that you will be prevalent. But in the study of English, the moral and mental science of literature, for which you have less taste, you will be less likely to take the highest place, and to take it, you will be obliged to labor with more diligence and with less pleasure than upon the scientific studies. Do not let yourself think this unimportant. To an engineer, a good style of writing his reports, forcible reasoning, a presentation of his views in good language, are almost as important as to a lawyer. An engineer is called upon to furnish the grounds of argument before courts in many cases, to make reports upon which depend the construction of great works involving millions of dollars and his own success. To him who is best at these things will be the prize of money and reputation, an implement which brings both.

Be careful of your language. Do not use slang terms. I find you never write a letter home in which you do not use the word "haved." This is slang. Avoid it. It begets a habit of inelegance and laziness of language. "Punished," "reported"—the same very good English word will express your meaning much more elegantly. Cultivate a taste from reading good books, history and letters. Avoid novels and tales which teach nothing but baseness and vice, which give false views of life and manners.

Keep up your Latin and Greek. They are of the greatest use in making your English pure. Moreover, the books that you study in these languages are full of references to the history of the published opinions of antiquity, to their manners, arts, literature, and religion. No man is an elegant or polished member of society without a good knowledge of this so fixed in his mind as to be a part of him. They influence his conversation at home and abroad and his writings, both on business and pleasure.

Study the languages which are taught at the Academy, and endeavor with your roommates to practice them so as to make them well. If you will make it a rule, for instance, to speak French or Spanish when they are taught, during a part of the day in your room, you will soon, by practice, be able to speak with ease and fluency.

At the same time that you thus attend to your studies, remember that it is of great importance to make it a rule to do everything thoroughly, well, to leave nothing half-done. Fencing, boxing, gymnastics—all these things should be well done. It is a characteristic of great men that they do all that they undertake with enthusiasm, put their whole hearts and minds into it, and thus they succeed. This life is full of competition, and only those who are strong can hope to outrun their competitors.

While they are yet young, you have now competitors who are much older than yourself. But remember that with each year, the value of this difference in age becomes less. A difference of two years at 40 is not such advantage as at 17.

I think you are disposed to be satisfied with partly doing a thing and that you are tempted to leave a lesson when you know it pretty well. I remember that I could never get you to learn a chapter perfunctorily. You always needed prompting. This habit I tried to correct, but I did not succeed. You are older now. The warning may be effectual. It ought to be so.

I send you a dollar, not, as you request, to purchase a chair. As this is only the returning to you of money you sent to me, I venture to do it. I would not send you a present of money, against regulations, but should send it to your credit at the treasury.

Ever your affectionate father, Meigs

Autograph letter signed, MCMC, MCMP, Ms. Div., LC.

Montgomery C. Meigs to John Rodgers Meigs (Shorthand Transcription)

My dear son: Washington, January 9, 1860

I have your letter to your mother this morning, giving your standing in mathematics, and I also have a letter from Major Kendrick, who was kind enough to write to me and give your standing in both mathematics and English. He says you have good men both above and below you to contend with and that I ought, under the circumstances, to be satisfied. He takes an interest in you, you see.

As I have lately written to you very fully upon the reasons for which I desire you to take the best possible standing, I will say no more upon that subject now. No father has a right to complain of his son for standing as well as you now do. It is a standing of which any father would be proud, and I and your mother are gratified and heartily congratulate you at your success. But the position in the Engineer Corps which I suppose you desire, and which I certainly desire you to have within your choice when you graduate, is not yet secured. It is a long race to the end of the fifth year, and in that time, you know, many accidents will happen to those who do not keep up their attention and determination.

I depend upon you not to lose any ground. You're very younger in mind than in years as yet. Some people develop early. This has been the case with Rowland. I have read one of his letters, and it would do credit to Washington Irving. He is undoubtedly a very talented author. But I presume that for mathematics, you have just as good a head as he has; and if you can take a stand in English so high as 4th, you have no reason to fear him in the end. Your standing in English is better than I expected from you. To English you ought to pay particular attention, for the reason that it is a study for which you have the least natural aptitude.

It will therefore be necessary to watch yourself carefully in this and the like branches. Mathematics and scientific studies are to your taste, and you will not find difficulty in getting knowledge upon them. You will find a difficulty in so learning each day's lesson as to be perfect in your recitation. Thus, keeping up your marks, you show them your stand in mathematics, which is according to your marks. No doubt each of the three who take on the highest place is well acquainted with his course, and the only criterion therefore left for the settlement of the standing has been the marks and the daily recitation.

Your friends were all proud to hear of your success. Your father and mother and Aunt Nannie will think that you might have, perhaps with more attention to daily work, done better.

God bless you, my son.

Ever your affectionate father, M.C. Meigs

Autograph letter signed, MCMC, MCMP, Ms. Div., LC.

Montgomery C. Meigs to John Rodgers Meigs (Shorthand Transcription)

My dear John: Washington, January 26, 1860

I have not written to you lately. I have been otherwise occupied by what you have seen in the newspapers.

In the first place, I received notice that the Secretary of War had directed that I should be sent to the Tortugas, giving up my employment upon the aqueduct, taking a place there which is not of sufficient importance to give me the kind of mental and bodily work which I have had for the last 7 years. Indeed, the fort at the Tortugas was begun, laid out, and left above water by a 2nd lieutenant to build it, to play the part a good master mechanic would play.

I succeeded in getting the Superintendent's attention to this matter, and I trust that I shall not be deprived of the honor of finishing the aqueduct, as I have been, apparently, of completing the Capitol.

You have seen by the papers which I have sent to you that the people have presented to me a tea set of silver, a very pleasant proof that I have not failed to do my duty towards them as well as towards the United States. Your mother is very pleased with the silver as well as with the compliment to me.

Your uncle Henry Meigs, from Columbus, arrived here on Saturday evening. He looks sad, poor fellow. While I do not see as much trouble at the state of political affairs, he thinks that there is great danger of a separation between the North and South. For my part, I believe that when the politicians have done talking, the wise people will speak out and show that they do not believe in the necessity of so foolish an act as a separation. In the meantime, I intend to endeavor to do my duty and to make myself ready to be a useful member of society, whether the two parts of the country keep together or not. We must all live, and there must be some government. Not one so good for the whole as the one under which we now live can possibly be constructed again if this falls apart from the folly of its statesmen. Oh, for another of your Washington, of Jefferson, of Madison, of Webster, or of any of the good, strong men who lived in former days. I see those who now figure on the stage, and I regret to say that among very few of them do I see any signs of greater wisdom, virtue or intellect than among ordinary men whom I meet in my business.

Your business is to acquire all the knowledge you can at West Point. Prepare for a useful part in life. It may be that trouble will arise in this country, that the two parts will separate, though I do not believe it. If they do, there will be employment for all the military knowledge of the country, for the inevitable result will be a civil war. Two such nations cannot separate and remain apart and in such close contact without differences arising which must end in bloodshed. Fortifications along the frontier, then extending inland for thousands of miles, now thought unnecessary, will give employment to engineers. Battle and [illegible] will employ the soldiers and officers of the nations. Fear, murder, robbery, and sudden death will be the fate of the mothers and sisters of the people. Could they avert such a fate?

I took your uncle up to the large culvert number 12 and showed him the bridges of the aqueduct. I send you a photograph of the appearance of Rock Creek bridge. You see that the centers are down and the bridge shows its true appearance, though the abutments are not yet built. Show this to Major Kendrick.

You see a description of the bridge which I wrote and which was published in the Philadelphia and Washington papers. It must be going the rounds. I have seen it in a Boston paper, which I send you.

I have not lately sent you papers, as you said that they brought many visitors into your room. This, I know, must interfere with the proper occupation of your time. I therefore ceased to send them, to have thus put an end to the nuisance. I now send you some, as I wish you to know what is going on in the country and in the world.

I wish you also, by practice, to learn to correct your own very bad style of writing, which can be done only by reading and studying good authors. When I was a boy, besides my lessons in school, I used to read a passage from a good writer and then, closing the book, write out all that I could remember of it, as nearly as possible in the words of the writer. By comparing what I had written to the original, I corrected and detected faults in my own style. That is a good practice and is much easier for you than it was for me. I had not the facility which is given by knowledge and vision. And it took much more labor and more paper to write them than it does now, with the aid of vision.

We are all well. Your Uncle Henry went to Philadelphia this morning. I hope his visit to me did him good. All unite in love to you and look to you to do us credit.

As you grow older, I doubt not that you will dominate the differences between you and the two above you in the class. Try to gain a good style of expression, both in speaking and writing. It is very important. I know a man of fine talents, formerly one of the most promising officers of engineers, who resigned and has now, after 40 years, gone back to a condition not as good as he began with upon one of the fortifications. I believe that the want of the power of expressing himself in good language was the principal cause of his failure. A good mathematician, he ought to have been a good and skillful civil engineer.

Ever your affectionate father, Meigs

Autograph letter signed, MCMC, MCMP, Ms. Div., LC.

Montgomery C. Meigs to John Rodgers Meigs (Shorthand Transcription)

My dear son: Washington, February 4, 1860

I am sorry to see by your letter of the __ that you have allowed your competitors in the class to beat you in marks even in a study such as geometry, in which you say that you are so well prepared that you have not work enough to keep you from idle hour.

Your marks give you the third rank in both English and mathematics. This is, I must think, the consequence of your carelessness, against which I have so often cautioned you—[against] being satisfied with partly and not perfectly knowing a thing.

You do not date your letter, I see. Do you not know the day of the month?

Your mother sent to the express office last night a package of curtains, etc., for which I enclose an express receipt of freight and lading thereto. I have written to Colonel Delafield to ask him to permit you to receive the package, telling him what it contains.

I saw Colonel Crozet last evening. He came to see us before going to New Orleans, for which city he set out this morning.

We have had very cold weather here at Grafton Cottage, Professor Page's country seat, 10° below 0. Our bathroom has been frozen up for several days, which is very inconvenient.

We are all well. I have been very busy, which has kept me from writing to you as often as I wished to. I have not so much work to do as before, but still I have things to attend to which take up my thoughts and attention and so completely occupy me that I have not been able to spare the few moments of leisure to write to you. Colonel Crozet tells us that you are looking very fat. Fat as an oyster, was his expression.

Your mother will write a few lines to you, to go with this.

I spent a few hours in Philadelphia the other day, Monday last. Your grandfather and grandmother are looking well. You have not yet, I believe, answered your grandfather's letter. You should not have left it so long unanswered.

Do your best, my dear son, to fulfill the hopes of the many persons who take a warm interest in your success at West Point and guaranteeing your success in life. Write frequently, and remember we are all looking to you.

Ever your affectionate father

Autograph letter signed, MCMC, MCMP, Ms. Div., LC.

Louisa Rodgers Meigs to John Rodgers Meigs

My dear John Thursday, February 9, 1860

I cannot let this anniversary of your birthday pass by without writing you a few lines to tell you that I have been thinking of you. To day, and of the time when I just saw you 18 years ago. It seems very difficult to me to realise that so many years have passed by since that eventful period, and that I have actually a son, so far advanced to manhood. The years are passing rapidly away for both of us. You are advancing up the steep hill of life. I am *descending* down the vale slowly but surely and yet it seems but a very few years since that I was young & fresh as you are. I was scarce 25 years old when you were born and yet I felt myself of very mature years, & I remember that I actually *blushed* at the inappropriateness of the expression when the Doctor spoke of the likelihood of your being a strong and *vigorous baby,* from the fact of your having a "*young* and healthy Mother." I thought myself quite advanced in life and that the Doctor's calling me *young* was quite a reflexion on my years.

We advance so rapidly from one stage of life to another that it all seems like a dream. In a few more years you will have arrived at all the dignities and privileges of manhood and the battle of life for you will commence. You must put on your Christian Armor and go forth into the strife.

You are as yet on the threshold of life, but you must polish your armor and brighten your weapons for the Contest. Do not let all your thoughts be directed to making you a good soldier for this world's warfare but remember that you once promised to become a soldier of Christ. It is a great comfort to me in my separation from you, knowing the temptations which beset young men, to remember that you have this obligation resting upon you, for I feel that it will prove a protection against many a snare and temptation. My dear boy I pray that God may so conduct you through this life that in the world to come you may obtain the life Everlasting.

Friday Morning

Dear John—your father has just called for my letter—he is going to the post office and I have no time to write you a longer letter as I wished to do this morning.

Ever your loving Mother

I hope you received your package containing the curtains. I have not heard from you to announce their arrival.

Autograph letter signed, LRMC, MCMP, Ms. Div., LC.

Mary Montgomery Meigs to John Rodgers Meigs

My dear John — Home, February 10, 1860

I received your letter on Wednesday, and was very much relieved on the receipt of it, having begun to fear that you, like my New York correspondent had forgotten me. . . .

Mary [Blagden] told me this morning that she really believes Hattie Coyle is *in love* with you. . . . although she has the greatest quantity of beaux she cares for none of them and has the stupidest time when talking to her most devoted admirers. She very seldom dares ask about you but never sees Mary without inquiring most particularly about *my* health etc. Now all the world knows she never cares a straw about *your* sister if anyone happens to mention your name she starts and blushes, and turns from her own partner or the person she is talking with, and listens for every word. Mary told me so sometime past but is more certain of it every time she sees Miss Hot Cakes. . . .

Yesterday you were 18. wern't you? We all intended to write you a Round Robin but someone came in and prevented it. . . .

Goodby ever your affectionate sister.

Mary M. Meigs

Autograph letter signed, JRMC, MCMP, Ms. Div., LC.

Montgomery C. Meigs to John Rodgers Meigs (Shorthand Transcription)

My dear John: Washington, February 24, 1860

I have received your letter to your mother, with one to me, in which you tell us that you have not been able to get permission to receive the curtains. I thought when you first asked for them that it was rather strange that the cadets were allowed to receive such things; but upon your statement that your inspecting officer approved, your mother bought the material and sent them to you.

When I was a cadet, the use of curtains was not allowed. I can see good reasons against them and none for them. In the first place, they encourage a violation of the regulations as to lights after taps, and in the next, they are expensive, mere luxuries, not necessities, and I think the fewer luxuries a soldier during his youth and education is supplied with, the better for him and the country he is taught to serve. Then, if supplied to some and not to others, it introduces a distinction of money and wealth among the boys, who should be taught that they are in all things, except talent, morality and industry, in which God makes them to differ, alike. I should, however, have preferred to see you desiring the comforts only of your convenience and not desiring to have anything better or more luxuriant than the poorest of them, with economy, can afford to buy out of his regular allowance. While I did not like to interpose, your mother, with the natural feeling of a mother for the comfort of her child, wished to send you the curtains. I cannot say that I feel much regret, therefore, if you, for now, have to practice the self-denial which I should have preferred to see you take willingly upon yourself.

The reasons of Colonel Delafield are satisfactory to him, I suppose. He is responsible for his own action, and I have no desire to quarrel with him about so small a matter; nor is it for your interest to succeed, if you could, by my aid in getting permission to receive these things against his wish. He is a friend of both of you and myself, and it would be an unfriendly act upon my part to endeavor, by permission of the Secretary of War, to overrule his opinion in this matter. So let the curtains stand. And do not trouble yourself with reasonings about the sufficiency of his reasons. They may be good and sufficient or bad and insufficient. That is nothing to you. In either case, he declines, for reasons satisfactory to himself, to let you have the curtains. If you never have any greater trial of discipline in your military career than this one, you will have an easy time of it.

On the 22nd we had an inauguration of the statue of Washington by Clark Mills. There was a great parade, thousands of people, and the militia from New York. A very well-drilled and good-looking body of men they were, to come on and march in the procession. The President dedicated the statue, Mr. Bocock made an oration, and the whole celebration in honor of Washington cost, I suppose, those engaged in it and the Congress of the United States not less than $40,000. This, in honor of Washington, is a small sum compared with the benefit of any celebration which will increase and keep alive the reverence which all Americans should feel for his honored name and example. But I could not help regretting that so much money could not be applied to the production of a better statue than the one which was set up day before yesterday. There is in *Harper's Weekly* a woodcut copy of the statue which is better than the statue itself. In the woodcut copy, the neck of the horse is unnaturally low. In the statue, it is out of all proportion. I would not give one of the plaster casts, which I sent to the Academy by Crawford, for the whole bronze statue. As a work of art, the least interesting of these casts is far above and beyond the bronze statue by Mills.

You have not told me how they look and how they are arranged in the hall of the Military Academy. I wish you would report on this. I believe I have asked you before.

Ever your affectionate father, Meigs

Autograph letter signed, MCMC, MCMP, Ms. Div., LC.

Louisa Rodgers Meigs to John Rodgers Meigs

My dear John — Saturday, March 10, 1860

. . . Mary looks anxiously over your father's mail in the morning, and was quite disgusted this morning when she found there was no letter from you. She says she thinks you are not doing as well as when you first went to W Point for you write much less frequently and get more demerit.

I was very sorry to find you had so many demerit last month. You are aware that from whatever cause they may proceed they will help to determine your standing which ought to make you very careful to try and have as few as possible. Mrs Roland came to see me yesterday. She seems much pleased with Tom's success and said she was quite unprepared to find him taking a place above you. She says he has always been a hard student, laborious and persevering, and to that she attributes his success. They have never considered him particularly talented at home, I believe and therefore are the more surprised and delighted to find him at the head of his class. . . .

If you are not to get your curtains you might as well desire the Quartermaster to return them to me. I can find some private opportunity of sending you the small articles, or I can send the collars by mail. They will only be rotted by keeping them. I will ask your father however to send you directions on this subject. I suppose Mary will write you to day.

Ever afft your Mother

Autograph letter signed, LRMC, MCMP, Ms. Div., LC.

Montgomery C. Meigs to John Rodgers Meigs (Shorthand Transcription)

My dear John: Washington, April 3, 1860

I send you the report for the month of January. I am sorry to see that you have been getting so many demerits. This demerit certainly you could avoid by care, and knowing as you do the importance of escaping them, you ought not to subject yourself to this injury. I have so much confidence in you that I do not doubt the subjects of these reports are small matters or what in your eyes appear to be small matters, but they count against you.

I hear from persons that have been at West Point that you have a good reputation among the officers and professors as a promising youth who can stand higher if he will be industrious and attentive.

I think you ought to come out head of your class, with the knowledge with which you start and the aid of the advice of your father, who has himself been through the institution and who is therefore able to point out to you dangers which injured him.

We are all well. The weather is warm as spring. I have been riding over to the insane asylum today.

Ever your father, Meigs

Autograph letter signed, MCMC, MCMP, Ms. Div., LC.

Montgomery C. Meigs to John Rodgers Meigs (Shorthand Transcription)

My dear son: Washington, April 7, 1860

I have sent to you from the Engineer Department copies of document numbers 20.29.22. printed by the Senate, all of which relate to the affairs of the public works under my charge and therefore will be of interest to you. Put them away, for hereafter they will be of still greater importance and interest as you are obliged, in your own course through life, to stand up for the right to favor a duty against power and authority.

It is difficult to do what is right in this position and at the same time keep within the bounds of military obedience and respect. I think I have done so. I have certainly shown by strong opposition that I did not much really respect the authority which I believe has been prostituted to bad purposes as I have kept within the technical bounds. So far, a court has not been called upon me.

We are all well again here. Monty Macomb has had the scarlet fever, but no other person in the family has thus far taken it, and Monty seems to be well again, though he has as yet kept in his room, for fear of taking cold. The weather grows warm. I have been all of 3 times lately up the aqueduct. It is now a very pleasant ride.

Having mounted your last 3 stereoscopic descriptive problems, I took them last night to the [Saturday] club at Mr. Peale's. Mr. Peale and Mr. Taylor were much pleased with them.

Captain Franklin, however, says: "Don't let John get into extra duty drawing for Major Delafield. It is a thing which will interfere with his studies and will equal more than his military duty or training." In this last, you know you are deficient for not being at the Academy during the encampment. Therefore, you should not do anything which will interfere with it.

If you draw 3 of these and give them to Professor Church, you will have done enough. It is, to be sure, good practice of the art or science of descriptive geometry, but if you practice it to the injury of other parts of the course, you will do wrong.

I fear sometimes that I send you too many newspapers, so as to interfere with the proper attention to your studies. If so, say so. Do not yield to temptation.

Ever your affectionate father, Meigs

Autograph letter signed, MCMC, MCMP, Ms. Div., LC.

Mary Montgomery Meigs to John Rodgers Meigs

Dear John, May 4, 1860

The enclosed as you may perceive are 7. collars and two stocks. I suppose you have given up all hope of ever getting them, but mother has been so much discouraged by the treatment of her last bundle that she did not feel very much inclined to make any more. She intended to write to you today but is now in bed suffering from a sick head-ache. You ought to write to *her* oftener than you do.

Father received from the Dept a copy of your monthly report. You are 1. Mathematics and 7 in English, how on earth did you get way down there? 7th why John what is the matter, you don't seem to be able to keep up in both your studies, when up in one you are down in the other, just like uneven scales. Besides you have 15 demerit, and 36 since Jan. Why can't you have as few as Rowland and Michie?

Father is very much displeased for he knows it is your own fault, and you ought to be ashamed really. You sit in your own room and draw those stereoscopes which to be sure are very pretty, but good lessons are much prettier. At first you [illegible] you had not the military way of reciting but there is no excuse now having been there almost eight months.

Please dont get anymore, for I have told every one you were third in all your lessons and when the June reports are published to have them all see you way down out of the stars will be rather hard for their faith in my word.

It would not require so very many more to turn you out remember that.

Before you left you used to laugh when anyone told you to be careful, there was not much use in saying anything about it to you if you don't choose to mind *what they say,* but being always as obstinate as a mule and just as headstrong there is no use in writing anymore on the subject.

I only hope you will try for your fathers sake if not for your own. He will not be satisfied till you are at the head in all the classes and have been without demerit long enough to have a *leave.* We are all well except mother. Capt Macomb has not yet returned.

yours Mary M. Meigs

Autograph letter signed, JRMC, MCMP, Ms. Div., LC.

Montgomery C. Meigs to John Rodgers Meigs (Shorthand Transcription)

My dear John: Washington, May 6, 1860

In your last letter to your mother, you say that Mary is the only person who writes to you frequently now. I think if you will refresh your memory, you will find that you do not write to your father quite as often as he writes to you. Do not get out of the habit of letting us hear from you. You know, I believe, that we are always anxious to hear of your well-being.

I was much gratified, and so was your mother, that the last monthly report placed you first in mathematics, but our pleasure was a little dashed at seeing that in English study, which I fear you neglect—and I have never been able to persuade you of its importance—you were only 7th. This was the report for the month of March.

Here we have summer heat suddenly upon us. I yesterday, Saturday, took Mr. Frederic Graff, the son of the engineer who built the Fairmount Water Works in Philadelphia, and himself, after his father's death, for a long time chief engineer of the works—he remained in charge of them until political influence was brought to bear upon them, and then he was turned out—I took him with me to see the Cabin John Bridge.

I met him in the street, looking at the Jackson statue with a view-meter, and, recognizing an old friend, spoke to him. He said that he had taken a view of the iron bridge over Rock Creek in the morning, and I invited hin to come to the club, which met here last night, and to go with me at 3 P.M. to see Cabin John. We rode out, a very pleasant ride in the country, beautiful, in the bursting green of the spring. He took his camera, one of [King's?], with him; but when he came, he found that one of the uncertainties of photography had got him. The glass was out of place, jolted out in the carriage, and the attempt was a failure.

He has been to Europe and made an extensive tour among the water works, bridges, and works of engineering. The Cabin John he has nowhere seen surpassed.

I have lately been several times to Baltimore when I am consulted in regard to their water works. I believe that I shall save them, by my advice, some $30,000. So the knowledge gained by so many hours of study and work is not altogether wasted, though the present government seems to place little value upon it.

I saw Colonel Crozet here a few days since he has returned from West Point. He did not see you, which I regret. It seems that they met at West Point and the 2nd day of their meeting received an order from the Secretary of War dissolving the board [appointed to revise the West Point instruction program,] of which he was a member. So all the labor and thought given to the revision of the course is thrown away.

We are all well. Monty and I took a walk today after dinner. Starting out at 4 P.M., we walked out on Bladgen Road until we struck the Piney Branch, and down that branch and Rock Creek to the Paper Mill Bridge, and so home. It makes, as you know, quite a long walk for a hot afternoon. We returned home at 7 ½, having, by pedometer, accomplished 8 ¾ miles. Monty is a good walker. He walked on Saturday up to the [illegible]. He grows stronger, gets his lessons well, and is a good boy. Your sister you will hardly know. She has grown to be a young woman and no longer a child.

I have sent you all the papers about the prizefight, for I do not doubt that a young man will find as much interest in them as all the old men of my acquaintance seem to and indeed as I, myself, have found in it.

I think that such an exhibition of courage and endurance is worth knowing, reading of, and indeed is worth something to the world, now perhaps no more luxurous than in the days of the Roman and Greek heroes. It must have been a luxurious people who built the Parthenon. Such refinement is not attended by a people engrossed only with severe study and occupation.

I have lately been looking over the plates in a small edition of *The Antiquities of Attica*. Of them all, [illegible] have the best and most tasteful architecture.

Ever your father, Meigs

Autograph letter signed, MCMC, MCMP, Ms. Div., LC.

Montgomery C. Meigs to John Rodgers Meigs (Shorthand Transcription)

My dear son: Washington, June 1, 1860

This day of examination begins. I pray God to give you such coolness and readiness as will enable you to do justice to the knowledge which I do not doubt you have obtained of the truths of your course of study. Should you come out in this important examination at the head of your class, it will give great pleasure and gratification to me, your mother, your grandmother, and to all those kind friends who look with so much interest to your success, first in the military academy and next in the career you will have to run in life.

I lost much standing when at West Point, at a single examination, by finding myself forgetful of a demonstration which I thought I knew. I have always since doubted whether that demonstration was not in fact an obscure one, not so clearly demonstrated as to give a clear idea of its truth and bearing. However, be that as it may, it was part of our course and I ought to have been able to demonstrate it as satisfactorily as it was in the book. But whether it was because I had not a clear understanding of it and whether it was pure embarrassment and forgetfulness, I do not know. I missed it at the blackboard in the examination hall, though I then went through a long and severe examination by questions upon the course generally, in which I did not hesitate nor buckle at a single question, for I understood and knew them all thoroughly. It put me down several files in my class. This was at the last general examination. I recovered most of these files at the June examination but not all of them. This may have been one reason for my losing a whole year's promotion.

I speak of these things to show you how important is a perfect familiarity with the whole course and over the most insignificant demonstration of a problem in [illegible].

We have, as your mother no doubt has informed you, been much interested in the foreign visitors who have lately been among us. They leave in a short time now. Though pagans and idolaters, they, by good manners and civility, patience and modesty, have won a friendly feeling from all that approached them. It is said that they will visit the military academy. If so, you will, of course, have the opportunity to see them yourself.

I have sent you many newspapers and a drawing of them which are faithful enough to give a good idea of their appearance. You, yourself, will recognize all the scenes in which they are represented, the President's house, etc.

I sent to them a portfolio filled with over 100 drawings of the aqueduct, Capitol, Post Office, etc., with which they were much pleased, saying that it should go to the tycoon, their great emperor.

I send you today a copy of the report of the committee of the House of Representatives, the committee on expenditures in the War Department, upon the action of the Secretary and the War Department upon the contracts for marble columns. I send you this that you may see that this committee, directed by vote of the House of Representatives to examine into this mess, declares that I have done my duty.

There will be at West Point presently a young man from North Carolina, a Mr. Craige, that is a son of the Honorable Burton Craige of the House of Representatives, from North Carolina. Mr. Craige has been a good friend of mine for some years past. He was at one time chairman of the Committee on

Public Building and supported me throughout. He said he wished to introduce his son to me before he went to West Point. I did not see him, however. He is, I believe, from what his father tells me, an honorable young fellow of [illegible] stature and one who will be properly advised in the beginning, be kept out of bad ways, and may do credit to his father and friends. You may assist in this and thus help to repay the obligations under which I stand to his father. Find him out and tell him who you are and that I requested you to make his acquaintance. Advise him. Keep him out of the scrapes of tormented plebes as much as you can.

I hope that you have too much good sense and too much good feeling and dignity, however, to join in this vulgar tormenting of plebes. It is an amusement for which I had no taste and from which I suffered no inconvenience when I was at West Point. I used to hear of it, but I do not think I ever saw any of it, except the passing from the positions of plebes to see whether, when on duty as sentinels, they would do their duty.

Wishing you every success and that God may have you in His holy keeping,

I am ever, your father, Meigs

Autograph letter signed, MCMC, MCMP, Ms. Div., LC.

On Sunday, June 24, 1860, Montgomery C. Meigs wrote in his diary: "Father I am very thankful to the Almighty."[1] He wrote this in gratitude for the fact that John was first in his class. More good news was forthcoming. On August 29, 1860, the board of medical officers reexamined John and determined that his medical condition did not prevent him from carrying out his responsibilities. Therefore, he was fully qualified to be a cadet. Of concern, however, was the ever increasing number of demerits. By July 1860, he had 46 demerits. At this rate he would have 138 in six months. One hundred in the same period would make him deficient in conduct. Montgomery noted this fact on a copy of the conduct report in September 1860. By the end of August, John had 16 more demerits. Of further concern to the elder Meigs was his own continuing disagreements with Secretary of War John B. Floyd. On September 20, 1860, Floyd removed Meigs from his work on the aqueduct and ordered him to duty in the Dry Tortugas.

1. MCMJ, MCMP, Ms. Div., LC.

Louisa Rodgers Meigs to John Rodgers Meigs

My dear John, Saturday, September 1, 1860

I sent you by Mrs. Roland a few days since some collars and pocket handkerchiefs and a few toilet articles that I thought you might not be able to procure at the Point. The myrrh is so excellent for a tooth and gum wash that you ought never to be without it. A few drops on your brush or in the water you use for your teeth is all that you require to use at a time. It is an excellent purifier of the mouth as is also the C soap.

The bottle *marked* Cologne is not cologne, but Lavender water from a larger bottle of some that I am using. It is a favorite perfume of mine. I thought when you were going to visit the young ladies you might like to have something of the kind.

The handkerchief I hope you will not lose. They are not particularly fine but very nice, and you should never go to call on any one, particularly ladies without having a fresh one in your pocket. Perhaps you have grown more attentive in these small particulars. You know how I used to have to *watch* you before you left me.

Who is to be your room mate this year, from something that Mary said in her last letter I inferred that Michie was to be the individual. I hope whoever he may be that he will prove an orderly good & studious companion & that you may mutually stimulate one another to renewed exertions for the highest honors of your class.

I suppose by this time you have struck your tents and bid adieu to encampment life for this year. I suppose it is rather difficult to get to work again, but I do most earnestly hope that you will determine to lose neither time nor opportunity of advancing yourself to the head of your class. You will of course see Mrs. Roland when she goes to the Point. I wish I could enjoy the same privilege that she does, but it is useless for me to think of it.

Your father will be glad to go and see you if he can but you must not rely too much on such a possibility. As his affairs stand at present he can make no plans for any thing. He is quite well and his spirits good as usual.

We have now some enjoyment of domestic comfort and I have more of your fathers society than I have done since we were at Rouse's Point. Loulie is the only member of the family who shares our society so that in the evenings we are entirely alone. Your father brings his drawings in the afternoon to a table in my room and I sometimes read to him whilst he is engaged upon his maps or plans. It has been a long time I think more than seven years since I had that pleasure before. Today he has gone up the line on horseback and will not be at home for dinner.

Capt Macomb is now engaged on light house duty. He is sometimes here for a few days & then goes off again on his travels. Your Aunt Nannie is still at Bailey's with the children your gd mother Aunt Kate and Jerusha. Your Aunt Nannie will probably go North to make a visit with her children to Capt Macomb's sister at Geneva, (Mrs Rose). Capt M paid a visit to Hammt last week. He offered to bring Mary and Monty home with him, but they were not quite ready to come. I think they will probably return next week or possibly the week after. I am rather at a loss what to do with Mary next year. I can scarcely realize that she is 17. She has a great deal yet to learn. I have determined not to send her to Miss Brooke again, but she ought to study for another year *at least* before she considers herself a *young lady.* I believe she is not very anxious to emerge from the chrysalis state, into the full grown *butterfly,* but I suppose if she once begins to enjoy *society* there will be an end of study. She enjoys flirtation as well as most people, though in a quiet way. She had quite an attachment I think for Mr HL but as he has outlived his sentiment for her, I hope she has done the same.

I must close this for I am taxing my eyes and yours with these crossed lines.

Ever affy Your Mother

Autograph letter signed, LRMC, MCMP, Ms. Div., LC.

Montgomery C. Meigs to John Rodgers Meigs (Shorthand Transcription)

My dear son: Washington, September 1, 1860

I was much pleased with the drawings of the problem of descriptive geometry. The lines are neatly and accurately drawn, so that the relief when looked at with the i's crossed is perfect. The idea is a good one, and you have worked it out well. I am glad you showed it to Professor Church, for it is a good thing, and he may have it carried [illegible] if he publishes an edition of the descriptive geometry.

I showed it to Mr. Schaeffer, Professor Henry, Captain Franklin, and Mr. Peale at the club last night, and they were much pleased with it. Mr. Taylor has got many stereoscopic drawings of lines and solids, and he was much interested with this one.

I took Senator Brown of Mississippi up to the Cabin John Bridge on Friday, and, on Saturday, Mr. Craige of North Carolina. He was chairman of the Committee on Public Building when I first came to Washington and supported me in those times when I needed support in the House of Representatives. They were both much pleased. Senator Brown is chairman of the District Committee in the Senate now and promises to do all that he can for the aqueduct. He only went as high as Cabin John Bridge, not being able to afford time to go farther. Mr. Craige went all the way up to the falls.

We took our lunches at the spring, using the bottle which we had there at that time to draw water in. The spring is clear as ever.

I have been very busy lately. Congress somehow seems to be the cause of work to all in Washington. And though I have not the Capitol and public building and have no money for the aqueduct, it seems to me that I am as much occupied as ever. I have not the same assistance that I used to have and have to do what is now to be done myself.

Colonel Crozet is going to start tomorrow morning for West Point and has promised to take a bundle for you. I send this letter by him and a small bundle of books which you have had before, I believe, but which you may find of interest to look into again.

Homer I hope that you will keep up a knowledge of. I have always regarded that I had not got far enough in Greek and Latin to be able to read with such ease as to keep up my knowledge of these languages. The little that I did know, going at 16 to West Point, after a single term only at college, has been of great use to me. You had advice [illegible] and can easily keep up your knowledge. A little reading every day will do this.

I also send you a [fun?] magazine. The *Spectator* you ought to read frequently. It is a model of good English; and nothing is more important for an engineer, a soldier, and a gentleman than a good knowledge or skill in using his own language. Officers and engineers are always being obliged to write reports that are extensively published, and upon the style of their reports depends very often their reputation and success in life and in the projects they have to advocate.

I must stop to go to church. Your mother is going off in a hurry, leaving me behind.

Ever your affectionate father, Meigs

Autograph letter signed, MCMC, MCMP, Ms. Div., LC.

John Rodgers Meigs to Louisa Rodgers Meigs

My dear Mother — West Point, September 14, 1860

Father has at last paid me his long expected visit. When I came here a year ago I was comforted by the thought that he would be frequently going on to N.Y. and could then come, but when Capt Franklin took the Capitol that hope was lost. Then your last courier made me hope for him in the last of August but as the time went by and he did not come I commenced to be afraid that I should not see him until furlough. Yesterday morning the officer of the day came up to my room and told me that Father was in the guard house waiting to see me. I hurried over there and found him in the Commandant's office where I did not have much chance to say any thing to him. Col Delafield excused me from recitation that I might have the chance to see him. If he had not done so I would only have had one half hour to spend with him. We went up to Fort Put and over the plain. His visit has done me a great deal of good, it has given me fresh energy to study hard when I saw how glad he was that I stood high. He introduced me to General Scott at the same time telling him that I was head in mathematics. The general remarked that "that was fine indeed."

It made me think how perfectly unselfish a parent is, to see his gratification. After sacrificing their own comfort for years to give it to their children and to educate them well, they seem to think themselves in some degree rewarded if they do well after leaving them. Father looks very well and in such good spirits. It made me happy to see his calm strong confidence in the result of this miserable contest. I don't think I shall ever be able to feel so if I am placed in such a position.

Gleaning nearly all my information upon the subject from the papers I have often felt anxious not knowing what they might be able to do with him there in W, but I don't think I shall feel so again for he told me, just how the case stood, so that no matter what the sec & President do it will be the better for him. General Scott told him yesterday that he was an "honor to his country" and that he took frequent occasion to say so.

Those two hours that he spent here will be much thought of by me until I see him again. I have to hurry with this to get it off.

Your affct Son J R Meigs

Autograph letter signed, LRMC, MCMP, Ms. Div., LC.

John Rodgers Meigs to Montgomery C. Meigs

My dear Father, West Point, September 22, 1860

I was startled yesterday by seeing in the papers that you had been again ordered to the Tortugas and to day I got a letter from Mary confirming it.

I hope the Sec has now put his last black stroke to his piece of villany and that Congress will not let him rest long in impunity.

It must take up the matter at an early date in self defense. I cannot see but that the Sec has been guilty of positive and wilful disobedience of *lawful* authority and for it I hope he will suffer. The papers say that capt Benham will hereafter make all disbursements on account of the aqueduct. What does this mean surely not that they are going to prosecute the work and spend the money Congress appropriated, without your superintendence. Unless they build it out of their own private funds I should think the disbursements must be very small indeed.

I have been over to the library to see what sort of a place Garden Key is. William's Florida 1837 says "The Tortugas are eleven in number, situated upon a coral reef which is about 15 miles in extent. It is about 90 miles west of Cape Sable 65 west of Key West and 100 miles NW from Havanna. Of that group a few of the largest are about 3 feet above the common tides. 7 of them are covered with mangrove bushes and verdant lignum vitae the smaller ones with herbs and grass. The SW Key though one of the smallest is most important to be known because it ends the chain of Florida Keys. A reef of coral rock extends a quarter of a mile SW from it. North of this is a long sandy Key. Under the lee of this there is good anchorage a quarter of a mile from shore. The best harbor is near Rush Key; this is sheltered from the sea by a large reef of rocks and a flat shoal and is quite smooth even in a gale. There is 18 feet water close to the bank. The channels among these keys abound with fish and turtle, but they afford neither fresh water nor wood except small bushes and are important only as a harbor. 5 or 6 miles west of Tortugas is a large bank of coral rock and white patches of sand." This is a pretty place to send talent, that has been entrusted by the Congress of the nation with the expenditure of millions of dollars, to kill turtles and fish, and be prevented from all communication with civilization. I should imagine that there can be very little communication between such a place as the book describes this to be and the main land. Congress however can easily require you as a witness ~~though I suppose~~ and bring you back from it. I have not taken that check to the quartermaster and shall not do so unless you tell me to again, for I can get along without the money, and you will probably want all you can get at this time.

I trust for the honor of Congress that they will not permit you [to] stay there long. I received ~~your~~ the report of my conduct. Remember there is no danger, for I do not expect to get as many demerit in a month in barracks as in a week in camp. Do not think that I am going to be careless on this account. I have 40 demerit to get before I have a hundred and can easily keep from doing it.

[Unsigned]

Autograph letter unsigned, MCMC, MCMP, Ms. Div., LC.

John Rodgers Meigs to Louisa Rodgers Meigs

My dear Mother — West Point, October 5, 1860

From the tone of your letter Father's return in a few months does not seem to be as certain as I thought it. What does Col Davis say about the order? Does he not feel sure that Congress will bring up the matter early?

A stay in Florida for the winter only will do Father unmixed good and give him a recess that he must have been wanting for years, but if he is to stay there during the summer it would be a different matter.

I am glad that you were not wholly unprepared for such an order. Father told me when he was here that such an order was not at all unlikely, but that anything they could do to him now would do him good for it would hasten the *denouement* and make it redound more to his credit when it came.

When is he going to start and will he go by sea or land?

I am rather sorry that Mary is going to a select school, for my stay at Mr Monroe's disgusted me with them. It is not fair to judge of all however from one and I hope that her teacher will be much more efficient than Mr M was.

Henley Smith was sent to a select school in Virginia you remember and he did not think much of them either.

Monty will do well at College I guess. And if he only studies hard this Winter and gets into a class a little above him rather than below him he will be able to enter the College itself next year. I expect that as good an education can be got there as at most colleges in the country and it is much nearer home. That long walk will do him good. My legs are better developed than any other part of my body and I believe it is principally on account of the 4 years that I walked to College.

[Unsigned]

Autograph letter unsigned, LRMC, MCMP, Ms. Div., LC.

Montgomery C. Meigs to John Rodgers Meigs (Shorthand Transcription)

My dear son: Washington City, October 9, 1860

I expect to leave this place for the South about the middle of the month. I hope to spend a few days with your Uncle Henry at Columbus.

I think that I will return before the winter is over, but I take with me books and the means of employing myself usefully, even should I be obliged to remain longer.

There is much of interest to one of scientific tastes in the new scenes of the Gulf of Mexico. The animals, the waters, the coral reefs, the shipping, are all new to me.

Be careful of your demerit, the only part of your discipline at West Point about which I have any anxiety. I know you will get knowledge and come out of the institution prepared to serve your country and repay the obligations of your education.

With fondest love, I am your father, Meigs

Autograph letter signed, MCMC, MCMP, Ms. Div., LC.

John Rodgers Meigs to Louisa Rodgers Meigs

My dear Mother, West Point, N.Y., October 29, 1860

An order of the Secs was read before the battalion this evening in relation to the Washington Acqueduct, appointing a board of officers to examine into the quality of the work already performed and estimate the cost of completing it. This shows a truly great confidence in the abilities of Capt Benham, is he not able to judge of the quality of work and estimate the cost of the completion. I dont suppose he is but I should not have thought the Sec would have published his disability in such a manner. He seems to be fond of boards of officers as he is continually forming them for testing guns and all such things. I hope they wont waste much money until Father comes back to finish it. Do tell Mary to write to tell me something about him. I only know from her last letter that he intended to go last Monday and was in good spirits. I do not even know whether he went all the way by sea or not, or what he took with him any thing about him in fact. I suppose dear Mother that you are very busy now with the household and on that account and because of your eyes I do not expect many letters from you. Still when they do come they are most thankfully received and carefully read. I feel a pain mingled with the pleasure however when I know what they cost you. I often think of you all at home and try to imagine to myself what you and Father are doing. My photographs help me in this and are often looked at. I like to go after supper to walk along flirtation and hum to myself scraps of the songs I have heard you and Mary sing. There is something pleasant too in sitting on the rocks about Lees Point to watch the vessels gliding along with the tide. One can just discern the sails of the more distant ones through the dusk and hear the beating of the pumps or the sound of the sailors voices. Flirtation is a great place in such delightful weather as this.

Many of the cadets are interested in politics and talk fiercely about disunion. Some of them have even gone so far as to write home for instructions in case of that event. The readiness with which the ten million treasury loan was taken up the other day did not seem to indicate that the bankers had great apprehensions of a dissolution. Riggs bid for over 2,000,000 $ of it. Do you not think that Sec Cobb ought to be hung as a traitor for the way he has been acting. They say he has made preparation for carrying the Government treasure into the Southern States at the first note of dissolution. I must go to my lessons.

Ever Your Affectionate Son John R. Meigs

Autograph letter signed, LRMC, MCMP, Ms. Div., LC.

Louisa Rodgers Meigs to John Rodgers Meigs

My dear John, Monday, November 12, 1860

This report was handed me at the breakfast table this morning, and was silently passed from one to another around the circle.

I cannot tell you how *shocked* and mortified I was to find the number of your demerit. Capt Macomb said he thought there must be some mistake and that he would take the report around to the Dept and make inquiries about it. There had been an omission of the name and we hoped some other report had been substituted for yours. He returned however a short time since, saying that he was very sorry but that the report was correct—76 demerit! He says there is only *one* in the class who has more, and he thinks it will be well to warn you that 24 more in three months will procure your *dismissal.* For the month of Sept. you have *14,* and for this month of October and November perhaps as many more.

I can scarcely tell you how pained I feel, at this intelligence. I feel it deeply enough but it disturbs me to think of your father's mortification. He has so much of a father's pride in your success, and will be humiliated to learn that you are falling so far short of his expectations. He was never very anxious for you to go into the service, and it was only on account of my oft repeated wish and your known desire to go to W. Point that he consented to use his influence for you.

You know how distasteful it was, with what *loathing* even that he consented to ask the appointment of this Administration, even compelling himself, to ask the Secy of War to withdraw his opposition. Is this to be his reward? Is he to see you dismissed or if not, is he to see you take a standing so low, in your class, that it will always be a humiliation to his pride to see you in the Army. I think your desire and expectation is to enter in the Engineer Corps. This of course you cannot do. You will have to go into the infantry regiments, which will place you out on some frontier post with no other occupation but drilling soldiers.

You will say perhaps these demerit are all for *trifles,* that may be, but that will not make the matter any better for you next January.

Your demerit has increased instead of lessening since last year, and your standing will of course be lower. You must have grown more indifferent. My dear boy, *think of what you are doing.* Your father and myself and all your friends have formed such high expectations for you. Are we to be so cruelly disappointed? Think with what a different feeling you will come home next summer if instead of coming triumpfull in success, you come rather in disgrace and defeat.

You have it in your power to amend, if you set diligently to work. You know I used to tell you that your small omissions would give you trouble at W. Point. You used to reply that when there was any strong necessity for a change you could do better. Capt Macomb says the most demerit he ever got in *a year* was *50.* A copy of this report was sent to your father from the Dept. He will probably not *say much* to you on the subject. It is not his way to do that. He thinks you are old enough and have intelligence enough to decide for yourself as to the course you intend to pursue. Your failure to do well will alienate his *love* and his *respect* and that I think would be hard for you to bear. I am sorry that he should have this anxiety added to his other troubles.

For the month of October November December, 8 demerit each month *will dismiss* you. You may *already* have that number. You seem to me to place all your reliance on your success in Math. Your standing in English is lower than it was last year, but I will say no more, but that I am much distressed and alarmed at the prospect. It would be crushing to your father to have you dismissed or graduate at the tail of your class, and such an expectation will not be a very cheering [illegible] to have in his exile.

Your aff Mother

Autograph letter signed, LRMC, MCMP, Ms. Div., LC.

Montgomery C. Meigs to John Rodgers Meigs

My dear son John Fort Jefferson, Florida, November 17, 1860

I have established myself here & am making myself as comfortable as I can while I feel so unsettled. There is much of interest forcing itself on my attention though I have thus far been so much occupied that I can not take time to look up the curiosities around me. Such odd animals a wonderful variety of crabs: stem crabs, spirit crabs, fiddler crabs, soldier crabs, crabs which scarce move, a crab which darts with the speed of spiders, crabs which live on the vertical face of a wall & jump like birds from one perch to another.

These keys are full of hermit crabs fellows with bright red & purple nippers with painted legs but with leather covered bodies & tails. Nature [illegible] will [illegible] & protected by a hard shell all their prehensil & locomotive organs has left it to their own prudence to protect their bodies so they ~~borrow~~ take up with the cast off shell of a neighbor, into this, they insert their tails choosing one which will receive every part of them except the ends of their nippers. They walk about actively, high on the tips of their toes, climb a tree or a bush or a window blind. It requires a smooth sided box to serve as a pen for them a bucket they walk out of as easily as a roach could.

I got a pocket full of them on the beach this evening most of them small occupying shells about from the size of a cherry stone to that of a plum.

They not only climbed out of my jacket but walked all over my coat & back before I got home.

One whose shell had got broken & for which I brought home a new shell as different as possible in form from that he had been living in. [He] readily accepted the new home. He examined it carefully while a crowd of his companions looked on & when satisfied the new shell was better than the old, he slipped out his tail from the broken one & slipped into the other to the great disappointment of the friends apparently waiting for a bite of the exposed tail. The vacant shell was the object of attentive examination to all the little fellows who no doubt hoped to find some meat left in it.

The spirit crab is of the color of the sand runs like a witch a short distance then stops short. So sudden are the starts & stops that it is almost impossible to keep sight of him. He comes and goes like a spirit & is well named. But this same spirit has two black eyes like glass beads sharp & shining which are set up on stems so that he can see all round & his ghostly nippers cut sharp into whatever they bite.

It takes an active boy to catch one on a smooth ocean beach.

I can see every day a dozen pelicans fishing. They sail over the shallow water, and every one drops at the same instant making a splash & scaring the fish which fleeing from one get into the reach of another. Strange instinct. It is a rapacious bird. The fireman here declares that one appealed to him for help & surgical attendance. He came up to him, examined his pouch, which was torn by the spine of a fish, extracted that fish, sewed up the rent pouch & the pelican appeared to understand. It became domesticated staid about until he got well again & finally departed when able to take care of himself.

There is always a tame one here. They are far more familiar than turkies or chickens. The present pet sits by the black boy who cleans the fish for breakfast & asks for his share fastidiously. I have killed a scorpion which is common. It is just like the picture on the map of the heavens. Its sting however is said to be about as bad as a bees only.

The schooner *The Tortugas* is a fine sailer, a good sailboat with a comfortable cabin & a trip to & from Key West over this beautiful phosphorescent sea is very pleasant.

The waves splash away in great maps of light.

Fort Jefferson I must describe to you some other time. You dress in summer clothes white pantaloons & find it just about pleasant.

Ever your affec father M C Meigs

Autograph letter signed, MCMC, MCMP, Ms. Div., LC.

Louisa Rodgers Meigs to John Rodgers Meigs

My dear John — Tuesday, November 20, 1860

I received your letter in answer to mine a few days since and although it has somewhat relieved my anxiety, it leaves a disappointed feeling in my mind of which I cannot rid myself.

To know and feel that you have the talent to take such a position as your father and I and all your friends could take pride and pleasure in, and to know, that your demerit for trifles will place you below men of inferior abilities and opportunities mortifies me very much. Your friends have all thought so well, so highly of you, that it is of course a disappointment to see you fail of their expectation. All those who know you are aware that you have not received so many demerit for *moral delinquencies,* but they also feel that such carelessness is of course very culpable. Capt Wright called to see me yesterday evening. I spoke of you and your standing at W. Point and regretted that you had received so many demerit lately. He said he had observed with regret that you had quite a large number, "76 I think" said he showing that he had examined the reports carefully.

He said it was a great pity if a young man had the mental ability to get though W. Point that he should fail to take a high rank, from mere inattention to trifles. He thought it a just and proper system of discipline. He evidently thought there was not much excuse for carelessness and inattention to duties. He said it was well worth trying for the head of a class, for even if a man were fortunate to graduate in the Engineers, the difference of position in that would make many years difference in his rank and promotion as an officer.

I cannot see why Roland should be your superior officer. Of the others I know nothing personally! I suppose R has scarcely received one demerit for the past term.

If you permit yourself to think lightly of so much demerit as you seem to do, in spite of the rank which your lessons would give you, you cannot fail to graduate low, for as you permit yourself to violate the regulations so often in trifles, you will cease to be troubled by them after a time, and you will be content just to escape dismissal. If you had not your *talent* for *Mathematics* to rely on, you would be at present in a position where I should be ashamed to see you. Major Lee and all the officers from W. Point, to whom I mention the subject, seem fully satisfied with the system of discipline at the Point, and say it is as fair and just as any system can be. They seem to think inattention to *trifles,* no *trifling* fault. Major Lee the other night talking of W Point, said he had no pity or sympathy for those whose abilities gave them preeminence but who would not exert themselfes, to conform to the rules and regulations of the Academy. He said he had a feeling of pity for those who had by hard and patient and toilsome labor to work their way through W Point. You know some such in your class, I dare say.

You think it is not just that we should expect *mediocrity* from you. If your father had expected that only from you, I do not think he would have applied for your appointment. If you do not graduate in the Engineer Corps I should be very sorry to have you an officer of the Army. I do not think your peculiar talents and tastes would find a legitimate sphere of action in any other Corps.

Those young men who are sent to those military outposts, lead a very idle dissipated life generally. I should be sorry to have my son subjected to such associations. Your father has pleasure I know in finding that your tastes assimilate so well with his own in many particulars, and hopes to see you following his own profession as a scientific man.

My dear boy you will think perhaps this is tiresome, but I must admonish you to reflect on your course. You think it tedious and wearying I dare say to be so constantly tied down to a system of formalities, but it is worth the trial, to get *no demerit.* If you fail in that endeavor, you will certainly come through with far less than you have been lately getting.

Your Cousin Raymond came to see us a little while ago. He is now stationed at Annapolis and has the second in command there. He says the system has been very lax at the Academy, but he is introducing a reformation. I have heard others say that he is a thorough disciplinarian, and that he is working a great change among the Midshipmen. He says there is no excuse for inattention to trifles nothing is trifling that infringes a regulation.

The Midshipmen now visit at the houses of the officers very freely. Your Cousin Eliza has had several little entertainments lately. Fanny and Bessy Higbee are there and make her house quite gay and pleasant.

I have had several letters from your father lately. He has arrived at the Tortugas and seems to be enjoying the change of scene and climate very much. He expects to have plenty of hunting and fishing. His letter from the Tortugas is very voluminous or I would send it to you. Mary wrote to you on Sunday and gave you all the news I suppose.

The political state of the country fills every mind with some anxiety. "Secession" is the subject of every one's conversation.

I think that the Union is in great danger, but I trust the Divine Wisdom will find some means that we do not know of to deliver us from the evils which threaten us.

I must hasten to a conclusion, for I have already written another letter and my eyes are giving out. Let me hear from you soon.

Ever your aff Mother

I suspect much of your demerit arises, from the want of being "up to time." You are not ready at the sound of the drum. You perhaps allow yourself to sleep in the morning a minute too late, and then have to rush wildly down, without being in proper trim for inspection. This *certainly* can be remedied.

Autograph letter signed, LRMC, MCMP, Ms. Div., LC.

Montgomery C. Meigs to John Rodgers Meigs (Shorthand Transcription)

Fort Jefferson, Tortugas, Florida, November 25, 1860

My dear John: Lat. 24 37 N Long 82 52 W

I have not heard from home for two weeks or more, because the last steamer coming in from New Orleans brought me no mail, but tomorrow I shall send up the schooner to Key West to get the mail, which I expect the *Isabel* steamer to bring from Charleston.

I had no idea of the delightful climate which I find here at this season. The thermometer keeps between 75 and 80°, the only exception being during the cold of [illegible], by which we have had one blowing for nearly 2 days. This is a strong wind, with a clear sky and flying clouds, and cold, with no rains such as storms in the north have with them. A single shower of rains for a half hour at the beginning of this storm is all that we have had. The thermometer has been 68° during the cold. The sea is white with wave crests which break all over the harbor.

I went fishing today on *Two Suns.* We caught in the schooner some 12 or 13 fish which would make about 70 pounds all together. Fine fishing here. Some of the fish here are beautiful. The kingfish which we caught is the first of the season, a single fish about 31 inches in length, a fellow which can leap 20 feet, and 6 feet in height from the water. Their jumps are like those of the deer.

Some fellow gave me a Sanibel's Island bait made of mother of pearl for this fish expressly. It was bound upon a silk guitar string covered with wire, and the kingfish which took hold of it walked away with it, cutting the guitar string with his front teeth.

I have just read a service and sermon. Last night I married a couple according to the forms of the Episcopalian Church. I am governor of the islands, United States Commissioner, commanding officer, etc., and thus all these things become my duty. I had a good congregation this morning.

Last night we had got the company assembling to the wedding, and after the wedding there was a party in which several damsels behaved with all the grace and eloquence of your sisters.

There are about 25 slaves here. All come from some [illegible], two of broad [illegible], both black as the ace of spades.

You would enjoy being here for some time. Such strange animals, [illegible] fish which look like lobsters, etc., etc.

Write to me and let me know how you get on with your studies. I am lonely and need letters from home, from those I love.

We have here now a United States steamer, "Mohawk," which has been in the harbor for a week past.

Ever your affectionate father, Meigs

Autograph letter signed, MCMC, MCMP, Ms. Div., LC.

Sometime after December 8, 1860, John Rodgers Meigs could report to his father that he had the highest marks in his class. He also improved his proficiency in the use of weapons. On December 15, 1860, the younger Meigs was ordered to join the first section of the fourth class in the use of small arms.

Montgomery C. Meigs to John Rodgers Meigs
(Shorthand Transcription)

My dear John: Fort Jefferson, December 17, 1860

I hope you will be enjoying a merry Christmas when this reaches you and that you will be free from all anxiety in regard to your position at West Point.

I am oppressed with a fear that the demerit which you were getting so often may have ruined you and me by losing you the place at West Point for which I worked so hard and which is of so much importance to you. When you lose this, you are a beggar.

I am not able to give you capital to set up in business. I have as much as I can do to support and educate your brother and sisters and keep your mother in comfort, upon my pay.

It is not a light matter to lose an education such as you can get nowhere else and the opportunity to adopt a profession which, though it does not make men rich, gives them position and honor and the opportunity to be useful to their fellow men and of good service to their country.

Here we are well, though we feel anxious about the condition of affairs in the country. These forts upon the Gulf are not armed and properly manned, and I should not be surprised, in the excited state of public feeling in the Gulf states, at a demand, from some Southern Confederacy or from the State of Florida, to give up to the state authorities this base. They have not the means of keeping up such an expensive station if they take it, but I have no means of defending it. I have no mission to expect mechanics and Negroes who are not hired to be shot at. I have no guns and no powder or other ammunition.

It is a position to be held by a great nation to exercise police over commerce of the South, and it should be garrisoned by a thousand men to build with warlike swords.

Remember me to Professor Kendrick and to all friends.

Ever your affectionate father, Meigs

Autograph letter signed, MCMC, MCMP, Ms. Div., LC.

Montgomery C. Meigs to John Rodgers Meigs
(Shorthand Transcription)

My dear son: Key West, Florida, December 27, 1860

I have been to Havana, having taken advantage of the schooner's coming up here for the mail, which arrives tomorrow, and gone over to Cuba on my way up here.

I enjoyed my visit to this first really foreign town I have seen. Everything there is so new and strange to me—the people, the way of living, the houses, the streets, the climate, the trees, the fortifications, the soldiers, the police. There is nothing like it in the United States.

The houses have thick walls, wires or grates but no glass in the windows. The doors are large enough to drive a carriage into, and the carriages are kept in the front halls of the houses. They are generally only 2 stories high, built around courts which have pillars and arches to support their interior piazzas. The style is very much like that of the houses in Pompei which you have seen pictured in the descriptions of the excavations by Gell and others.

Bright colors adorn the walls inside and out of houses which often seem built of crude stone or plastered; and the surfaces of the walls, both inside and out, are painted, with the exterior in bright blue, reds, yellows, etc. All this makes the streets very gay. They are like the pictures of Canaletto of Venice. The Moorish style of building shrine Alhambra, the mosques, is the same general style as that employed in Havana, at least so far as the arrangements and plans of the houses goes.

The streets are full of soldiers, and the town is surrounded by forts and by a rampart, ditch, and [illegible], having a scarp and counterscarp wall built of masonry. So there are many forts occupying all the positions of importance, so that the harbor is perfectly commanded from many points and by many guns.

10 or 12,000 soldiers and police keep this city in order. It has about 180,000 inhabitants, I am told.

Rows of palms decorate and shade the public walks and drives. Many handsome villas belong to the government, used for the residence of the governor and kept generally for the generals commanding the artillery, engineers, etc., etc. All these, too, appear in Venetian or Moorish style. Arches rest upon pillars and columns. The whole scene is like a panorama of gorgeous pictures.

I hope you are prepared to pass a good examination, and I hope that you have escaped the penalty of demerit. God preserve you from this misfortune. All [illegible] enable you to preserve yourself.

I have not heard for 2 weeks from the United States, and I have no news from home at this distance to give you. I am well and try to do my duty as contentedly as I can, hoping soon again to see all my dear loved ones at home. Tomorrow I expect letters from home.

Ever your affectionate father, Meigs

Autograph letter signed, MCMC, MCMP, Ms. Div., LC.

Mary Montgomery Meigs to John Rodgers Meigs

My dear brother, Home, December 30, 1860

You will hear before this is received of old Floyd's resignation, is'nt it the best thing in the world, to think of that old scamp being "*out.*" Oh' dear! it is too good, it must bring father home very soon, then what will become of us all go north or south? which will you go for? with father or exercise your own mature judgement and decide for yourself?

I had a great mind to ask Capt Franklin yesterday what he thought of the good news, but my bashfulness restrained me, maybe he dont think it good, he has always got along so easily with him. He cant be a friend of that nasty old hog!

I have had my picture taken on a "carte de visite," as they are called. Mollie is going to send you one with a letter as soon as they arrive. I have only seen the negatives but flatter myself that they will be quite good looking.

Why hav'nt you written an account of your proceedings on Christmas day? We got your report yesterday. You have three demerit for Nov and 92. since the commencement of the academic half year My dear old brother I hope you will have mercy upon our feelings and never as long as you live give us such a scare as we have suffered from this winter. Linus Cunningham joined me in the street the other day and said he had been at the point and sent word to you but was obliged to leave before he expected. He gave a glowing discription of your attainments received from the Proffs. and cadets. Do astonish them more by coming out some month without a demerit can't you? Sue heard Mr C. talking about you and as soon as he had gone said that is good I knew John would do well. I always had a great respect for him, because he is so talented and *good!*

Hattie Coyle told me to remember her very kindly to you and say she wanted very much to see you.

Mollie and aunt Nannie send much love.

Breakfast is ready and I must hurry down. Goodby Happy and Many New Years to you

Yours untruly Mary

Autograph letter signed, JRMC, MCMP, Ms. Div., LC.

It is January 1861. It is the advent of the Civil War, and John Rodgers Meigs is at West Point, his father is at Fort Jefferson, and his mother, sisters, and brother are in Washington, D.C. In February, Montgomery would return home; by the middle of the year, both Montgomery and John would have the opportunity to serve their country.

Mary Montgomery Meigs to John Rodgers Meigs

My dear brother, Home, January 6, 1861

You have of course heard that the southern states are seizing the forts. As far as we know Fort Jefferson has not as yet been attacked, but Genl Totten told mother that father was expected to stay where he was and defend the station even tho' he had not force, or be *disgraced.* It seems to me that the Genl. must be of a most fool hardy temper if he advises an officer to risk his life and that of his men, in such an attempt as this would be. Genl Scott is now here, uncle John saw him this morning. He said Capt Meigs has not been forgotten, reinforcements have been sent to Ft. Jefferson, and he is to defend the fort if attacked. Mother has written very violent letters to father, but as the mail from the north is being opened in S Carolina she is trying to get them returned if they have not already been sent south; they may do much harm as she abuses the sect. and whole concern in no very measured terms. Genl Scott has requested that all the Military companies in the city shall be obliged to keep their muskets and 40 cartridges in their houses to be ready at a moments notice. The National Rifles have sworn to do thus but to keep the peace here on the 4th. of March.

Think of father's being placed in such a position as that now held by Mj. Anderson when Florida goes out he will be—can he be called upon to shed the first blood in civil war? God grant that it may not be!

It makes me so angry to hear people talk as Fanny Bailey does saying Oh there is not going to be any war and laugh at the very idea, if they heard the southerners talk and knew as much of the feeling there and among army officers, they might change their opinions.

Uncle John and others have gone so far as to engage carriages to take their families from town on the 3d. Aunt Annie and Willie are going out into the country. I suppose we shall stay here. I hope so, for I dont like the idea of running away from danger. FB. says this will be the safest place on that day as there will be a sufficient number of US troops to keep order, and with Genl Scott as commander, I think against such forces the Wide-awakes will fight to little advantage. Mr B will take care that this place be well protected as his precious old head would otherwise be in great danger. Mr Lincoln will without doubt be inaugurated and safely installed in the White house, but what must be his feelings at taking the reins of such a government into his hand, a broken and divided union and empty treasury, and above all he is hated by half the people he means to protect with a hatred which cannot be described. I wish we were all all together at the north again. I hope you did not get demerited for that supper, you can't afford to run such risks and in such a time as this you ought not to let mother have more to bear than what is now weighing upon her mind.

I have to be careful, you may not have long to stay at the Point but while you are there *do* credit to your father by your conduct. We are all well. Mollie goes home tomorrow.

Yours Mary

Autograph letter signed, JRMC, MCMP, Ms. Div., LC.

John Rodgers Meigs to Louisa Rodgers Meigs

My dear Mother, West Point, January 9, 1861

Mary's letter rather startled me by showing how serious matters have become in Washington.

We never realize any of these great national evils until they affect us personally. I wish you had engaged carriages to take you out of the city on the 3d of March. I suppose that [there] can be no danger of a serious row on the day of the inauguration, for Genl Scott has shown his usual wisdom in taking all necessary precautions. It is said that 3 companies of flying artillery have been ordered on already, one left this place this morning. All the militia in Virginia could not do anything against them assisted by a little cavalry. They could dash along from one end of the avenue to the other at a gallop drop into battery and sweep the street while the militia are moving a couple of blocks. Just think of a cloud of over 70 horses and men and 4 heavy guns with their caissons flying along a street. The very sound would be enough to scare some of the state troops. Why even out on our soft and dust cushioned plain a battery makes so much noise that you have to shout with all your might into the ear of the man on the same seat with you in order to be heard, and over the stones of Pensylvania Avenue I should think it would be horrible. Napoleon tried the effect of cannon in the streets of Paris once and a single discharge was all that was needed. Wouldn't it be glorious if a detachment of cadets should be ordered on, I would like nothing better than to belong to one of those batteries for the day. It would be fun to see the people run.

The only danger is that some ruffianly lunatic might shoot Mr. Lincoln with a revolver while in some exposed situation. No military force could prevent this if there should be a man reckless enough to try. Mr. Lincoln ought certainly to wear a mail shirt.

I pity poor Father the suspense he feels must be very trying. Here we who get the news each day cannot tell what they will be on the next and he who has to wait 2 weeks at a time cannot imagine.

I do not think that there is the least need of anxiety about him, because Fort Jefferson could not possibly be held by any force with which the State could afford to garrison it against a war vessel and they would not therefore risk its attack especially after it has been reinforced. But as it is, the Key of the Gulf it will always be well supplied with troops by the US. I received a letter from him yesterday describing his visit to Havana. Please send me all that you can spare from your mail.

Two more jokes have been gotten off upon old Del by one of the officers here. A few days ago he was officer of the day and old Del told him "Now Mr. Hascall I dont want you to allow any member of the examining board whatever to receive his papers while the board is in session for the news is so exciting that I am afraid he would not be able to attend to his duty." So Mr. H. told all the officers the orders with which they probably felt a little disposed to grumble. In the course of the morning in walked old Del's orderly dumped a big batch of mail down upon his desk and retired. The old gentleman seized one of the papers first thing, smoothed it out, and laid back in his chair oblivious of his duties, adjusted his specks and got a glance at the very interesting telegraph headings. Up stepped Mr. Hascall, Colonel Delafield: Your paper if you please! Contrary to orders to read it! He delivered it up without a word a martyr to his own discipline and Mr. Hascall carried it off amidst the smiles of the board.

The Colonel is not liked by the officers on the Point and seems to reciprocate the feeling excepting to members of the Engineer Corps to which he belongs himself, to them he is very cordial.

The other day at muster after several of the officers had marched in and had their names marked out, Mr. Weitzel stepped up. The Colonel received him in his blandest manner with Ah! Our Young Engineer friend Mr Weitzel! The next man was this Mr. Hascall. Old Del said not a word to him, so he spoke for himself, Ah! Your young artillery friend Mr Hascall!

I got through my examination in math yesterday better than any man in the two highest sections. I did not hesitate upon or miss any thing. We will be up in English and French in a couple of days. Love to all

Ever your Affct Son J. R. Meigs

Autograph letter signed, LRMC, MCMP, Ms. Div., LC.

Mary Craythorne Meigs to John Rodgers Meigs

My very dear John, Philadelphia, January 12, 1861

It was really very odd that your letter arrived just when it did. I had determined to write to you this afternoon, as soon as I had written my German letters. When I had half finished that, yours of the 11th was handed me. I am glad you like Nookie's picture. I do not think it does her justice, she is a fine looking girl, but is not stiff as there represented. Did you observe the pile of coats hats etc. at her feet? Henley Smith was with us when the picture was taken, & wanted to put *himself* at Nookie's feet, but I would not allow that. Oh Johnnie dear I have had such a nice time in Washington. I liked so much being with my sweet

Sister Lou & Nookie again. As for the rest of the family I am in love with them all, they were so kind to me. Is not Mrs Macomb one of the most amusing companions, & dearest little thing on the face of the earth? I have heard all my life how agreeable she was, but her charms surpass my most sanguine expectations. As for Capt. Macomb! Well, I hope he will be my friend forever.

You ask about Henley Smith. I was delighted with him. He is not I should think very brilliant, but he is young. That which I like most about him is that he is a perfect gentleman, as far as I can judge, in thought as well as in conduct. I supposed from Nookie's & Henley's behavior that they were really fond of each other, but if I am to believe all Nookie says, it is but on one side, & that she does not care much for him. Henley told me one day he had made up his mind that Nookie only cared for him when no other gentleman was by. I have given her several severe lectures on the subject of flirting, & have tried to make her believe that flirting is not only very foolish but very wrong. She has not much respect for my wisdom, so I fear such good advice will come to naught. But indeed John, I am sorry to see Nookie inclined to be a flirt, for you know a woman or a man who flirts is generally heartless, so I hope & pray that dear girl may never become such a miserable thing. I saw Susie Ledyard several times, & like her very much. The precociousness of the young men in Washington surprises me. Just think of your old Auntie talking earnestly with a youth in a *dress* coat, of fourteen summers! Mr Frank Turnbull by name. I could scarcely help laughing when I thought of his few years, but he is a very nice little chap.

Sister Lou gave me a photograph of you which I have put into a beautiful book for the purpose Emlen gave me at Christmas. It always does me good to look at your picture, but I wish I had a better one of you.

How shall I begin the disagreeable subject, upon which Mother & father & brother John, said I must write you? I hope you will not be hurt at what I have to tell you. Oh dear John, if you only know how much we all love you, & how proud we are all to have you such a fine fellow. Indeed I sometimes think I love you too much, so you must remember that if it were not so I could not say to you that which I must needs do. Here it is John. We all heard of the narrow escape you had from being dismissed during the last six months, and regretted exceedingly the proximity of such a shocking thing. Just imagine how *we* should as well as yourself been disgraced by your dismissal. Although we might know that it was for matters of small importance, mere carelessness in very little things of no importance, the world would not ask whether you were dismissed for leaving your wardrobe in disorder, your coat unbuttoned and such like, or whether it was for dishonesty, drunkenness, or bad morals, you would be disgraced *for life,* and my dear brother Mont who had so much to think of, & so much to bear, how would he feel about it, Sister Lou too. God knows she has enough to try her, in her separation and anxiety about her husband.

I want you now to make up your mind, that you will never go to another of those *suppers* you described in a letter to Washington, & which has caused all this alarm about your position at West Point. It will only be for a little more than three years, & then you will graduate I hope very high. If you were caught at a supper it would cost you as much as twenty demerits, would it not? & surely you know by this time that it is hard to spare so many at once. I can understand it is almost impossible not to get demerits for little negligences, but you can certainly, avoid a supper which might give you so many. ~~at once~~ Brother John who loves you dearly and is very proud of you, told me to say to you, that he never could forgive you if you were to be dismissed.

They have none of them seen your letter to me, but I know they will be very glad to hear of your high standing in mathematics & French. I congratulate you with all my heart. Do you have numbers given you in your class at this examination? If so do let me know how you stand, also Mr Michie & Mr. Rowland. Please write to me at once and tell me you are not hurt at what I have said to you, for I would not offend you for anything. I shall look anxiously for a letter every day.

While I was in Washington I heard Mr. Dolittle & Mr. Benjamin speak. Both very well, but both extremists in, the first republican sentiments & the other, all for secession. Mr. Baker from Oregon made a beautiful reply to Mr Benjamin, which I hope you have read. Today Mr Seward is speaking, & everyone looks to him for something conciliatory. Father who was reading it at dinner time was not much pleased with it. I have not yet seen it. Is not our country in a frightful condition? However I cannot but feel that all this present trouble will be in the end, good for the *whole* U.S. & that we shall be more united than ever under an amended Constitution, & that affairs will be so settled that the "Irrepressible Conflict" will be a thing no longer in every one's heart and on everyone's tongue, but forever more in the dim past.

Your story [illegible] is as you say, too good to be last. I suspect her poor fellow has paid dearly for his verdancy. Thank you for *intending* to send me your Cadet seal, but it was not in the letter. Are you sure you put it in. I always wear the button you gave me, to my watch chain. I am sure you are dolefully tired of this long letter, & note me a bore no doubt, but as I do not trouble you often you must not think me so this time. Remember I shall look anxiously for some news of you in a few days.

With great love from all, (Don't tell Nookie what I told you about her) believe me my dear John

most affec. yours Mary C. Meigs

Autograph letter signed, JRMC, MCMP, Ms. Div., LC.

Mary Montgomery Meigs to John Rodgers Meigs

My dear brother, Home, January 13, 1861

Many thanks for your kind letter, I prize them almost as much as if they were not my due. I wish with all my heart that a detachment of cadets would be ordered on here, it would be quite as pleasant for the girls as for you. It cant be however. So there is [no] use in longing. I asked Croix Porter yesterday something about W Point, and he immediately began to laugh and wanted to know if I could explain why young ladies were always so anxious about Cadets and Middies. I did'nt think they were, and as he insisted upon an answer gave brass buttons. He staid with me all the time we were at the capitol and requested me to make him a Union cockade which I am going to do. Any objections?

I met Willie Gwin in the street yesterday. The snitch only bowed and gave me a very mournful look. I wish I had spoken and asked him why he behaved like such a goose the last time he was here, he has'nt been to see me and I suppose he has gone back to college, is'nt it horrid in him? You dont know, but it is so don't you get offended at young ladies and thinking the offence was intentional never ask "why," it is very jackassical in your sex, and very common. All's well, here as far as war is concerned. People are however leaving the city as they fear trouble on the 16. of February. It is very foolish I think there can be no danger, when there are so many US troops to keep the peace.

Mr Sewards speech yesterday was very fine they say, we could'nt hear it, he did not however please either party, tho' he recommended concessions, and in all probability spoke Mr Lincolns views, as he is Sec. of State elect. Mr Sumner says if these are Sewards sentiments he will leave the party, the very best thing possible. It will show the south that Mr L. views are not such as Sumner [and] Hale and the whole tribe preach.

There have been no letters from father very lately. In the last he tells, or hopes we all stand by the stars and stripes. He says he received a letter from Maj. Lee probably advising him to turn traitor and give up the fort to the first man who demands immediate surrender. Maj Lee says that the single sargeant who defended Ft Pike ought to be hung. That is no way for officers who are eating the bread given by the *United* States to talk. If he is treated so badly by the north why not resign? I *hate* to hear them talk. Mr. Lee is always violent and on that subject he raves.

I am glad you like the picture. I dont no matter in what light or position I regard it. If it suits you cut any[?].

I had a long letter (12 pages) from John Macomb, a very nice boylike and hearty epistle. He says it has been the darling wish of his heart to correspond with me and hopes an intimacy may be kept up by writing till we meet again.

Blarney, I know but it is not unpleasant to hear such sentiments from young *men,* if 17. ans. You need not fear any impropriety in the correspondence. As soon as he takes a more intimate *stand* than that now held I will let him know that we are "*quits.*"

We are very glad to hear you passed such a good examination. I hope in English and French you stand as high. How are the demerit, very bad. You have not spoken of them for sometime, let us know in your next.

Mollie wants you to send her a button seal. I received a short note from her yesterday, she is very glad to get home as are her friends to receive her. She's a jolly old thing every one was delighted with her. Henley is desperately smitten. He shows it by not having been here but once since her departure, before he came two or three times during the day. Polly likes him very much too, but of course considers him a perfect [illegible] and a very green one at that.

Aunt Kate goes home in the morning, we shall miss her very much. You have no idea how quiet she has become, even more so than formerly. When there is anyone here she never says a word, it is very natural that she should feel like retiring within herself for no one can fully sympathize with her, in the great union which came almost as soon as her true happiness had begun. I have to answer Johns and Mollies letters tonight and must bid you adieu.

Mother is going, or is writing to you now writing is about all she does nowadays besides talking politics with any man she can get hold of, and she does it well particularly when Aunt Nannie is here to put in her oar, which as you know is bright and smooth. But good night.

with much love yours ever Mary M. Meigs

Autograph letter signed, JRMC, MCMP, Ms. Div., LC.

Montgomery C. Meigs Jr. to John Rodgers Meigs

Dear John — Washington, January 17, 1861

As I have not written to you for some time I thought I would send you a few lines containing an account of the accident that happened out at the College yesterday. I believe I told you in a former letter that since Dr. Samson has been President they have had a steam heating apparatus by which the College is entirely heated. On Thursday evening a new engineer arrived who was to attend to the engine and keep water in the boiler. Yesterday morning about ½ 8 oclock the engineer had gone to his breakfast and left the pump that supplies the boiler with water going. Some how or other the pump stopped and of course the boiler got red-hot without a drop of water in it. When the man came from his breakfast he saw the boiler in that condition ran to the pump and turned the steam on so that it would work, and then ran for the door. Just as he reached it the boiler burst with a great noise, tore the engine-house to pieces knocked down the walls and, half covered him with bricks from off the wall. He poor fellow draged himself out with a broken leg and burnt body and leaned against a tree. They took him in the College and did all they could for him but they sent him down to the Infirmary where I hope he will get well soon

Mother Capt Macomb Mary and I went down to the Theatre last night to see the performance of some genuine French Zouaves who were at the siege of Sebastapol where they formed a dramatic corps and acted in the theatre of the camp. The bill states that once while they were enacting a piece, the Russians made a sortie and interrupted the performance. They (the Zouaves) were obliged to seize their muskets and go out to fight in their theatrical dresses during which one of them was killed in his female attire and several wounded. At this time one of the banners being seized and carried off by one the Russians one of the zouaves fought with him and killed his opponent and retook the standard, which fight was acted just as it occurred, the Zouave with a gun and bayonet and the other with a sword. The breakfast bell has just rung and I must leave off. We are all well here. Good bye. . . .

Your affectionate brother Montgomery Meigs

Autograph letter signed, JRMC, MCMP, Ms. Div., LC.

Edward T. Fristoe to John Rodgers Meigs

Dear Meigs Columbia, Boone Co. Missouri, January 25, 1861

I suppose you are aware I have left the Old Columbian College and am now a Prof in the State Uni of Missouri. I received your very kind letter before I left Washington and was much pleased to hear from you, and learn that you were doing so well. You ought not to be satisfied with anything short of 1st rank. You have such tastes & talents for Scientific Studies that by ordinary application you may very soon become distinguished in any branch. Only be just to yourself and you may anticipate a distinguished and honorable Career, and leave behind you, perhaps valuable additions to Science. I shall watch with pride your future Course, and only wish I could have many pupils like yourself and our friend Hatcher.

I wish you would send me a Catalogue of West-Point, and also a Copy of its present Laws. I wish to know what text books are used in the various departments etc. Also anything else you think of interest will be most thankfully received.

I wish also you would send me a list of the best *Works* on Military Science especially Infantry & Cavalry, such books as are not to voluminous. Also the best works in Military Engineering. I would like also to get the name of some work which gives a good account of the Zouave drill, and any new methods of drilling, and send me word where these works can be found. You see I am using you to help me, and I hope you will excuse me.

Let me hear from you soon, and believe always your True friend

Edwd T Fristoe

Autograph letter signed, JRMC, MCMP, Ms. Div., LC.

Montgomery C. Meigs to John Rodgers Meigs (Shorthand Transcription)

My dear son: Fort Taylor, Key West, Florida, February 6, 1861

I have come here to get some more guns and am now embarking [on] a barge which I brought up from the Tortugas for the purpose, 7 8-inch columbiads, which were those which I carried down sometime ago, and the smaller guns for flanking purposes, which I have got from one source and now make up a total of 25 guns, which will in a few days be mounted in the casemates of Fort Jefferson. With these we can, though our garrison is small, bid defiance to a strong attacking force. The fact that we are strong in guns will, I doubt not, prevent any attack.

We had a narrow escape before the arrival of the company of artillery from Boston in the Steamer *Joseph Whitney.* The people of Florida and Louisiana were very free in their assertion of their intention to seize this place and Fort Jefferson. But thanks to good preparation and the expulsion of the traitors from the cabinet, we got ready in time and prevented the attack being made.

I was not displeased to recognize, upon the officer who sat in the boat sent in by the *Whitney,* the uniform of the artillery. For until I saw this, it was with me a heavy guess as to whether the steamer in the offing brought friends or foes.

We are now comfortable and have our mess arranged so as to be pleasant, much pleasanter than it was to live in the solitary style in which I passed the few previous months.

I hope you are attentive to your studies and to your conduct and that you have taken a good stand in your class in spite of the carelessness which has given me so much trouble about you.

The country, it is probable, will have need of her faithful sons, and you, who are in her care, I hope will do your best to perfect your [illegible] branch of science, which may enable you to repay her for your education and mine.

God bless you.

Ever your father, Meigs

Autograph letter signed, MCMC, MCMP, Ms. Div., LC.

John Rodgers Meigs to Lousia Rodgers Meigs

My dear Mother, West Point, February 9, 1861

I ought to write to you today of all days in the year for as perhaps you have remembered, it was just 19 years ago that you first saw me as a little baby by your side.

I have been thinking as you used to predict I would do of your watchfulness and tenderness without which I could not now be alive, and for which I have never made any return. They say that the only reward that parents expect, and one with which they are well satisfied, is to see their children make good use of the lessons they have been taught at home after they have left it. It is this that makes me and I guess most other men in the Corps work to stand high. But how selfish it seems that we should suppose that a thing which is of use to ourselves alone is in any degree a repayment to our parents for their care.

I have been surprised to see how seldom I look back over the life I have already spent, although it now amounts to a considerable portion of the three score years and ten upon which are all so apt to count. We seem to keep constantly in our minds the present and the future almost to the exclusion of the past which only comes in fitful flashes of small duration. They say though that one constantly looks towards the great central part of life and when that is past retrospect, becomes the habit of the mind.

I have been trying to see what is the earliest thing which I can recollect but cannot decide, for every thing is mixed up without any reference to dates. It is not the great things of my life that I recollect best, but little trivial ones. I remember being in the yard at the old Morans house gathering tomatoes with the cook and being told that we were soon to go to Washington to see Grandmother and Great Grandmother. I remember talking to her and asking her what made those *big blue wrinkles* in her hand.

I recollect riding round the room on Cousin Raymonds shoulder and watering flowers with you and Uncle Gus in the garden. About the next sojourn at Detroit I recollect a great deal. I can see you now quite plainly as you were walking down the hall carrying something into the kitchen, when you turned and saw Father come in the front door from his six months absence. I remember how soon after we moved down to the Fort where we had a long and happy time in spite of Mr Wilson's copy book and the sicknesses that would come. For myself I used rather to like them. It was so pleasant to have you and Aunty Trowbridge sitting, sewing in the room, while I lay abed watching you and perhaps listening to some story. Then the blanc mange and the quails Father shot for me were so nice, I was willing to be sick for the pleasure of eating them. How awful it would be to be sick here, it would be so desolate and unhomelike down at the old hospital. There is a poor fellow there now, named Cuyler, perhaps you know some of the family, who is very ill of the typhus fever, with inflammation of the bowels and the lungs. I pity him very much and hope I shall never be in the same case.

Yes there would be but little pleasure in being sick now even if I could get blanc mange and game. Our capabilities for enjoyment seem constantly to decrease and those for suffering increase as we grow older. That makes me think that more should be spent on children when young than is generally done. Ten cents now would give Loulie more pleasure than ten dollars when she is as old as I am. What a pity then it is to throw away the many opportunities of giving her enjoyment, opportunities that can never come again when once they are lost.

Those curtains that you meant for my birth day present one year ago have become my birth day present for to day. I was sent for by the Quartermaster to see if there was nothing in that bundle that came for me sometime ago that would do instead of the things furnished at the commissary department. Hard times had made them change their policy. I got everything. The curtains are beautiful, I had no idea they were any thing like so nice as you have made them. The room now looks so cosy and comfortable and homelike that it is pleasant to be in it. I wish I could show it to you through one of those wonderful glasses told of in the Arabian Nights. I am sure you would be pleased with its looks. They look so pretty that I cannot help looking up at them every now and then to admire and thinking of the trouble they cost you. That trouble, which I know you took with them, puts an additional beauty in them, for me, which old Lester cannot see, but he too keeps remarking, "Yes, Meigs, Those curtains *air* a *pretty nice thing* they make the room look first rate." It is altogether the best room in the division excepting one, and that is nice, more through the painstaking neatness of the occupants, than the excellence of their furniture. It is no trouble for us to keep the room nice now, because it looks well even when not fixed up. I would not live in such a cheerless dreary looking room as I had last year for any consideration. If I could only draw well I would send you a picture of the apartment. Old Peter Michie is scribbling away opposite to me, he has permission to visit in here on account of the stinking gas that comes up out of the furnace in his room. The heating of barracks is very bad, being entirely insufficient on cold days like this. It is one of the very coldest we have had this winter, and the wind is blowing like a young hurricane besides. Give my love and thanks to Aunt Nannie for her window fixings. They set off the curtains beautifully. I am receiving the benefit of those nice thick socks now too. I had begun to need them very much as I had to manage to keep warm by wearing two or three pair of cotton ones.

Love to all

ever your affct son John R. Meigs

Autograph letter signed, LRMC, MCMP, Ms. Div., LC.

LOUISA RODGERS MEIGS TO JOHN RODGERS MEIGS

My dear John, Saturday, February 16, 1861

I received your letter of the 9. Feby some days since. It gave me pleasure my dear boy to find that you begin to realize, what I have told you, many times, that you would do, that a Mothers love and tenderness, are never felt more sensibly than, when you are deprived of her presence. You begin to think and to feel as a man. I can scarcely realize the fact that I am the Mother of two grown up children, but I must begin to accustom myself to the idea. If you could see me seated at the desk with specs adjusted and the silver of years, shining through my hair, you would have no difficulty in supposing that I was quite old enough to assume the dignity of such a position. It seems to me a very short time since, that you were a little white haired curly headed boy, standing at my knee. I have had more contests with you, than with any other of my children for you were the most determined and rebellious of them all. I can remember with what sadness and almost despair, I have turned away from you sometimes, when I have seen what little effect all my counsels and reproofs seemed to have made upon you. I feared that you would never love and honor me as you ought to do, and that filial affection is such a safeguard for young men. I am truly thankful that the time is come when you begin to remember with pleasure the many evidences which I know you possess of your parents love and care. Those periods of severe illness which you remember at the Fort, how many sleepless nights and anxious days they cost me and your dear father. That whole summer was one of constant anxiety and fatigue. Auntie Trowbridge is here now. I read your mention of her in your letter to me, she seemed pleased to be remembered by you. She always inquires of you with kind interest, and says you were always a great favorite of hers. She desired to be remembered to you when I wrote. She was looking at the two little goldfish in your little aquarium in my room. She was delighted to watch them. How she would have been interested in your large Aquarium. She has the same taste for Natural History that you have. . . .

This city has assumed a very military appearance. The sound of the bugle is heard and soldiers are seen at every point. The Southern people are very indignant at such a display by Genl Scott of Federal troops. They think it aggravating to the people and totally unnecessary. Genl Scott is not one however to do things by halves. Old Abe is on his onward march to W. I wish he would not make public speeches. A gentleman said here to day that his speeches were retarding the progress of the Peace Convention. It is supposed that he intends to support the extreme wing of his party, and put aside all those who are disposed to favor compromise. I hope & trust this may not be true. If so of course we must give up hope of *Union*. We must divide into two distinct Confederacies.

I do not think I should like to go with the extreme North. The Greelys & the Sumners, & such fraternity. I saw your Uncle John and Aunt Nannie last night, they are quite well, and were glad to hear that you were well and doing well. All your friends here seem to be interested in your success. You have now a reputation of your own, as well as your father's to work for. I have received no official report of your standing at the last examination. I look forward with happiness to the pleasure of seeing you next summer, and having you at home once more. Do try my dear boy to have fewer demerit than you had the last term. I hope you continue faithful to your religious duties, "Remember thy Creator in the days of thy Youth—" it will bring its own blessing. We require the constant help, and aid of prayer in all transactions of life.

I hope you take care of and preserve for me your father's letters.

Do not advance political opinions, it is unwise for you and dangerous.

I must conclude this letter, I have been writing to Grand mother Meigs and must now write a few lines to your Father though I hope he will have left Ft Jefferson before they reach him.

Ever your aff Mother

Autograph letter signed, LRMC, MCMP, Ms. Div., LC.

Although John Rodgers Meigs would learn of it only later, Montgomery C. Meigs's tour in the Dry Tortugas came to an end on February 13, 1861. Back in Washington by the twentieth of the month, he received a warm welcome from citizens and officials alike. On April 7, 1861, he and others left New York on a secret mission for Key West and Fort Pickens to secure the island fortifications for the Union. With the accomplishment of his mission, Meigs earned the respect and gratitude of the president and members of the cabinet. Following his return to Washington, he became an unofficial military secretary to Secretary of State William H. Seward. Accordingly, Meigs prepared reports and gave advice as the occasion presented itself to both Seward and Abraham Lincoln. For his experience and service, Meigs was appointed brigadier general and quartermaster general of the United States in June 1861.

Louisa Rodgers Meigs to John Rodgers Meigs

My dear John, April 7, 1861

I have not heard from you for some time and am sorry that you do not write home punctually. I think you ought to be able to write home once a week. If you established a habit of that kind you would not find it difficult to gather materials for a letter. We are all well here, but the aspect of our public affairs make us all feel anxious & dispirited. There is a restless feeling of uncertainty, as to our future, and to what termination, [illegible] this present excitement is carrying us. The wisest man among us cannot foresee or foreshadow when & where our difficulties will end. In cases of great public difficulty & emergency men have often been found who seem equal to the occasion, whose great elevation of character united to wisdom and prudence, such as is not generally accorded to mortals has been able to grapple with the danger and who by controling and influencing the minds of other men have guided their country safely through great trials & perils & victories to peace & prosperity. In the great struggle for our Independence as a Nation, We had a Washington! Oh that his mantle could fall upon the shoulders of some one of our political rulers, who could guide us through the difficulties which now threaten the existence of the Government which our fathers bled & struggled to create.

It is not to be concealed that matters are rapidly approaching a crisis.

The policy of this administration does not seem to have made itself intelligible to anyone. The country looks anxiously and expectantly for its declaration.

I believe that it is about to speak in one way or other, but in what manner I cannot say. The papers have probably informed you of the sudden activity in fitting out & sending ships to sea, of the mustering of troops, in N'York, and the entire mystery which is presented in regard to their proceedings. You have probably seen that your father also has been detailed for secret service. He left us last Wednesday or Thursday evening and proceeded immediately to N.Y. Last Sunday which was Easter, I preceded him to church and was surprised that he did not follow me. He did not come to church at all nor did we see him again until the evening. A gentleman whose name & face Bridget did not know had called to see him directly after I left the house, and he had gone out with him. That was all the intelligence I could gather on the subject. Your father would not, and could not, he said give us any account of his proceedings for the day. He had been deeply engaged all day on matters of great public interest. From what he told me however I am fully aware that he has been employed on a service of high trust & importance. I could not form the least idea of its nature. He told me that I must ask not questions, that he could answer none, and that I must not believe any thing I saw in the papers or any thing that I heard. He could not tell me even when he would write to me, or when he would return perhaps in three weeks, perhaps not for six. He was bright and animated and full of excitement, seemed to anticipate a pleasant & successful excursion of some sort. Some think a fleet has been sent to St. Domingo, others think an attack on Pickens, or rather a reinforcement of troops, but everyone is completely mystified.

For once, they seem to have baffled the efforts of the correspondents of the press. They do not even pretend to know the object of the expedition. The papers informed me that a force of carpenters had gone on with your father, but they slipped away so quietly that no one knew that they were going. Thomas went with your father. He took him without any notice. He drove up the carriage to take your father to the cars. He had not the least idea that your father intended to take him with him, but we found the next morning that he had gone.

The whole affair has been managed very successfully, so far as secrecy is concerned. I shall write to you as soon as I hear any thing from your father. He told me not to excite myself by conjectures for I could not foresee the least idea of his mission, but to wait patiently until I heard from him. This, I shall try to do. I was very glad my dear boy to find by your last report that you had so much less demerit for the last two months. It is a great comfort to me and will be I know to your father to find you improving in this respect. I received a report yesterday.

I suppose Mary has told you that Capt Macomb has been ordered back to Detroit. You know what a shock that must have given us & for though in many respects it is pleasant to Capt. M, it is a very great loss to us, a breaking up of our family circle which has been so happily united for several years. Your Aunt N. and the children will not go I think before the fall, and Capt M will not I hope be obliged to leave before your father's return. Being military people who ought to be living always in expectation of a move we have been very fortunate to remain so long together.

Monday morning—Mary had a letter from you this morning. I have had no letter from your father since he left. He told me he thought he should not write to me at present. I had a newspaper yesterday which was enveloped, and [illegible] by him which was intended I suppose as an assurance that he was well. The paper mentions that he has sailed in the *Atlantic*. Her destination remains a matter of conjecture.

Your Uncle John adheres to his affection for his "native state" but says he will go wherever he is sent at this time.

Ever your afft Mother

Autograph letter signed, LRMC, MCMP, Ms. Div., LC.

Montgomery C. Meigs to John Rodgers Meigs (Shorthand Transcription)

My dear son, U.S. Ship *Atlantic* At Sea, Gulf of Mexico, April 24, 1861

I suppose you will feel some curiosity to know where I have been. Your mother no doubt has informed you that I disappeared from Washington suddenly and without letting her know that I was to go. Poor Mother heard [illegible] all sorts of dangers and she [illegible] and behaved with reasonable firmness in this trial.

Well, I went to New York and got this steamship, the *Atlantic*, of the Collins Line, and I have to perform my promise to the President. I have, with the means he placed at my disposal, relieved Fort Pickens and reinforced it so they [illegible] in the garrison. About 1,000 men do their duty [illegible] take it out of the hands of Uncle Sam.

We had a very rough and weary passage, lost 4 of our horses by the boat that we carried to Pensacola, 69 out of 73. In landing, we drowned one, one died of exposure, having got loose and swam 3 times around the steamer, and one was turned head over heels in the surf and broke his neck. We took with us six months' provisions and stores, large quantities of munitions and ordnance and grain, enough to carry the garrison up to about 1,000.

Colonel Brown, an old soldier and a brave man, commands and will no doubt give a good account of himself and his country.

I hear [illegible] bad news of things at Washington, and I hasten back to look after my family and my country. God grant that I may be of use to it hereafter, as I trust that in this relief of a beleaguered fortress I have been already.

Study, my son, learning what I fear is to be the most important of arts in this country for some years to come, the military art. Prepare yourself by study and temperance for the watchful, laborious life of a soldier. Be ready, if need be, in the cause of your country, to give up your life at a command. Devoted to right the cause of liberty and justice, the Lord may look upon you as one worthy to be descended from those revolutionary sources from whom we draw our blood.

From your father, Meigs

I have made some sketches from [illegible] paper. Look at them as [illegible], although I presume they will make good pictures out of [illegible] and hasty sketches. Of landing of horses, of stores, et cetera.

Autograph letter signed, MCMC, MCMP, Ms. Div., LC.

On July 2, 1861, John Rodgers Meigs, after spending almost two years at West Point, came home to Washington, D.C., on furlough until the twenty-eighth of August. To his father he looked "stout and well,—the life evidently HAS agreed with him."[1] His father could be further pleased with him. The young Meigs stood head of his class in mathematics and third in general merit. Unbeknownst to John Rodgers Meigs or his family, he would have the opportunity to take part in the first major battle of the Civil War—the First Battle of Bull Run.

On July 18, 1861, he joined Major Henry Hunt's battery, part of Brigadier General Irvin McDowell's army, as a volunteer aide. After-battle reports contained praises of Meigs's participation in the fighting. Israel B. Richardson reported that during the battle Meigs "carried my orders promptly, and a braver and more gallant young man was never in service." According to Hunt's report, the young soldier "performed his duties until the close of the action, with spirit and intelligence, and was very useful . . . in conveying orders, observing the enemy, and rallying our troops."[2]

On July 22, 1861, young Meigs returned home at 8 A.M. to tell his father of the Union defeat as well as his own actions, including commanding a regiment. As further proof of his actions, Richardson wrote to Montgomery C. Meigs, about the "gallant conduct of your son, and his exertions in carrying orders for me in the field." So impressed was Richardson that he "recommended him to be immediately made a Lieut. in the regular Army."[3]

Following his first participation in battle, John Rodgers Meigs remained with family and friends and kept up with his correspondence. He visited both sets of his grandparents, at Havre de Grace, Maryland, and in the Philadelphia area. His days at home were shared with another participant at the First Battle of Bull Run, James S. Brisbin, who was wounded twice in the fighting and who was befriended by Montgomery C. Meigs.

1. MCMJ, July 2, 1861, MCMP, Ms. Div., LC.
2. Report of Israel B. Richardson, July 25, 1861, and Extract from Report of Bvt. Maj. Henry J. Hunt, Second U.S. Artillery at Battle of Bull Run, July 21, 1861, in *OR,* series 1, vol. 2, pp. 376–77, 380.
3. Israel B. Richardson to MCM, July 28, 1861, MCMC, MCMP, Ms. Div., LC.

Montgomery C. Meigs Journal Entry

July 22, 1861

This Morning it rained. John came in and waked me up at 8 A.M. telling me that they had fought a great battle and been defeated; it seems that he played aid de Camp to all the Generals and Colonels in the neighborhood; had been very usefully employed. In fact he took command of a regt. No one else seeming to do it, and they obeyed his commands.

They were attacked by a strong force of Cavalry and infantry but they were repulsed by a few rounds of canister and musketry.

He thinks they made a great slaughter; the enemy repeated the attempt once or twice with the same result. Finally, Richardson backed down on Centreville and there joined McDowell & [illegible] & [illegible]; they remained some time in this position and then retreated to the lines over the river, covered the retreat.

This was done in pretty good order. John about 2 A.M. left them to ride into town and tell Gen. Scott to send a remount to Hunt's battery whose horses were getting worn out, and to send fresh troops to reinforce—when they would hold Fairfax. I took him to report to the General who was much pleased with him and asked what was his position etc. Orders however had been given by telegraph to come into the old lines of the Army, and this was done; so that though I sent out the remount it did not find Hunt until late, and was sent back, he preferring to keep his horses, now rested. It was a bad defeat; the papers of the day give full accounts of it, and here I leave it. Our Army is much demoralized and we must form a new one. Beauregard ought to drop down on us before we get into order again. Troops are offered in great numbers, but 5000 men could have taken this City on Monday morning I believe.[1]

1. Transcription copy, John George Nicolay Papers, Ms. Div., LC.

James S. Brisbin to John Rodgers Meigs

Dr Sir & friend Boalsburg, Centre Co., Pennsylvania, August 18, 1861

On my return home from Washington I found a letter from you on my table waiting my arrival but have been too sick to answer it. I have been very sick since I left Washington the last time. I had a severe attack of fever and at one time the Doctor despaired of my life. I am now able to be up and walk about a little but am still very weak. I expect soon to recover here among the Mountains. I wish you would come up John and spend some time with me. We have game and fish in abundance. Minerals of many kinds abound here. Then we have lizards snakes toads and all manner of creeping things. The celebrated Bear Meadows are only four miles from here. There are no Meadows like them in the world. Plants are to be found in them the names of which are unknown to the most learned Botanists. We have many Caves in the vicinity and nearly all our hills contain fossil remains. I think if you would come up you could spend a few weeks or if not so long at least some days right pleasantly. I would delight to have a few days in the mountains with you.

About the horse—it appears I will loose him. Your father writes under date of August 1st that he had the Govt Stamp on. I looked him over carefully but could find none. Bull Run was a pretty dear day to me. My own horse which had been brought from my fathers farm and was a fine animal I lost on the field. I saved nothing but one small pocket Revolver.

I have one of the nicest presents for your Mother she ever saw, something both useful and beautiful. Mother was anxious I should present her some token of remembrance and it will afford me real pleasure to do so. My parents feel truly greatful to Mr. & Mrs. Meigs for their kindness to their child. Both father and Mother join me in the wish that you will visit us & partake of our mountain hospitality which is always cheerfully given. I see your name almost every day in print and I can assure you it gives me genuine delight to read your praise.

Be prudent now and you have a prestige that will carry you high. If you conclude to visit Centre Co you must write me many days before you set out for we live some 30 miles from Rail Road and I desire to have a carriage at the Depo to convey you over the Mountains. I have already written more than I had intended and will close. Write me soon & Believe me as ever your

Sincere friend & Most obt Servt

Jas S Brisbin

Autograph letter signed, JRMC, MCMP, Ms. Div., LC.

John Rodgers Meigs was appointed sergeant in the Battalion of Cadets on August 29, 1861, and two days later he was detailed to the math department at West Point. He was to report to Professor Albert E. Church upon his return in early September. But the month of August was not to be a quiet one. In mid-August, Meigs was called as a witness at the court of inquiry of the conduct of Colonel Dixon S. Miles at the First Battle of Bull Run.

During the battle, Miles's Fifth Division was the U.S. Army reserve, but one brigade and an attached brigade commanded by Colonel Israel B. Richardson were forward along the run guarding Blackburn's and Mitchell's Fords. After the defeat of Brigadier General Irvin McDowell's main forces, the units commanded by Miles and Colonel Israel B. Richardson covered the retreat. A dispute arose between the two, and McDowell relieved Miles for alleged drunkenness, as charged by Richardson. Miles subsequently requested, and McDowell ordered, that a court of inquiry examine the allegations raised by Richardson, who asked and was granted permission to appear as accuser. The court was headed by Brigadier General William B. Franklin and included Colonel John Sedgwick and Captain Charles Griffin. The court began its hearings on August 12, 1861, and continued in daily sessions until August 19; it then suspended its sessions until October 18–29, 1861. It concluded that Richardson's charges were justified but that sufficient evidence could not be found to sustain a conviction of drunkenness, and that no further proceedings were necessary, a decision confirmed on November 6, 1861, by Major General George B. McClellan, who had succeeded McDowell as commander of the army.

The court of inquiry summoned John Rodgers Meigs to testify as a witness for the prosecution. He testified on August 19, 1861.

As part of his defense, Colonel Miles and some of the witnesses who appeared on his behalf alleged that part of the confusion and disagreement between the two colonels may have resulted from persons' giving conflicting orders and disrupting Miles's tactical disposition of his troops. Some witnesses charged "Cadet Meigs" as one of the persons responsible for this confusion.

On August 19, 1861, Oliver Duff Greene, first lieutenant in command of a battery of the second artillery that supported Richardson and Colonel Thomas A. Davies along Bull Run, testified that "I received an order twice from a young gentleman that I was told afterwards was Cadet Meigs." When asked, "What reason did Cadet Meigs give you for issuing orders?" Colonel Richardson or his counsel objected, which was sustained by the Court and so went unanswered. On October 18, another witness to Meigs's activities—James McMillan, who served as inspector general of Colonel Miles's division on July 21, 1861—testified: "There was confusion caused by unauthorized persons giving orders; one instance was General Meigs' son." Another witness, Colonel Thomas A. Davies, testified that John Rodgers Meigs acted "as crazy as a loon." In spite of the testimony of the witnesses, Meigs denied giving orders during the battle, and his superior officers praised his service.

In the following record, "the Accuser" was Colonel Israel B. Richardson or his counsel, Lieutenant Colonel Sylvester Larned. "The Accused" was Colonel Dixon S. Miles or his counsel, Reverdy Johnson.

Excerpt from John Rodgers Meigs's Testimony in the Case of Colonel Dixon S. Miles's Court of Inquiry, August 19, 1861

Cadet John R. Meigs, a witness on part of the prosecution, being duly sworn:

Question by Accuser. Where were you and did you see Colonel Miles the defendant, about sunset of July 21st, last?

Answer. I was on the hill near Centreville, and saw Colonel Miles.

Question by Accuser. Was he intoxicated?

Answer. I think he was.

Question by Accuser. State fully your reasons for so thinking.

Answer. I recollect very little of Col. Miles's appearance; but I do recollect of being perfectly convinced he was intoxicated, and of telling General McDowell so.

Question by Accuser. Did you convey or give any orders, and if so under what authority, about sunset?

Answer. I gave no orders after seeing General McDowell for about an hour before that time. I had been making suggestions to some of the colonels of the regiments; I do not recollect giving any orders, and I had no authority for giving any orders.

Question by Accused. Did you ever see the accused before the 21st July, or have you ever seen him since?

Answer. Never before the 21st of July, nor since, except to-day, that I recollect of.

Question by Accused. On whose staff, if any one, were you acting on the 21st July?

Answer. As volunteer aid to Major Hunt. I carried some orders for Colonel Richardson, also Colonel Miles during the day.

Question by Accused. Did you, or did you not, on Centreville Heights suggest or order a colonel of a volunteer regiment to deploy his close columns into line?

Answer. I did not.

Transcription copy, RG 153: Records of the Office of the Judge Advocate General (Army), Court Martial Case Files, 1809–94, Case II 498, NA.

There appears in the Montgomery C. Meigs Papers in the Library of Congress, in the handwriting of John Rodgers Meigs, a "Review of the proceedings of the court of inquiry in the case of Colonel Miles." In this review, young Meigs writes an extract from the testimonies of Captain James H. Butler, Lieutenant James McMillan, and Captain Duff Greene. Meigs concludes, "But this is not material in the determination of the question whether Col. Miles had from drunkenness occasioned a confusion showing him unfit for command." Another related document in the Montgomery C. Meigs Papers is a letter dated August 23, 1861, from Miles to the elder Meigs in which Miles writes "to offer you an apology for my abrupt conduct to you at Centreville. I was informed I misunderstood your remark, which under the pressure of great excitment at our unfortunate condition . . . led me into language to you I regret and which I very respectfully request you will not only forgive but forget."[1]

1. JRM Other Papers, MCMC, MCMP, Ms. Div., LC.

Charles D. Meigs to John Rodgers Meigs

My dear John Hamanassett, Saturday night, August 31, 1861

Your uncle Frank, who as usual came to us by the 4. P.M. Train, brought me your very agreeable letter for which, you see that I now make haste to return my hearty & sincere thanks.

I do not know whether you are old enough to appreciate what I regard as a very consoling truth & which is this, that the best and most abiding comfort in this world is to be loved by good & virtuous & true hearts.

A bellyful of bread & meat, & a glass of good wine are undeniably most excellent creature comforts, & they are of a sort we can not well do without, but the day will come, if you should live long enough when you will have hold of some fair blue eyed girls hand, & hear her say with a sigh & whispered love that sounds as if it came straight down from heaven "Dear Johnnie of my soul, how I love you!" When you hear it, then you will know, what now you only believe, that to be loved is one of the consolations of man's life.

Now my dear John I do not pretend that I love you as that girl will, but this I can say, that I do love you dearly, and that my affection for you seems to increase with the lapse of days and the increase of distance. I miss you, here, every day, & often every day & there is hardly any thing in this world I much more strongly desire than the pleasure of your presence & conversations; except indeed it might be the public beheading or hanging of Jeff Davis & his reb butchers Beauregard & Johnson—both of whom merit death for their perjuries—let alone their insults to the nation & their murders.

I believe that my mind, or heart is peculiar in this, that in the general I love my friends very much, but when I do so I am never well satisfied with them.

My attachments to different people make me afraid for their reputation, their probity, their honor or their fortune, for, when I love a person I would have that person "all right" not for a day only, but for ever & ever.

You seem to be an exception to this rule of my mind, for in loving you I am quite at rest and contented, & I suppose this is because I am confident in your virtue & fortune. Should you fail in either before my death, which can not now be far off you will break your poor old Grandfather's heart you Dog, & I know you would not do that if you could help it, for I am sure you love me too.

But as I might spoil you by praise I will quit, lest I shd. bring about the very end, which now I fear not.

It is a satisfaction to me to know that you got to the Point in good time, and that you are now promoted to be Sergeant of a company. May you live to be a Lieutenant General, & you will if you don't die young. Do you make it your business & your aim to reach the highest military rank & power in your country, for unless Jeff destroys our nation as he is now trying hard to do, the day is at hand in which we shall require the greatest Military Science & Skill, because our nation was designed by God to be the model & the refuge of the oppressed in all lands.

If we extinguish this civil war, we shall be more potent than Assyria or Rome, & we shall go forth breaking the chains of mankind perhaps as far as Thibet or China & Japan. Prepare your mind & soul then for the conquest & Christianization of Japan, but let that soul be ever in absolute subjection to the law of truth justice & philanthropy. Never be a king, or serve a kingdom, but set before your eyes the beautiful, the Godlike character of Gen Washington, who knowing that a man is a mortal & perishing creature, knew, as well, that virtue is the best & most durable reward of life. There never was & there never will be a truly virtuous man, who would not rather perish a thousand times than oppress his own country men.

To be patriotic, to love ones country is not an abstraction, for ones country is not an idea of the mind, it is the material thing on which the contentment & pleasure & safety of millions of hearts reposes & trusts.

When I think of this wicked war I rejoice I am old. When I remember you, then I lament that I am more than 21. years of age, for if I were young again I might hope to follow you with my [illegible] & wherever you shall live. But after all, when I do die, why may I not hope to gaze at you from out the face of some summer Moon, or peep at you behind a gorgeous evening cloud in heaven, & sympathetically move you ever more safely onwards in the march for Truth & Honor.

Hamtt looks as green as June.

Nathan's arm gives him no more pain & is going on finely.

Auntie Mollie often says "Dear old John what a dear fellow he is" etc.

Your grandmother loves you as well. So does Jimmy Mandy & black Charlie & Mary Ann & Issy.

Your mother & sister & brother are well. Gen. M. says "We are [illegible] to save the Union to put down the rebellion & to punish the Traitors." This, in his letter to me of Thursday.

I hope he is a prophet.

Pray dear John [illegible] to send me a line to show you do not forget.

your affectionate CDM

Autograph letter initialed, CDMC, MCMP, Ms. Div., LC.

Though well thought of by his family members and others for his actions in battle, John Rodgers Meigs learned from his mother that he made a different impression on a young family member. Also, she felt that he should not receive too many letters similar to the one Charles D. Meigs sent to him, as his grandfather "is not as discerning as he was when your father was your age." John was to listen to his mother's admonitions.

Lousia Rodgers Meigs to John Rodgers Meigs

My dear John, Safe Moorings, September 4, 1861

I received your letter from WP. last evening and was very glad to hear of your safe arrival as it is the first intimation of the kind I have had. I have been expecting a letter from you every day since your departure and have consequently been *every day* disappointed. I should have written to you sooner, but forbare to do it, on principle, as I thought it your duty to write to me, at least to let me know that you had arrived at your place of destination. I have been obliged to control myself with the old adage of "no news is good news." I suppose you must have had a pleasant meeting with your friends in New York.

I am glad to hear you have a pleasant room mate, and that you are once more comfortably established in quarters. Now my dear son you have another year, before you and the opportunity offered of honorable distinction, do not fail to use all your energies for that purpose so as to gratify the high hopes entertained of you by your friends.

Do *for my sake* as well as your own, cultivate a little more of the "Suaviter in modo," so essential to every man. There are some little *peculiarities* of yours which I wish most earnestly that you could get rid of. I want you to tell me in your next if you bought that powdered aloe, and if you use it. When you write me word that your nails are of a proper & becoming length I shall feel as if you had accomplished a victory. I have one favor to ask of you. Will you put away entirely the use of the *tea spoon,* that also is conspicuous. Do you know little Minnie, broke out at the table the other day suddenly with "Mama do you know how Cousin John eats with his spoon, this way," and then she began to take little short sups with a loud noise for which she was immediately checked of course, but it shows how your manner impressed her. Now *do* my dear boy, listen to what I say. You know I love you very much and in all these admonitions have only at heart *your welfare.* Do not get vexed when you read this and think I am boring you. There is no one else in the world that would take the liberty of speaking to you as freely as I do, however much they might be impressed by any peculiarities of manner. I do it, at the risk of making myself exceedingly disagreeable, but I would rather do that than have you allow your faults to get the victory over you.

Great talents & great virtues are sometimes allied with such repulsive personal habits, & such rude address, as in the case of Dr Johnson, that men as they grow older, become unbearable almost to their friends. A little *self restraint* & control early in life would have remedied these defects. Now I have done. Do not be vexed but listen to your Mothers admonition. I want you to set about the reform *at once.* If you do not you will be the *slave* of habit. Tell me in your next whether you were provoked or whether you are edified by my remarks and whether you design to profit by them. I know this is quite a different letter from one which your *Grandfather* wrote you a few days ago, but it would not do for you to receive many such as he writes you. He is not as discerning as he was when your father was your age. . . . Mary is going to write today and will give you all the news. I have not had a letter from your father for many days. I suppose he has been too much occupied. Did you not rejoice to hear of Stringham's victory. What an ignominious position for *Barron* I am told that Com S. turned on his heel, when Barron extended his hand upon meeting him saying, I recollect Sir to have met you before.

I must close this so as not to inflict crossed lines. Loulie stands by me and sends her love.

Ever your afft. Mother

Autograph letter signed, LRMC, MCMP, Ms. Div., LC.

Frank B. Meigs to John Rodgers Meigs

Dear John — Philadelphia, September 11, 1861

I enclose you Adams' Express Cos Recpt for one bundle, contg the rules you left at 1208 Walnut.[1] I have packed them up so that they should carry without breaking or niching the edges. Your sister Mary is in town was at dinner to day & is very well. Your Mother was in town yesterday well & in good spirits. Aunt Molly is expected in tomorrow for the season. There is a letter at home to day from your Father dated Baltimore yesterday. He was detained there by an accident to the train from Washington, the engine ran over one cow & broke the legs of another smashing up the baggage car & damaging [the] engine, but not injuring any person beyond scratches & bruises. He says he thought they were putting on the brakes harder than usual. [He] is going west to look after some expenditures, says all was quiet at Washington yesterday morning when they left.

I was at Ham' on Sunday. Your Grandfather is in much better spirits, the victory at [Fort] Hatteras has cheered him up. He thinks B. regard, Jeff & Co will have to fall back from before Washington. If they dont, they will probably be driven back before long. It is remarkable how quiet the movements of troops are kept, since McClellan took charge. Our papers never publish the arrival & departure of men now, so we have no means of estimating the force at Washton. I hope they will touch the rebels up, along the coast & keep them uneasy, we were all glad to hear of your promotion & hope you will be able to keep clear of the demerits.

Very truly Yours Frank B Meigs

Autograph letter signed, JRMC, MCMP, Ms. Div., LC.

1. The home of John Rodgers Meigs's uncle, Franklin Bache Meigs, was at 1208 Walnut Street, Philadelphia.

Montgomery C. Meigs to John Rodgers Meigs (Shorthand Transcription)

My dear son: Washington, September 20, 1861

I am glad to find that you have been honored by the appointment of instructor, but be careful not to let these duties too much interfere with your studies of your own lessons.

I have heard officers of the army lament that they had [illegible] this costly when it had made them lose rank in their classes, and this was losing rank in the army on graduating.

I have not been able to find the blanket, India rubber and wool, both of which you took to Bull Run and I believe brought back. Where did you leave them? They say that you did not return them to the stable, though the horse blanket belonging to the stable and the saddle itself are there. I wanted to use them, so write at once and let me know where they are.

I have been to St. Louis, though on a disagreeable duty [to investigate corruption in the quartermaster office of General John C. Frémont's command]. I have seen the President and cabinet since my return, and I believe they are satisfied with the way in which I have performed my part.

Letters from home. All well.

The enemy is strong, but I believe that General McClellan is confident of beating them if they attack; and he will, I am sure, not attack until he is sure that he can beat them in such an attack.

Napoleon says that no man wins in war except by employing the means of success which are somewhere at the point of conflict that your adversary brings.

I sent the sword and your note to Mrs. Hunt.

[illegible] is well—the mare I have not used for some time. I have been trying to buy a new horse, but he is [illegible], though very strong. I probably will buy him.

Study faithfully. You have now seen enough to feel the importance of applying yourself to your studies and of doing faithfully your duty as a cadet, avoiding demerit and all breaches of discipline, which makes men brave and successful soldiers capable of fighting the battles of their country.

Ever your father, Meigs

I gave your note to [illegible] by mail, the proper person. If you do not get an answer soon, send me one such and I will try again.

Autograph letter signed, MCMC, MCMP, Ms. Div., LC.

The following letter was written by Mary B. Hunt, sister of Second Lieutenant Presley O. Craig, who was killed at Blackburn's Ford, at the First Battle of Bull Run.

Mary B. Hunt to John Rodgers Meigs

My dear Mr. Meigs October 1861

I owe you many apologies for not having answered your letter but it was not that I forgot it, but that we have had so many stampedes on the other side and my husband has been obliged to go over when anything has occured, that I have been kept in a constant dread of something happening so that I could not compose my mind sufficiently to write.

I think my dear brother could not be known without leaving the same impression on one's mind as he has left on yours. He was always so gentle and considerate. We all remember you with affection, you were nearer his age than any who went, and he was so glad when he heard you were going, I shall always think of you in connection with him. I should be so glad to hear from you whenever you have the time or inclination to write. I send you a ~~photograph~~ *carte de visite* of my noble brother. It is not as good as might be expected, but it was the only likeness that we had. We are having some large photographs taken from it and I will give one to your Mother to give to you. I do not think we will ever be able to get one with his expression.

My husband desired to be remembered to you when I wrote. I begin to dispair of our ever having peace. The family desire to be affectionately remembered to you, and believe me.

Truly Yours Mary B. Hunt

Autograph letter signed, JRMC, MCMP, Ms. Div., LC.

Louisa Rodgers Meigs to John Rodgers Meigs

My dear John Germantown, Pennsylvania, October 4, 1861

If the N York papers have failed to give you the intelligence that QrMaster General Meigs & "lady" were stopping at the Astor House last Tuesday, I shall be the first probably to communicate the fact. We only spent a night and part of Tuesday in the city. It was quite tantalizing to me to be so near you and not be able to run up and make you a visit. I wished at least to write you a letter from the city but was afflicted with a headache which compelled me to forego that pleasure. You will feel curious to know how I came to make so flying a visit to the Great City. So I must tell you about it. On Friday last, Major Macomb came on to spend a couple of days with your Aunt Nannie and attend to some business which he had in Phila. He arrived at night in the midst of a heavy gale. Don't you think he proved himself a good topographical Engr when he piloted a hackman out to Mrs McDowells over a road he had never travelled from Phila. When the hackman arrived in Germantown he was very anxious to put his passenger down in some of the by roads, insisting that there were no houses in this vicinity. The night was intensely dark and it was raining furiously. Macomb persisted however. He had a little card map of Germantown which Mrs McDowell had given Nannie and with this he found his way safe to our door. This is however just an introduction to your father's visit. On Saturday night about 1. oclock I was aroused by hearing someone try the front door after opening the gate. This noise had been preceded by a whistle which I suppose first aroused me. You know I am always on the *qui vive* for robbers, and concluded that this was a highwayman, bent on plunder. I sprang to the window, and looking out found, fearful to relate, a man standing under my window. He did not speak nor did I remaining in breathless suspense. After standing for a moment looking up at my window he walked out into the street and took a survey of the house. He threw up his arm with I knew not what determination. I thought he was going to take aim at my window. At this critical moment I ventured to break the silence by a loud "hem" which was intended as an admonition that his proceedings were observed though I did not care to interrogate him concerning them.

To my exceeding astonishment & joy the figure responded by calling me in a most familiar tone, and I rushed down to admit this midnight prowler to my own room. He came in so quietly that not a soul in the house knew of his entrance except Miss McDowell and her sister. I had intended a great surprise the next morning at breakfast to the rest of the family, but Minnie & Loulie, betrayed my secret. Your father made up his mind on Saturday to come on by the 2.30 train, and arrived in Phil too late for the 10 train. I think he came out in the last train which leaves Phila. He spent Sunday with us, and as he was obliged to go to N York on Monday Eveng, I concluded to accompany him. I thought I should like to take a peep at N York which I have not seen for the last 8 years. I might have a chance of seeing you, and at all events I should have the pleasure of spending so many more hours with your father, which as he declines to let us go to W at present is something of an object with me.

We left Phil in the six-evening train, and on arriving in N Y, proceeded at once to the Astor which looked quite familiar although I think I have not been there since you were five years old. The time when I attempted one day to punish Mary for misconduct, and you rushed up the hall and knocked me down, as I was kneeling on the floor, saying "now you are going to whip my sister, take that." Do you remember it? I was very distressed at your conduct, but shutting the door to keep out the noise, administered prompt punishment to both. The offence was never repeated. I was much in hopes that your father would remain in N York long enough to allow me time to make you a visit, but he could not tell me in the morning what hour he would decide to leave, "just as soon as he could get through his business" was all that he could say. Of course it was impossible for me to venture on a visit to you particularly as your father told me that you might be engaged at the time of my arrival and I might not have an opportunity of seeing you for some time. I felt quite lost in N York. What a great Babel of a place it is.

Your father took me over by my own request to Barnum's Museum in the morning. I have carried away recollections enough to last me for some time.

I could not be induced to look in there again as I said to your father "It is very fillin for the price." Were you in the least in doubt the instant you put your eyes upon the "What is it" to declare his species. He is of the "Genus homo" being nothing more or less than an idiot black boy, that I discovered the instant that I looked at him. His head has been shorn, & with his deformed brutish scull & long ape like limbs, he might possibly deceive some persons. He has not the intelligence of a monkey and it seems difficult to believe that such a creature has a soul, the germ of immortality in that wretched looking body. Yet I believe that it is so. I am persuaded that he is a human being, though in a most degraded form. We paid a visit to the Hippopotomus who is a great curiosity, but a disgusting looking one, to the Seal & the Sea Lion who seemed to me to concentrate in his person the essence of all fishiness. He was a most offensive brute, & my visit to him made me sick for the rest of the day. It is a shame to place those pretty aquaria in the same room with such a beast. I could not examine them though I tried to do so with handkerchief applied to my nostrils.

Mr C C Trowbridge called to see us with Mr Sibley during the day. He is always very affectionate and glad to visit us. He told me if I could remain in N York a day longer he would take me up to W Point to see you. Capt Saxton came in and sat during dinner with us. He is very busy in N York. I believe he expects to be promoted very soon. He is a fine officer. A number of gentlemen called to see your father during the day. We left on the six train for Phil. Your father dropped me at 1208 Walnut & went on to Washington.

I must hasten to close the messenger waits.

Ever your afft Mother

You say nothing on the subject of my *last letter* in your reply.

Autograph letter signed, LRMC, MCMP, Ms. Div., LC.

Louisa Rodgers Meigs to John Rodgers Meigs

My dear John, Germantown, Sunday, October 6, 1861

I find that your father has just been writing to you & will enclose with his letter a few lines only to let you know that, I do not intend you should have any excuse for your sins of omission in writing to me. We are all well here. Your Aunt Nannie & I were both very much delighted by the arrival of your father & Col. Macomb on Friday night. Your Aunt Nannie had some expectation of seeing her husband, but your father's visit was a happy surprise. He leaves here tomorrow for Boston and has invited me to accompany him. I feel very much tempted to do so, but do not like to leave your Aunt Nannie, just now. Your Grandmother & Aunt Jerusha, have gone to Havre de Grace.

Your father has not been quite well since he has been home. I think he had something like one of the aguish attacks he had when he used to be going up to the Aqueduct. He has never gotten entirely free from the effects of that miserable disease chill & fever, which he contracted first at the Fort in Detroit. It appears sometimes after long intervals when least expected. If it had not been for this little attack of indisposition I think we should have left here yesterday evening, and made an effort to spend this Sunday with you. Your father says he cannot possibly go to see you during the week. He has to hasten as much as possible to get back to Washington. I shall make an effort to see you before leaving here for home.

I hope we may return there before many weeks. Mary laments very much the loss of her rides on horseback. I am afraid her mare does not get exercise enough to keep her in good condition. Mary came down to see me on Friday and brought your letter to her which she had just received, for me to read. My dear boy, if you only knew what a relief to my mind that letter was, you would not have deferred writing so long on the subject. I have been really unhappy for she had taken up the idea that you had perhaps had the boyish folly to make some revelation of sentiment to Miss H. C and that you hesitated to tell me of it. The idea haunted me like an incubus. I woked up at night and thought of it, & it was many times in my mind during the day. The idea of your separating yourself from me & so many that loves you of fettering and clogging your young life with such an attachment, which could bring nothing but unhappiness, was dreadful. I wrote to give you some idea of what I felt. I did not like to say all that I thought of it, for fear of exciting your old spirit of opposition. Why did you never write me a word in answer to my letter. If you knew how pained I was at your silence you would I am sure have treated me with more confidence. I hope you will never indulge a sentiment for any woman until you are fully arrived at years of discretion. Your knowledge of the fair sex is so limited that you might very easily be deceived in your estimate of a woman. There is plenty of time for you to think of falling in love when you leave W Point. It always makes me feel sad to see a young fellow who has just entered upon the toils & struggles of active life, burdened with the care of a wife and family. For a military man it is especially to be deplored. I think it an equal injustice both to man & woman. You have only to look at the young wives of some of our young officers to see the truth of this. Your father has given you his opinion on the subject of your remaining at the Point. I hope you will be guided by his judgment. I will not write any more at present. I fear you cannot read these crossed lines.

Ever afft your Mother

Autograph letter signed, LRMC, MCMP, Ms. Div., LC.

John Rodgers Meigs to Louisa Rodgers Meigs

My dear Mother, West Point, October 9, 1861

I did see in the NY papers that you and Father were at the Astor house and was told of it by several cadets. I thought there was a very good chance of seeing you as this place is only two hours from N.Y. I waited and hoped but when you did not come I concluded that you must be paying only a flying visit of a day and could not come. I was of course very much disappointed. The papers keep me pretty well informed of Fathers movements and doings now a days so that makes up in some sort for not hearing from him. I feel quite disappointed when I do not hear of him in this way.

That was quite a proclamation of his about the blankets the other day was it not. I expect he will get plenty in that way. Would it not be awful for those poor soldiers to have to winter it out without blankets. Things do not appear to be in the position we hoped for at this time do they.

Our troops instead of being in possession of Richmond and on the road to Charleston are lurking behind entrenchments, our armies seem everywhere inferior to theirs in numbers though we have 20,000,000 against 12. To be sure our men seem superior to theirs individually. In Missouri Gen Price is retreating before an inferior army in point of numbers because it is composed of better and more perfectly equipped men. I hope our naval expeditions will be of more advantage than they have been so far. It seemed so impolitic did it not to let the Hatteras victory stand by itself and merely give warning and time for preparation to the rebels against the expeditions that were to follow. They say there is a tremendous expedition fitting out in N Y of about 15000 men.

I saw that our friend Benjamin Corlies Stationer of Brooklyn was confined in Fort Lafayette. If he had any sympathies for the rebels what was he doing about our house last summer. Cousin Fanny has become involved in a law suit the details of which are in the papers on account of those letters that Mrs Lear left her of Gen Washingtons.

I find that I will have to resign my office as sergeant. They want to make those officers of the battalion who are acting assistant professors go to do duty with the battalion an unheard of thing and I am going to resign. Good Night. Love to all.

Your affct Son J R Meigs

Autograph letter signed, LRMC, MCMP, Ms. Div., LC.

Louisa Rodgers Meigs to John Rodgers Meigs

My dear Son, Germantown, Tuesday, October 29, 1861

I had a letter from you this morning the first which I have received for I cannot tell how many weeks. I have written to you *twice* without receiving any answer, and as the last letter which I received contained no allusion to the principal subject of a previous letter I began to feel that these had been unacceptable to you and that my admonitions had not been understood or appreciated as I hoped they would be.

If you knew how constantly you are in my thoughts I think you would not have been so long without writing to me. You say in your letter to me to day that "I have punished you enough," for what I do not know. I might say that you have punished me enough by your long silence to make me anxious and unhappy. I know you are not punctual in such matters, to most persons but I think your love and your duty might *compel* you to write me a few lines every week. There are in our large family circle so many who love you and who watch your career with kind interest that it is your own fault that you do not get many letters from home. You know how many you used to receive when you first went to W Point. You do not take the trouble to write to me, not even to Mary. Of course you can not have many correspondents. In all the eight years that your Uncle Gus was at California before his marriage he never allowed a single mail to leave S. Francisco without sending his Mother a letter, he allowed *nothing* to interfere with this duty. I have reason to blame myself somewhat for not writing to you more punctually. I have had so much headache that I have often excused myself on that account, but I shall not let it be my fault if we do not hear from each other, more frequently for the future.

I have been to Washington lately. Your father came out late on Saturday night to spend Sunday with me. I got his permission with some difficulty to let me accompany him home on Sunday night at 11 oclock. There were many things that we required from home as your father & Col Macomb have decided that it is best for us to remain here for the present. Your father seemed nervously anxious to get me away from Washington. I was there the day of the attack at "Balls Bluff" and he seemed to think that a General engagement would probably follow. That engagement, the loss of Col Baker, so many other gallant officers wounded or taken prisoners, besides the loss of our troops & the moral effects of the repulse are very disheartening & saddening. My visit to W was not calculated to enliven & cheer me. The house was so lonely & deserted. Your father & Col Macomb coming in only late in the afternoon to dinner, that it made me feel sad, and the knowledge that we were to be *exiles from home* for an indefinite period of time was dispiriting enough. We see the effect of these dreadful times everywhere in W. Your cousin Hen I found in the midst of packing and arranging her household effects. They have rented their house and she and Mr Smith and all the family are going to lodgings in *Baltimore.* I suppose the sentiment of the people there is more congenial to the feelings of the family, than any place further North would be. Your Aunt Annie is now at her father's & will I suppose spend the winter there. She left here with your Uncle John some weeks since. He came in from Missouri & reported himself immediately for duty at W. He has gone out with the Southern fleet which sailed from Hampton Roads. He went on board of the *Wabash,* which is under command of your Cousin Raymond. Your Uncle J is to take command of the *Flag,* a small *steamer* I think when he gets to their place of destination. He was not willing to wait until a larger vessel was ready, but another will be given him before long. He told your father he had seen a man from Norfolk who had just made his escape, being a Union man & unwilling longer to remain with those people. He says they have raised the *Merrimac* & fitted her for sea. They have converted her into an iron clad steamer, & that we have no vessel in our service which can compete with her. She can run down our largest ships and has powerful batteries on board. She was not quite completed but they were confident that they could get her to sea in a short time. They expected to run the blockade without difficulty. I hope they may be able to take Norfolk before that is effected. We must all wait in breathless anxiety to hear of the movements of the fleet. I think none of the officers excepting one or two know, where they are going to make a strike. It seems as if we were a long way off, from "conquering a peace."

I hope my dear boy that you will make yourself content at W Point for the present at least. It would be a great pity that you should lose so much of the course of study. I should think it must of course improve your efficiency as an officer. Of course your father is your proper adviser on this subject. We have already so many dear ones engaged in this dreadful contest, that I am distressed even at the anticipation of having your life exposed to its dangers.

Your father is much absorbed, and finds time to write me a few lines only occasionally. He has always been so faithful a correspondent that his silence alone would be enough to show me how fully his time is occupied. I thought he looked weary & harassed when I was last at home, but he feels so deeply the state of our public affairs, that it seems as if he could not relieve his mind of them for a moment.

We have moved into Germantown near the Depot since you were here. The house is pleasant and comfortable & more cheerful than the other. Your Aunt Nannie, Loulie & Minnie constitute our family circle. Your Grandmother and Aunt Jerusha go tomorrow to Sion Hill for a visit. My eyes begin to ache so I must conclude. Direct your letter to Germantown. Write to me soon my dear John. Write to your father also, and do not be surprised if he does not write to you very soon. Have you written to your Grandfather lately? He loves you so much that I wonder you do not write to him frequently. We all love & think of you & talk of you & look anxiously forward to the time when we shall see you again. May God bless & keep you.

Ever in the prayer of Your loving

Mother

Autograph letter signed, LRMC, MCMP, Ms. Div., LC.

Mary Montgomery Meigs to John Rodgers Meigs

My dear brother — Philadelphia, November 3, 1861

I received your letter several days ago, you can't imagine what a relief it was. From your having maintained such a determined silence to mother in regard to H.C. we had made up our minds that you had in some way committed yourself to her, and knowing it would not be approved of, had determined to say nothing. Mother spoke or wrote so often about it that she thought you were angry and so did not answer her letters.

I can't tell you how glad I am that [you] don't care for Miss C. except as a friend, but don't talk in such a silly way about never liking another girl as you did her. All boys feel so till they grow to years of discretion. . . .

Goodbye my dear brother and believe me ever your affectionate sister
Mary

Autograph letter signed, JRMC, MCMP, Ms. Div., LC.

Montgomery C. Meigs to John Rodgers Meigs

My dear Son, Germantown, Sunday, November 3, 1861

Your mother sent to me your last letter to her. I am sorry to see that you show some sign of discontent. Do not think that I love you less that I seldom write. I love you dearly my boy. I am proud & gratified at the good fortune which attended your good conduct on that fatal day when our soldiers patriots ill led were repulsed by beaten traitors from the first great battlefield of the rebellion. Out of that strife you brought my name unstained.

Go on in the service deliberately & carefully worked out for you, prepare yourself to act well your part in the conflict of which no man can see the end. The great fight for liberty & human rights for law & order & civil government against misrule & anarchy & oppression. The end is certain. God fights on the side of liberty & justice & in his own good time will give us deliverance but our nation has sinned, has not valued as it should have done the great privileges won by our fathers' toil & blood. Our people have left the choice of the Governors to the vilest of the low, have been absorbed in money making till they declined to serve their country & now he lays a heavy hand on men & men's purse, & blood & money are to be pressed out like water in defense of what is better than both, but which we have brought into peril by our own unwatchfulness by our selfish devotion to self.

What matters it that this or that cadet, intelligent in drill resigns, & is made aide to some unknown general, or colonel of some volunteer regiment.

Fit yourself for the strife so that when you step forth you may be armed for the conflict & be able to show yourself the soldier of Christ & of your country.

Thousands are now above you in military rank, millions were born & died before you. For you as for all, the great thing needful, is to do your duty, honestly faithfully, bread and clothing you will receive. What more do rank or pay give, a good conscience, a new spirit the consciousness of usefulness, these are the gifts which make men happy. They are not dependent upon Tom Dick or Harry or other positions, rank or pay but upon ones self.

Finish your course at West Point. The country now calls upon you for this as your duty. Do not plunge unprepared into the conflict & add another anxiety to your Mother & to me.

When you have completed your proper course we are willing to give you to the country in the hands of God, a taught soldier able to do good service.

I am here on my way to Boston to look into the means of clothing our army fast approaching five hundred thousand men & exhausting all materials for clothing.

Loulie who lies in the bed near me sends her love.

Your Mother & Aunt Nannie are comfortably established here away from war & war's alarms.

On the Potomac we held the rebel in check but as yet strike no great blow at him. McClellan succeeds Gen Scott in the chief direction of all our armies & I trust will be directed by the Almighty to great success. We fight the battle of mankind. Freedom against oppression. Law against license. Liberty against slavery.

Ever your loving father MCMeigs

Autograph letter signed, MCMC, MCMP, Ms. Div., LC.

Montgomery C. Meigs to John Rodgers Meigs

My dear John — Washington, November 8, 1861

I have just come from the review of the 100 guns & 3500 cavalry, a formidable force in numbers & formidable in fact to such troops as it will probably be called to meet. The material & materiel are admirable. That is the live & dead components of the mass but the men except the regulars want drill & discipline. Of the regulars many are recruits but they have old soldiers on each elbow & will do well. Still I think they are as effective as men that can be brought against them & they are rapidly improving.

The army has been manouvered for years which [illegible] so completely the right to expect devotion from all its members. There are no conscripts no drafts but freemen voluntarily putting on harness & discipline to maintain their govt & save their country.

Rightly headed I think they would do well.

God direct our leaders.

The plan of campaign is a good one. Have we yet the material? That is the question which McClellan & those with him must decide. Material, men, horses, wagons, forage, food, money, ammunition, guns.

The march of an army is a pretty & an easy thing on a map, or as described in history. 60 miles by the light division in 26 hours, but the history tells us nothing of the broken wagons, the worn out men, the abandoned horses, stores, munitions, which strew the route of such a march. To move 50000 men thro Ky. & Tennessee into South West Va., over a sparsely settled country full of mountains & hills & bad roads & no hay or grass or corn or pigs or poultry is not like a pleasant excursion of half a dozen on foot or in carriages finding food enough at every village or even farm house. The great part of a General's labor history does not record. The fighting, the direction, even the planning of the battles occupies in the whole seconds only to the hours of labor involved in the preparation & execution of marches.

I sent $300,000 to the Quartermaster ordered to [Robert] Anderson the moment I heard he moved into Kentucky.

This is but a flea bite, a small sum to organize an army such as he will need.

There are [illegible] in these days & the earth shakes & Capital shrinks at their tread.

We have over 300000 men in camps in these loyal states. The rebels claim 350000 in the border states & 300000 drilling at home. The whole population drills in some parts of the South. Well we have in N.Y. more men fit for soldiers than 2/3 of the whole southern states. NY. should give us, 300000 soldiers. Love to all.

Ever Yours MCMeigs

Autograph letter signed, MCMC, MCMP, Ms. Div., LC.

Louisa Rodgers Meigs to John Rodgers Meigs

My dear John, Germantown, November 9, 1861

I have not much time to write to you this evening. I have just written five letters and my eyes are fatigued. I must acknowledge the receipt of your last letter giving me an account of the Secy's visit to W Point.

I feel much gratified as I know your father will do at the flattering notice taken of you, by him as well as other officers.

Your conduct at Bull Run seems to have given entire satisfaction. That sort of *Eclat* is very fascinating to *women.* If there is any one manly attribute that particularly gains their favor it is "courage." I suppose, because it is opposed to their own timerity, and because they naturally look up to a man's strength & courage as their protection in danger.

I have no doubt the young ladies to whom you were introduced on that occasion, knowing that you had acted well & nobly your part on the field of battle, felt much more interest in your society than if you had never been exposed to such dangers. I am glad you had an opportunity of making some pleasant acquaintances and hope you will continue to cultivate ladies' society, more than you have hitherto done. It gives a refinement and polish that you will acquire in no other way. I was amused at the idea of your dancing the polka, thinking of your attempts at Warren. I hope you will take dancing lessons this winter, and improve yourself in polite accomplishments & break yourself of bad habits. *Tell me in your next* if you have broken yourself of *biting* your *nails,*

& *spooning tea.* Do not forget to tell me, you know how it worried me. I wished to ask you several times lately but feared to vex you. You know it is only my love and my interest in you that makes me anxious about these particulars.

I must not omit to tell you of the arrival of a very charming young lady in our house. Miss Christine Livingston Macomb, a young lady who will make a noise in the world, one of these days if the present efforts of her lungs in that way are any indication. She is really a very nice little baby however, and promises to be very pretty, though her complexion is rather too pink at present to make her decidedly beautiful. Miss Macomb is three days old. I wish you could have seen Minnie and Loulie's astonishment when they first caught sight of it reposing in its little basket. If it had suddenly dropped from the skies at their feet they could not have looked more astonished. Your Aunt Nannie is pretty well and would send her love if she knew I were writing. I read your letter to her giving an account of the Sec's visit. She was pleased at the notice taken of you and entertained by your description. It grows dark and I must conclude.

Ever your afft Mother

Autograph letter signed, LRMC, MCMP, Ms. Div., LC.

Montgomery C. Meigs to John Rodgers Meigs

My dear John — Washington, D.C., November 17, 1861

I send you a copy of Major Hunt's report printed by the Sanitary Commission.

Preserve it.

Col Richardsons report of which I have at length seen a copy uses in reference to yourself very much the same language.

It is on file in the Adjt Gen Office & recommends that you be made a Lieut in the regular Army at once.

It may hereafter be of use to you to know where to find it.

I have not heard from Phil for some days. They were all well at the last accounts.

Ever your affec father MCMeigs

I am glad to see that you have little demerit so far this year, though you will have more than can be necessary.

You took the first place in philosophy and drawing, but why are you 5th in chemistry? You certainly know more chemistry than most of those with you in the class.[1]

Autograph letter signed, MCMC, MCMP, Ms. Div., LC.

1. Shorthand transcription.

James S. Brisbin to John Rodgers Meigs

Head Quarters Recruiting Rendevous
My Dear John — Galion, Ohio, November 18, 1861

It is a long time since I heard from you, nevertheless I have not forgotten you. I recd your last letter after you had reentered upon your studies at West Point. I now write you to say that I have lately learned your father will in all probability be asigned an active Command. In the event of his taking the field I wish to join him and I will join my entreaty with yours that you may also be permitted to go on his staff. Would it not be glorious John for us to get into the field under such a General as your father. I do hope your father will take the field, he is doing a great work where he is, but his position is one of labor, with little renown. He may toil through this war and get but little credit for his great labor. We are making History every day. History that will never die. The men who are now the actors in this Awful Drama are stamping the impress of their characters upon this age in which they live. They are making their names some of "the few immortal names that were ~~never~~ not born to die." I tell you this war is the greatest the world ever saw. There is more at stake in this struggle than ever was involved in any previous contest. The final issue will affect the entire world. Men are now recording their names side by side with the heros of the Revolution. This—the second war of Revolution is of no less importance than was the first war. We cannot fully realize the importance of the time in which we live. We are making History *not* for ourselves but for the world, *not* for our day and generation but for all future time. Is this then a time to be idle, the moments are passing swiftly by, the golden opportunity will soon be gone and we will have done nothing to distinguish our names from the "Common Millions of Earth." Who was Quarter Master during the Revolution? Who in 1812? Can any one tell? And if so who cares. History at least has not recorded the names.

Your father is a cool calm and far seeing man. My Judgment is weak but I think he is Just the Man to Conduct & Campaign successfully. I do hope he will not remain longer in the Quarter Masters Depart. True he is invaluable to the Govt where he is and I question if any one could be found who would be able to conduct the affairs of his office with so much ability. But the field is the place. It is dangerous but its glories will reward the dangers. We have tried one battlefield John and I hope & pray we may yet be so fortunate as to try more. Write your father on this subject. I think it is a duty you owe him to contribute all you can, to his reputation. To the field then. May your father command. May you & I ~~we~~ have *one tent* and may we side by side share the dangers of the battle and contribute our mite to the glory of our Country.

Very Respectfully Your friend Jas S Brisbin

PS Tell me the name of the captain (now Col) who lived with your family when I was there. Write Me immediately "Dont fail"

Autograph letter signed, JRMC, MCMP, Ms. Div., LC.

Louisa Rodgers Meigs to John Rodgers Meigs

My dear John, Philadelphia, November 27, 1861

I received a letter from you a few days since, and have been intending to reply to it sooner but not having made up my mind in regard to our return home, I thought it best to wait until I had come to some decision in the matter.

I am sorry that you seem so confidently & have expected a visit from me for I fear I shall not be able to go to see you at present. When I allowed myself to hope that I might be able to go and see you, I thought it most probable that we should be detained here all winter. I do not now give up the hope of seeing you, some time during the winter if your father should be going to N.York for a few days and felt willing to take me with him.

The truth is however that it is rather an expensive thing making these expeditions. It costs quite as much to make the journey to spend two days, as if I should spend a month, & with a very much smaller amount of gratification. If I had *funds* enough, I should however make an effort, to spend a couple of days with you, but I really think I shall be obliged for the present to forego the enjoyment. I find that the expenses of another lady's dress added to my own, makes a considerable addition to my expenditure. Your father has been very kind in furnishing me with money, but somehow or other, it does not seem to afford me the means of carrying out all my desires and intentions. I have spent quite a sum for Mary, but she is still, like Oliver Twist asking for more. I have not been guilty of any extravagance. I am very sorry my dear boy to disappoint you, but I have found it a necessity all my life to practice a certain degree of self denial. If any opportunity should occur this winter which I can avail myself of I shall not fail to embrace it.

I suppose there are very few of the cadets who can enjoy the pleasure of seeing their friends very often. I hope you will be able to make a visit to your friends here this winter. It will give them a great deal of pleasure as well as yourself. I think I shall go home on Saturday. Mary is very anxious to get back to W. I think she will not find it as pleasant there as she anticipates. The opportunity which she desires for equestrian exercise, will be less frequent than formerly, as the winter is getting so cold.

I sent you a letter which I received a few days since from Brisbin. He is a devoted follower of your father & would I suppose be willing to follow wherever he would lead. There is some truth in what he says. I well know and I should never, seek to interfere with any decision your father might make in regard to taking a command on the field. Yet I cannot help feeling happy to have him where he now is. Knowing as I do that the benefit of his services to the Govt are incalcuable.

I am writing rather in haste as I am obliged to go out. I will send you a letter I recd from your father yesterday. It was sent down to your Grandfather, or I would enclose it now. Send me Brisbins letter when you have read it. I had no idea he could write so well. His letter quite touched me. It is written with so much feeling. In haste

Ever your afft Mother

Autograph letter signed, LRMC, MCMP, Ms. Div., LC.

James S. Brisbin to John Rodgers Meigs

Head Quarters Recruiting Rendezvous

Dr John, Cleveland, Ohio, November 27, 1861

Since writing you I have recd. a letter from your excellent father in which he communicated to me your good success at West Point. He says Genl Cameron even complimented you on your proficiency. Good my boy. I am delighted with you. You have proved yourself as good a student as soldier. My congratulation is not as important as a Secretary's of War but it comes from a heart that loves you with a true friendship. I have not time to write more. Think all manner of fine compliments and credit me with your thoughts I feel there. You know I do therefore I need not say much. My Department here is very laborious. Write me John, don't forget, soon and always remember,

Your friend & obt sevt James S. Brisbin

Autograph letter signed, JRMC, MCMP, Ms. Div., LC.

James S. Brisbin to John Rodgers Meigs

Head Quarters, Recruting Rendezvous

My Dr Lieutenant, Cleveland, Ohio, December 2, 1861

For I will call you Lieutenant & if the law will not allow you the title let me bestow it at least in my connection with you, so far as I can, you doubly earned it on one of the bloodiest fields within the age, and notwithstanding your Cadetship I shall always continue to address you as Lieutenant until you have passed from that to a higher title. I sympathize deeply with you in your present situation. I know how your young heart burns for the contest but be patient John your day is coming. You are very young and can wait for renown. If God in his providence will lengthen but your life to the alloted span of man's existence you will find many opportunities to be useful and grow great before you die. Husband the moments, lay now the solid foundation upon which all your future greatness must rest. America needs surface men who can act on the Moment but she needs far more deeply men learned in the science of war. Look for instance at me, poor devil. I must trust to the occasion and hard knocks to make my reputation while you skilled in the science of Arms can acquire for yourself the reputation of an accomplished soldier and wise leader. In a word *I* must be commanded. *You* can command. You can plan I must execute. *You* can be a man of genius, I must be the machine with which you work your purpose. Unless your noble father (God Bless him I shall never forget him) takes the field do not upon any consideration leave West Point. It would grieve your father and it is your duty to obey him, besides that he is older and wiser than either you or I and although we might not see it there is wisdom in his counsels. Hereafter we are to be a Military people and the nation must have many men of great wisdom to direct her affairs. To such men as you, men who can methodize everything, America looks for her future greatness. The unwise can serve but the wise must rule. Stick to your Books John. Graduate first, fight afterwards. Your father your mother we are all proud of you and I predict for you a brilliant and useful life. Be faithful be zealous and studious, learn to love your country more every day, and America will yet rank and thank you as one of her noblest sons. Remember there is no excellence without labor. Gather now while you can the golden sands of Knowledge for

> Were you so tall to reach the pole
> Or grasp the lightenings with a span
> You must be measured by your soul
> The *Mind's* the measure of the man[1]

Write me often. I shall always be glad to hear from you and will try to answer your letters promptly. In haste,

Very Truly Your friend and Most Obt Sevt

James S. Brisbin

NB, I am sorry you sent my letter to your father. Perhaps he will be displeased with me for speaking so freely and think I am persuading you to leave West Point against his wishes. You had better send this letter to him so he will understand me.

I sent your father your letter that he might see what a dutiful son you were & how gracefully you yielded your wishes to his desires JSB

Autograph letter signed, JRMC, MCMP, Ms. Div., LC.

1. The quatrain, somewhat misquoted, is the conclusion of the poem "False Greatness," by the English clergyman, hymnist, and poet Isaac Watts (1674–1748).

Simon Cameron to John Rodgers Meigs

My dear Sir: War Department, December 15, 1861

The pressure of public business has hitherto prevented me from expressing my appreciation of your gallant conduct at the battle of Bull Run.

It was my determination, after reading those portions of the reports of General Richardson and of Major Hunt, which refer to the part you took in that engagement, to promote you at once by a commission in the Army of the United States. On consultation with your father, however, I learned that you preferred remaining at the Military Academy, in order that you might have the advantage of a full course of Study, and thus better qualify yourself for useful service to your country. This determination meets my approval, and evinces a good sense and a sound discretion not often displayed by gentlemen of your age when promotion is tendered. It convinces me that this brilliant action of your boyhood, is but the commencement of a successful and an honorable career.

I should be glad, indeed, if the War Department furnished any means by which I could confer upon you a stronger mark of its approbation than the mere personal sentiments of its present Head. I trust however, that, in connection with this communication, the accompanying certified extracts from the official reports above alluded to, will serve as a testimonial of your gallantry on the 21st July.

Very truly Your friend

Simon Cameron

Letter signed, JRMC, MCMP, Ms. Div., LC.

John Rodgers Meigs to Montgomery C. Meigs

My dear Father, West Point, Christmas, December 25, 1861

I was very much surprised the other day to find upon my table a big official envelope marked Simon Cameron. I could not imagine what it was until I opened it and then my delight as you may suppose quite equaled my surprise.

I showed it to my class mates and they were all very much pleased and congratulated me right heartily, said they were proud to have such a letter sent to any one in the class. I could not resist the temptation to show it to Professor Church. He seemed very glad that I had received such a highly honorable testimonial from the Sec of War and I know he was too for he likes me as much as I do him and that is a great deal. He is a very kind hearted Gentleman though he makes some of the plebes hate him by sending them off in January. It seems the strangest thing that we should have such a collection of fine officers and noble gentlemen on the Point as we have now when before we have been afflicted with such mean spirited ones.

There are some things open to improvement yet though. They have got lately to giving punishments in addition to the demerit for the most trifling offences so that it is almost impossible for cadets to avoid getting them.

The regulations are almost as bad as the laws of Draco. There were about 200 extra hours of Saturday guard duty and confinement given out for 18 days in last month. One man had to spend all Saturday afternoon in pacing up and down in the area for being late at breakfast roll call, an offence that only gives one demerit. Another had to do the same thing for a little bit of rust in the concave head of his gun rammer.

Many of the cadets have already about as many extras against them as they have a reasonable prospect of walking off.

Do you think Congress is really going to refuse to increase our corps on the ground of retrenchment. It seems as if that would be very much like saving the dime to lose the dollar.

Everybody seems to recognize now the great use that the academy has been to the army and the country. It is famous for the fine officers it has turned out. I should think the argument that one Senator [Zachariah Chandler] used, namely that West Point, had turned out nearly as many traitors as true men might be as well applied to the abolition of Congress itself as to the academy. He may be sure that there are no traitors here now and that is more than can certainly be said of Congress.

Ever Your Affct Son, JRM

P.S. I suppose it would not be proper for me to write to the Sec to express my thanks for his very kind letter but you may assure him that I am fully aware of its great value and shall keep it to the end of my life as my provident trophy.

Autograph letter signed, MCMC, MCMP, Ms. Div., LC.

Charles D. Meigs to John Rodgers Meigs

710 Walnut Street
Philadelphia, Pennsylvania, December 31, 1861

My dear Johnnie,

I must put the *nie* to it!

I should vainly search all over & through the records of my brain to find the words [that] might fitly interpret the full sense & totality of my love for you, you *Darling* boy.

Now what do you say to that for *highfalutin!* Your letter came to me about 11 ½ A.M. to day, but *opened,* with a pencil endorsement by Mollie asking pardon for the rudeness, but "I wanted so much to hear from Dear John you know." So she opened it at 1208 and then got into the carriage that took her to the Depot, for Washington, where she now is, I hope in safety. We all read your letter, & all of us, with great pleasure, and I truly believe we shall all always have true joy & happiness when we hear from you, provided it shall always assure us of your health & safety. If there is a prime element in human happiness that arises from the inward conviction that our friends do truly love us & trust us & glorify themselves in us then you ought to be a happy youth, for it is true as death that one and all we are looking to your career for our honor & for a large part of our satisfaction. This is universal as to *our* people. We have no doubts nor any fears.

Tonight Uncle Frank brought me the beautiful map and as you know that Hamtt. is my fixed idea, I need not tell you how much pleasure it gave me, nor need I say a word about the beauty of its execution. I am to have it put into a nice frame with a glass to keep it safe & of course shall not forget who made it, nor the days when I saw you so earnestly laboring for the sole kind intent to give a great pleasure to your old Grandfather. Let the whole of that [illegible] my debt as "good works" & put good St. Peter in mind of his Tablets when you come to knock at Heaven's gate. He is to be sure to say yes, come in, for you have fulfilled the commandment, "honor thy father & mother," I am only your father once removed. But enough of this after saying my dear John I sincerely thank you for your kindness.

I am sorry we could not see you here for Christmas, never mind, the day must come in which we shall see you and you know how welcome you will be.

If you had come however, you would have found me very melancholy, for I am sure perplexed about the war, & my soul sickens with the thought that the entire remainder of my life, which I thought to play away so peacefully uneventfully & happily at Ham is to wear out amid continual alarms for my country. I don't live without hope, for I hope in my son, and I can look forward to the not distant day when you will give your courage and science & moralities to the support of those inestimable rights & privileges, which those dreadful men are trying to overthrow forever. I pray God to inspire your heart with [illegible] to your native land, to make you sink your own selfish interests in the presence of the rights & happiness of man, everywhere & raise your arm to strike down every assassin of these blessed liberties & franchises that your father upholds & for which your venerable grandfather & your uncle John so often offered their blood if the pouring of it out might be good for the safety the welfare & the Honor of this country. If we can keep clear of the threatened dangers from England, I believe we can win back our erring southern brethren, & I pray God enable our people to do so without the loss of another American man. If England were either wise, or honorable, it would give us assurance of nonintervention & leaving us to our own resources, I believe we could have a *cessation* of *war* within 6 months. But she hates us, & if she should attack us, I trust [illegible] was enough to hear of your leading our people to Ireland to [illegible] forever out of that land. Yes I would gladly hear of your entry into London at the head of the real nation. I meant the democracy of England, *they* do not hate us but only the detested aristocrats. But I must stop. Emlen is in full progress. He has to be quarter master & is to supply the wants of [illegible] I have entire confidence in his probity and ability. All our people are well. Christine is growing up & will become a beautiful and an admirable woman. So good night my dear boy. I read your battle of Bull Run this very evening. [illegible]

CDM

Autograph letter initialed, CDMC, MCMP, Ms. Div., LC.

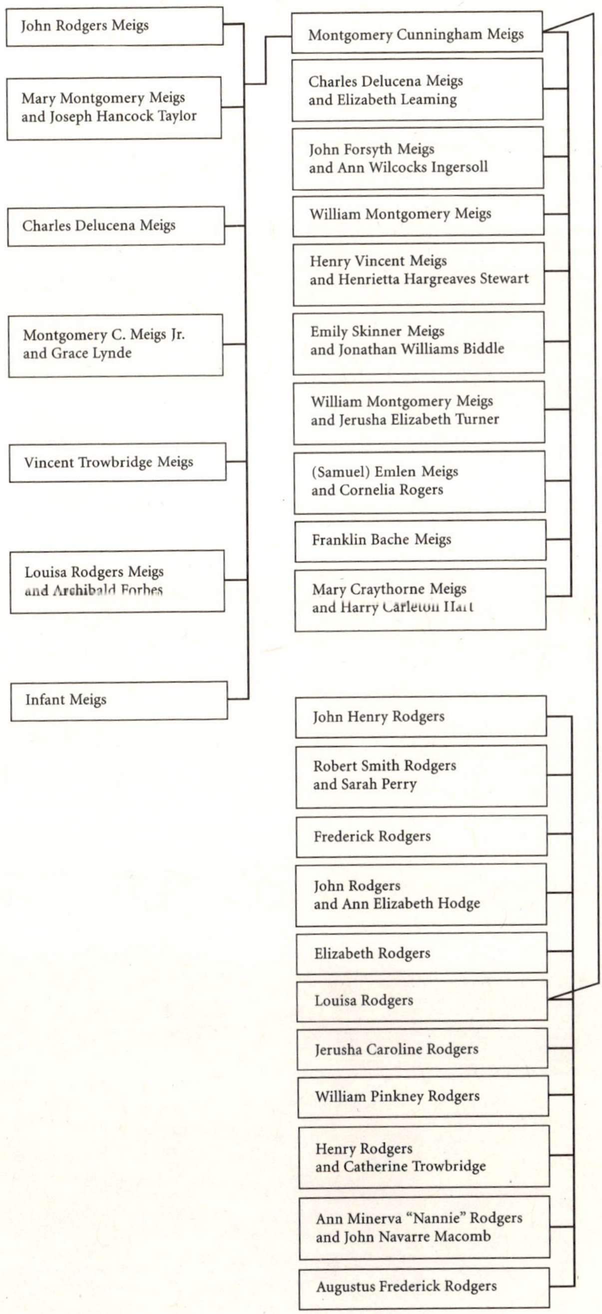

Family chart of John Rodgers Meigs

Montgomery C. Meigs, architect and engineer, served as quartermaster general of the Union Army during the Civil War. He was the primary role model for his son, John Rodgers Meigs. Courtesy of Library of Congress.
LC-DIG-CWPBH-03111

Louisa Rodgers Meigs, daughter of Commodore John Rodgers, head of the famous naval family, married Montgomery C. Meigs in 1841. She provided her family, and in particular her son, John Rodgers Meigs, with social and spiritual guidance. Courtesy of Library of Congress.
LC-USZ62-120546

John Rodgers Meigs, who was interested in science, is pictured examining a specimen prior to entering West Point. Courtesy of Library of Congress. LC-USZ62-120545

Mary Montgomery Meigs Taylor, younger sister of John Rodgers Meigs and one of his most frequent correspondents, is shown reading in the garden at Hamanassett, the home of her grandfather Charles D. Meigs. Source: Montgomery C. Meigs Journal, 1857–60, Montgomery C. Meigs Papers. Courtesy of Library of Congress.

Mary Montgomery Meigs Taylor, as a young woman. She named her first-born child after her brother, John Rodgers Meigs. Source: Montgomery C. Meigs Papers. Courtesy of Library of Congress.

Ann Minerva "Nannie" Rodgers Macomb, the favorite aunt of John Rodgers Meigs. Courtesy of Library of Congress. LC-USZ62-2029

Charles D. Meigs, a prominent Philadelphia medical doctor, was the paternal grandfather of John Rodgers Meigs. Courtesy of Library of Congress. LC-USZ62-116544

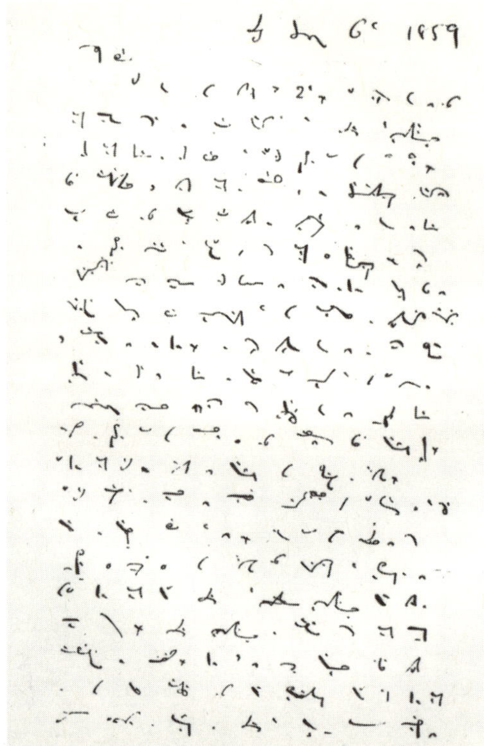

Photograph of page of letter written in Pitman shorthand used by John Rodgers Meigs and Montgomery C. Meigs. Source: Montgomery C. Meigs Papers, Correspondence, 1855–1864. Courtesy of Library of Congress.

(*below*) West Point Class of 1863. John Rodgers Meigs is pictured in the first row, sixth from left. Somewhat conscious of his height, he may have taken this position to minimize his self-perceived short stature. Courtesy of Library of Congress. LC-USZ62-39352

John Rodgers Meigs in West Point dress uniform. From a glass-plate photograph made in January 1862. Courtesy of the Honorable John Rodgers Meigs Alger, great-great nephew of John Rodgers Meigs.

Philip H. Sheridan was Union commander of the Army of the Shenandoah and of the Middle Military Division in 1864. He ordered the town of Dayton, Virginia, and all homes and barns in the surrounding area burned in retaliation for the death of John Rodgers Meigs. Courtesy of Library of Congress.
LC-DIG-CWPBH-01010

Benjamin Franklin "Frank" Shaver, a member of Company I of the First Virginia Cavalry, claimed it was he who shot John Rodgers Meigs. Courtesy of Betty Martin, Harrisonburg, Virginia.

George W. "Josh" Martin, a member of Company H (the Black Horse Troop) of the Fourth Virginia Cavalry, held himself responsible for the death of John Rodgers Meigs. Courtesy of Russell Hicks, Altavista, Virginia.

John Rodgers Meigs's sarcophagus at Arlington National Cemetery. Montgomery C. Meigs designed the bronze relief to show his son's fallen body. The revolver has an empty chamber to reveal that young Meigs had not surrendered but fought against Confederate soldiers who attempted to capture him. Photo by Mary A. Giunta.

Montgomery C. Meigs Family photographed at the Washington, D.C., family home. John Rodgers Meigs Taylor, the namesake of John Rodgers Meigs, is pictured seated to the left at the feet of his mother, Mary Montgomery Meigs Taylor, younger sister of John Rodgers Meigs. Source: Montgomery C. Meigs Journal, 1857–60, Montgomery C. Meigs Papers. Courtesy of Library of Congress.

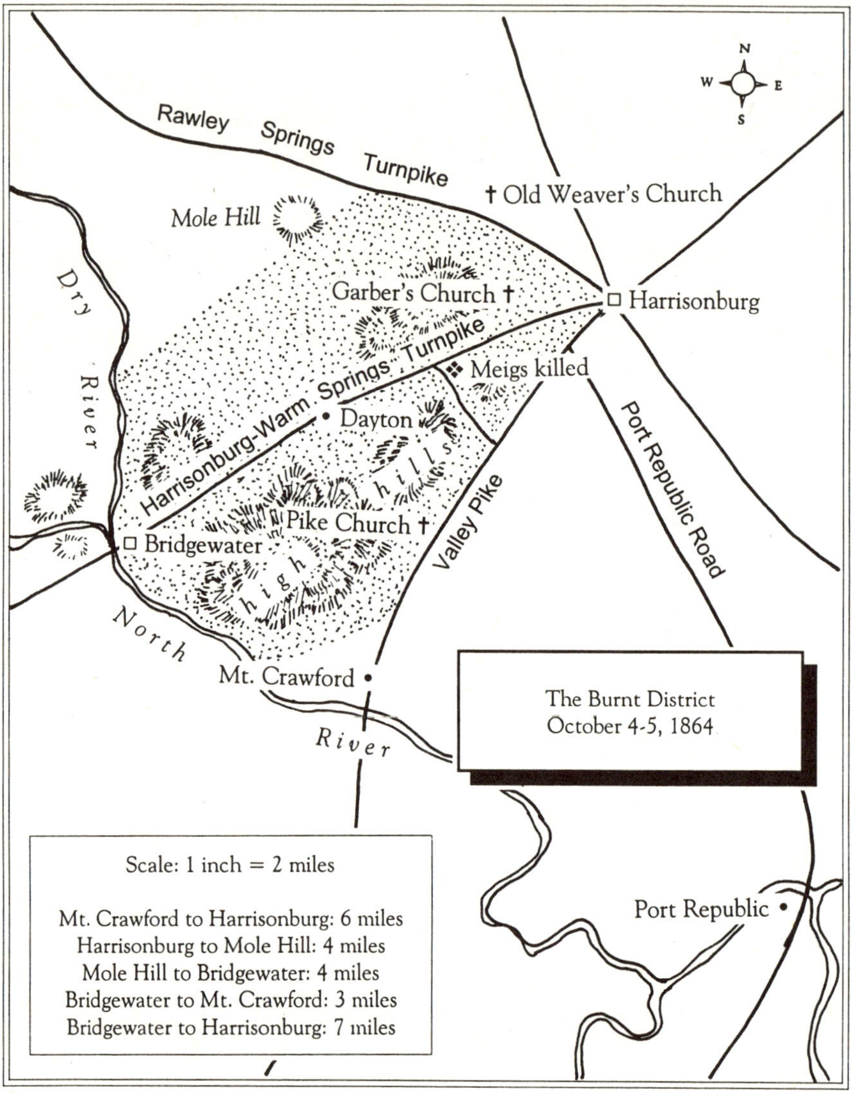

Map showing site of the death of John Rodgers Meigs and area where Union soldiers burned homes and barns. Reproduced courtesy of Katherine Tennery Bennett.

CHAPTER 3

A Cadet in Trouble

While some remained loyal to the Union when the war began, many Southern cadets, including those scheduled to graduate in June 1861, resigned from West Point to serve in the Confederate cause. With the attack on Fort Sumter on April 12, 1861, the remaining members of the class of 1861 petitioned the secretary of war and were allowed to change the date of their graduation to May. Members of the class of 1862 were granted their petition for early graduation in June 1861. John Rodgers Meigs and most of the class of 1863 stayed on to continue their studies.

By the beginning of 1862, some of the glitter and romance of the war was fading; by the end of the year, thousands of Americans on both sides were war casualties. For young Meigs, the year would be eventful at the Point. He would be challenged to maintain self-discipline under the rules and regulations of the academy and to remain there after being placed under arrest; he was also given fifteen demerits for speaking to Abraham Lincoln about his punishment. On a happier note, he enjoyed visits from his mother, sister, and other family members during the summer encampment with its accompanying social activities. In September, he returned to his classes and teaching responsibilities, and by the end of the year, he was looking forward to the Christmas holidays and a visit to New York City.

John Wood to John Rodgers Meigs

Dear Friend, Washington, D.C. January 3, 1862

I have long been anxious to find some subject of more than common interest that I might send you copies. The inclosed will prove to you that my search as promised [was] successfull. The Genl has at last found time to give me a sitting. And I hasten to send you one of the proofs.

I called on your esteemed Mother, on the 1st Inst., and then met for the first time your Aunt from Phil. The conversation turned naturally on yourself, and if you could have witnessed the expressions of pride and thankfulness for the past and hopefull expectations for your future, it would I am sure be a reward and an incentive, if any was needed in your case, to continued exertion to excell.

Your esteemed Father continues to grow in favour both in and out of the Army. In Congress his opinions are looked for and depended on. Indeed, many have expressed to me that he is the man for the times.

It is a common saying that Circumstances have made most great men, but in my opinion a man must be truly great in himself, first. Circumstances may be the means of making known his abilities, while at the same time the same circumstances would only prove the worthlessness of an inferior mind.

"Honor and fame from no condition rise. Act well your part there all the honor lies."

I thank you for your kind inquiry about my Wife and Boy. They wish to be remembered to you. Tho late we wish you a happy New Year. Yours truly,

John Wood

P.S. I find I have not mentioned the news here that you had lately recd a very Complimentary letter from the Sect of War, it is proposed by your Aunt that you furnish us with Copies, and for the purpose that you send it to me to be Pho., but perhaps this would not be proper.

Yr. J.W.

Autograph letter signed, JRMC, MCMP, Ms. Div., LC.

John Rodgers Meigs to Louisa Rodgers Meigs

My dear Mother, West Point, January 15, 1862

Mr Wood sent me what I thought a first rate picture of Father, the other day, in his General's uniform. I do not think it is quite so clear as that splendid one he took of Capt Macomb just before I came away but still it is a decided success and is much superior to that horrible one of him in his dress as captain of Engineers.

I have made a neat little frame for it and put it up on the wall just where I sit to study so that I can see it all this time. It seems almost as if he was looking at me himself sometimes. The cadets who come in admire it but to my sorrow they can see no resemblance at all to me.

I wish that I was tall like Father. I do not like the idea of being a short man especially as I am to stay in the army. My encouragement is that you all said I had grown some when I came back on furlough while I thought I had not grown at all. I am so glad that Mr Stanton is Sec of War. Father could not have a person whom he desired more I suppose unless it was Holt but perhaps he will be in yet in Welles' place, the people want it to be so, he has proved himself such as a staunch patriot and brave capable man. I am sorry for Mr Cameron but perhaps he will make quite as good a minister as cabinet officer.

Can any thing be more admirable than the way in which the President manages his affairs without fear or favor of any party or any body. Who thought that he would be such a splendid president. Mr Greely did not I guess or he would never have helped him to his place.

Why is it that Uncle John does not have a higher place given him and a more important position. Blockading the *Fingal* with the *Flag* is I would think rather small business for him. I wish as Aunt Annie does he could catch her. There is Cousin Raymond, whom I always supposed would rather look up to Uncle John as his superior in other things as well as rank making a fame for himself that will probably end in getting him a Commodore's appointment. I am sure Uncle John would if he had the chance offered him. I give you the marks as they were just before examinations. I had lost in Philosophy 38 tenths Michie 91 Rabb 111 McKee 136 King 192 Howell 295 Twining 362. In chem Michie had lost 3 tenths Rabb 13 Meigs 9 Townsend 16 Twining 6. We came out thus

Philosophy	Chemistry	Drawing
Meigs	Michie	Meigs
Michie	Meigs	Holgate
Rabb	Twining	Townsend
McKee	Rabb	Michie
King	Townsend	Rabb
Howell		

Professor Kendrick told us today in the section room that he had been in doubt whether to place Michie or myself at the head and he was not sure whether it ought not to be Townsend and Rabb instead of Rabb and Townsend. He said he did not suppose Mr Michie would object if he should say that he thought Mr Meigs knew as much or more than he did about the course but his marks were not so good and so he did not put him head. Michie is the most generous fellow I ever knew, he always seems to think he is standing higher than he deserves when the truth is often that he is standing lower so he said that he thought I ought to be ahead of him.

It would be a gross piece of injustice to put me there with the marks I have and I am very pleased the Professor did not do it. I expected to come out 3d instead of 2d in Chemistry.

I am sorry Aunt Mollie is going to leave you so soon just as she has begun to get comfortable and to know the ways of Washington City. It is a pity people can not afford more parties now adays. I should think they were quite necessary to keep society stirring. If they were not so expensive when they do have them there might be more of them.

What a horrid beastly picture that is in Harper's Weekly of poor Miss Matilde Slidell. I did not think such a journal as that is would disgrace itself by descending to such miserable attempts at wit as that. I should think any one might see who feels in the way the maker of that picture did upon the subject that the more silent we were the more graceful the surrender would be. It will spoil all the effect of Mr. Sewards nicely written letters if the journals of the country are filled with such foul things as that. What better satire could Punch have than to copy that picture and say where he found it.

Love to all. Ever your affct Son JR Meigs

Autograph letter signed, LRMC, MCMP, Ms. Div., LC.

John Rodgers Meigs to Louisa Rodgers Meigs

My dear Mother, February 9, 1862

This makes the third birth day I have spent away from you. I am 20 years old. I would rather be younger until I graduate. I used to pride myself when at College about being a little ahead of my years in education but I am not so now. I have always thought that ⅓ of a mans life is a good deal to spend in getting ready to spend the rest and one third has gone by if I live to be 60 years old. I hope you are still keeping up the idea of coming on here next summer, it wont do at all to put it by. I hope to have a nicer time than I have yet spent at West Point which has been a dull place to me. Perhaps it is better so as I have had nothing to interest me but studies. I hear continually evidences of the value of this education to men who are engaged in the present war. There was a very fine high spirited little fellow here last year who stood at the head of the present 1st class in math. He unluckily got into disfavor with some of the officers who used to persecute him as we thought and showed their dislike in various ways. Two or three of them were perfect fools any way and Wetmore got the better of them in two or three scrapes which increased their bitterness towards him. I heard an officer acknowledge a few days ago that he had never seen such a bitter feeling as one of those officers displayed towards Wetmore. Well after having been courtmartialed two or three times he was dismissed or resigned I forget which and left last summer while our class was on furlough. He went out to Ohio and took command of a battery which was raised somewhere near his native town and was engaged in the battle at Somerset. His battery was the most or one of the most effective on the field and turned the tide of battle, as they say. He is looked on as quite a little hero, just past 20. It was miscalled Wheatmore's or Whitmore's battery in the papers. Of course we are all delighted at Wetmore's success, such things are good for the credit of the institution.

Good night. Love to all.

Ever your affct Son

PS I have received all your letters and am very much obliged to you. I send you a letter from Monty, how nicely he writes.

Autograph letter signed, LRMC, MCMP, Ms. Div., LC.

Montgomery C. Meigs to John Rodgers Meigs (Shorthand Transcription)

My dear John: Sunday, Washington, February 9, 1862

This is your birthday. We are sitting at the table in the back parlor by the light of the bright gas. Your mother is reading the story of Mount Vernon, of the relics of Washington, most of which have lately been taken from the house at Arlington that the traitor Lee left, these relics of the father of the country which he has deserted.

God grant to us, my dear son, the time to see you do some great good to your country, which I think will, in the rest of my life and during the beginning as well, be in need of the talents, ability, courage, and patriotism of all our sons. Here we have a great army, well equipped, in good spirit, full of patriotism, courage, and zeal, waiting, I suppose—remember, I am not in the secrets of General McClellan—for the drying up of the most muddy roads that I ever saw.

I see that one of the boys that lately resigned from West Point has had the opportunity to win such a name of credit for himself as you did at Bull Run. He resigned, got command of a battery based at Cleveland, Ohio, and caught up with Thomas' army just in time for the battle. He had spent 3 years at the academy.

I am glad to think that you are at the academy, learning that which will make you hereafter more useful to our country than you could be now. At your present age, you could be only a lieutenant of a company. You would not have the chance perhaps in a long time of being as useful as you were at Bull Run. I hope to see you someday an engineer or perhaps a successful general. What I would most like, however, would be a successful completion of this year before you leave the academy.

It is trying the strength of the nation. Men enough full of patriotism are to be found to supply our armies, but money is [illegible] hard to get. Taxes, taxes, these are what we need. The Congress is taking scolding about the exaggerated tales they, themselves, spread about profiteering and cheating in contracts, keeping out of sight and out of mind the fact that millions of dollars are honestly earned by men who have worked for the government and given thousands of their property to it and who wait patiently for payment until the Congress raises the means. We owe immense sums of money to these contractors. To some companies we owe as much as $400,000. To some single persons we owe $200,000. We wait, they wait, do not complain, while Congress talks.

Why don't some of those who are able and direct the war and criticize those who stand ready whenever the proper command comes to go out, taking their lives in their hands, to fight the battle of the country—why do not some of those go into the field themselves? Some of them would make good private soldiers who are now very bad legislators. They lost the battle in Kentucky, so [illegible] won. The fight was won by the bayonet. Some of the [illegible] boys were killed with Bowie knives, several with bayonets, and they took the muskets out of the hands of the Mississippi Tigers. Thank God our men do not take such bonehead and bloody-bones names. They are not pirates.

We are all well at home. I have not written to you for some time because I am so busy that I really do not find a moment from one week's end to another in which I can sit down to write, except a business letter.

I need not tell you how much I am gratified by your good standing at Jan examination. I trust that you will take the head of the class, after all your demerit. This is the only thing I [illegible] with you about, except that I wish you would be more careful and learn every lesson thoroughly. You sometimes get bad marks in the recitation. This can only be from not having well studied the lesson of that day. The practice which you will get by teaching the calculus to the class under your charge will be of great use to you in making you more ready in its implementation.

God bless you, my dear son. I am proud of you and look to you for much comfort hereafter. If in this war I am struck down, I shall leave your mother with confidence that you will take care that she never suffers from want. I have not been able to make much saving from my pay, having always found that she was ready to spend the whole of it. Even now, when it has become larger, I have the greatest difficulty in making her believe that she ought to put away a part of it, as it will not last forever, and it is wicked to spend money uselessly in such a time as this.

I wish to live as we have heretofore and put all that I do not find it necessary to spend into the loans, so as to contribute my might, both in money and work, to the saving of our country. I think that the rebellion is now beginning to go backwards. With good roads to successfully fight the war, its back and neck will both be broken. God give us both in His own good time.

Ever your loving father, Meigs

Autograph letter signed, MCMC, MCMP, Ms. Div., LC.

John Rodgers Meigs to Charles D. Meigs

My dear Grandfather, West Point, February 17, 1862

I received another of your kind letters and one from Monty to day. I have sadly neglected my duty in writing to you and to Monty indeed to every body. Even Mary complains. The only reason my dear Grandfather is that I am generally a little ashamed of my letters. You I suppose whose thoughts and words flow out so smoothly and naturally as water from a fountain cannot appreciate the sufficiency of such an excuse. I do not believe you have felt your tongue tied up or your pen constrained to move through tortuous ungraceful sentences, of whose defects you were fully conscious but which you could only improve by the labor of revision, very proper for matter which is to be published but not at all so for a natural familiar letter to a friend. I have felt this and do every time I write and it is the reason I write so seldom. It is very dispiriting is it not, to have the knowledge of a deficiency of that kind pressing upon you in conversation and letter writing and is too apt to make one shun society or feel constrained while in it. Do not let any one see this dear Grandfather for I should not like them to read it. I know you are so charitably disposed towards me that you will not think the worse of me for it and may perhaps suggest a plan for getting over it.

I am always flattered by your letters more than is good for me but you make me feel like living for higher and nobler motives than I do. I see the same thing in Fathers letters, when he does find time to write to me, he urges me to work and study and improve that I may do good not to myself but to my fellows and countrymen. I am convinced that these are the principles that rule his own actions but I feel that they are very far from being mine.

I was delighted to hear of Uncle Em's appointment though I doubt whether it was for his own interest to give up his business for service in a short war. It is a great thing however to have had an influence for good in controling the great events fraught with good and evil for the human race that are happening every day. I suppose that in the days to come the actors in this war will be looked back to with the same feeling that is held towards the heroes of the Revolution. It is a struggle for principles as great and perhaps greater than formed the issue in those days and it is certainly on a far more magnificent scale.

I hope McClellan is to be proved the greatest General of the age by the success of his plans. If he can carry out that grand idea of encircling half a continent with a cordon of armies and by one tremendous advance compelling it back to its allegiance, he *will* be the greatest general of the age.

He has the best opportunities for knowing and from his conduct in stopping the pursuit of a flying enemy when we had obtained a great advantage. We can only infer that he is *sure* that when he does strike, the blow will carry all before it and that he gives up these local advantages because he is certain of making the crash universal and simultaneous.

The only fear is that that pitiful mud will make all this too late so that even if the victory be gained its effect will be lost by the action of Europe.

If that disinterested holy and impartial England undertakes to interfere with us as she has with Mexico she may get herself knocked flatter than Spain has been on her first advance in Mexico. She has caught a Tartar there has she not that she little expected to find. Perhaps that wicked attack on a weak nation by three great robbers will be the means of uniting her again under one strong government of home manufacture, but as the great object of the allied powers was "to cure the dissentions of unhappy Mexico" they may conclude that their end has been obtained and retire satisfied.

I read General Beauregards interesting account of the battle of Bull Run. He seems to tell a straight story and one that renders a good many things intelligible that were not so before. His statements are sometimes very onesided however. In one place he speaks of the attack made on the position of our battery and says nothing of the ignominious repulse sustained by two regiments from the fire of 6 guns but on the contrary compliments them highly for their courage. I believe that there were four regiments instead of two that attacked us but he only complimented the two. We had a couple of regiments behind us but they did not fire a single musket. The 6 guns drove back about 30,000 men.

In another place he says that General Bonham thought proper on account of darkness to direct his brigades back to Bull Run but says nothing of the cannon balls he met coming down the road as soon as his force came in sight of our position. This was the time that that body of cavalry appeared coming along the road from behind a piece of woods and were shot at from our battery on the hill at Centreville. It seems from the report that our position narrowly escaped being the most bloody one on the whole field, for he had ordered an advance of his whole army upon it but the order miscarried. This fully justifies McDowells leaving those heavy reserves inactive, for which he has been so much criticised. If that order of Beauregards had not miscarried he would have had full need of them to escape destruction, but with them there I believe such an attack would have ensured the victory to our side as our whole force would have been brought into play then. Lest you have not seen it I send you the report. Love to all.

Ever your affct Grandson, JRM

I will write to Monty soon.

Autograph letter initialed, CDMC, MCMP, Ms. Div., LC.

John Rodgers Meigs to Louisa Rodgers Meigs

My dear Mother, West Point, February 24, 1862

We have today one of the hardest winds that has blown at West Point since I have been here. It roars over the chimney like a great organ pipe and it is almost impossible to make head against it. I had to laugh this evening at little Mordecai, as I came up from supper, I saw him with his martial cloak wrapped tight, and his arms folded about him with his head bent down to the front, boring his way along through the blast. Pretty soon he got onto a sleety piece of Ground, just then the wind redoubled its efforts and he came to a halt, then commenced sliding backwards until his feet came in disagreeable proximity to a gutter full of water, so he turned around concluding if he must go in to go in forwards. His feet reached the water, his hat blew off and went rolling away over the snow and he brought up against the snowbank with his two arms thrust down into it about a foot. Then to cover his mortification and add to the effect his cloak blew quite over his head. The last I saw of him he was remaining in this position with his feet in about 2 inches of water apparently cogitating as to what he had better do. I would have got blown over a bank about 20 feet high this evening if it had not been for a fence that stopped me. The wind is ______

We have had a great excitement. As I was writing I heard a hollering out in the area and soon made out that every cadet that had a throat was bellowing firyin! firyin! We rushed out and down into the engine house, each seized a bucket and off we started for Professor French's which was burning down, as we panted across the plain we could see the sparks flying like long sheets of fire in the wind.

Just as I reached the gate I saw Mrs French open the door, I rushed at her, tumbled head over heels over a snow drift and up the steps onto the porch where I put out one foot in a masterly manner to stop myself but out it slid and I laid full length before her. I picked myself up in a hurry however and was surprised to see that she did not look excited only surprised at my want of ceremony. What's the matter Mrs French? said I. I don't know, said she. What's all this excitement about? Nothing only we understand your house is burning down. Why its the first I knew of it perhaps its only the chimney won't you go up into the garret and see. So Lieut Adams and I went up into the garret, an *Irishwoman* brought *a candle* to help see the fire with if there was one and we discovered that it was the chimney "only that and nothing more." So the crowd was pretty badly cold and somewhat disappointed, not that they wanted the house burnt down but then they would have liked just a little out building merely to have some fun and get off lessons tomorrow. If it had been a real fire the wind would have swept it over nearly all the houses on the Point, so it was especially lucky that it was only a chimney. It added another proof however to the necessity for good fire engines here. We have two useless old tubs with bursted hose and squeaking pumps whose worthlessness is so well recognized that we did not attempt to haul them out. I can spit higher than they can throw. If Professor Frenchs had been really afire 60,000 dollars worth might have been destroyed.

Just think what an enthusiastic fire company we would make. All living in a great hive this way we could turn out on a moment's notice. I suppose *economy* prevents our having one though.

I have enjoyed to night the reading of Mr Cox's analysis of Mr Gurley and his military plans among which is that of leaving the battlefield by the help of an oxtail and shouting Bull! run!, Bull! run! I expect McClellan enjoyed the speech next best to Mr. Gurley. You must have read it.

I was sorry to see a statement in the paper today, that Mr Stanton did *not* at that railroad meeting pay a high tribute to General McClellan as previously reported. I should think that on account of the rumors floating about concerning him and McClellan it would have been a valuable opportunity to contradict them.

I hope and I suppose most army men do, that Mr. Stanton is not going to allow the credit of our victories to be fastened onto him by the politicians of the *Tribune* and *Post.* Or even if he denies that he is entitled to it, I hope that he will go farther and show who is. In that letter to the *Tribune* he speaks rather contemptuously of the idea of "*organizing victory*" as if the honor of a victory was due solely and entirely to the men engaged in the fight and not at all to the great chief who has collected and disposed the armies and planned the whole campaign. General McClellan does not seem to be getting his full share of credit for our glorious successes and yet it seems only envy and politics that prevents him from getting it. At the Point here we all look up to him as we used to General Scott as the perfection of every martial and noble quality.

Does it not do you good to read of the way the old General is treated in New York by the citizens whenever he is seen. At the great mass meeting at the Cooper institute at every allusion to the hero of Mexico the whole audience of thousand's of people rose and gave hearty cheers for him and they were only representing the feeling of the whole country towards him.

We all think we know that he lost the battle of Bull Run for us but it does not seem to have altered his place in the hearts of our people.

How do army officers in W feel towards McClellan, do they stick to him as at first or has jealousy crept in and made them side partly with the abolitionists.

I often think as you do that Father might now be gaining the place in the hearts of the people that Halleck is if he had taken command of the department of Kentucky but he has done what he always has done, and what has elevated him so high already and that is what he thought was *right.*

We had no extra holy day on last Saturday but were assembled in the chapel to hear "Washington's Farewell Address." What a wonderful paper it is. I had never read the whole of it before. A salute was fired and the band played Hail Columbia, Washingtons March and the Star Spangled banner with quite stirring effect.

It is a great pity that Uncle John did not keep the place he had on the Missisipi. It seems to have been impossible for any honest officer who was not a fawning sycophant to keep on good terms with Fremont. I was glad when he was relieved.

I am afraid that Mrs Reales family has not a great addition in Gibson but he may make a more sensible officer than he did cadet. I have noticed that some fine cadets have turned out to be worthless officers. Good night. Love to all.

Don't let Father see this letter or he will say I am conceited and so I would be if I were not only talking to you and those in the family.

Ever your affct Son J. R. Meigs

P.S. That printing is done with waxed paper made by melting wax on a slate and soaking the under side of the paper on it. I send you a piece.

Haven't I written a long letter now, though I sat down thinking I could not scrape together more than a dozen lines.

Autograph letter signed, LRMC, MCMP, Ms. Div., LC.

John Rodgers Meigs to Montgomery C. Meigs

My dear Father, West Point, March 11, 1862

I was made happy to day by the receipt of 5 letters half as many as I had during the whole of last month.

You had an excited time in Washington over the news of the *Merrimac's* exploits. It is certainly nothing more than the Navy department deserved after all the warning they have had. It seems as if anybody who knew that iron was harder than wood ought to have known what would happen when she got out. It is now proved to those who know nothing about the construction of vessels that iron clad ones are altogether superior to wooden ones.

I hope Congress will now be induced to finish the Stevens battery. It is a disgrace to the wisdom of the nation to have sunk half a million in a project which is decreed worthless without a trial.

Do you now think that the superiority of our rams over cannon balls is also shown by the way in which she finished the *Cumberland.* She could not have been sunk in so short a time by the fire of a whole fleet of frigates.

They will soon have to return to the old Grecian method of fighting by running vessels into each other for two such vessels as the *Monitor* seem to be able to fight with balls for an indefinite length of time without injuring each other. I should think the *Minnesota* might have destroyed that vessel if she could have got off the shore by running right over her. She is so low in the water and her roof slopes up in such a manner that a large vessel like the M ought to be able to mount right on top of her and press her down by sheer weight.

We heard this evening of that grand order of the Presidents decreeing an advance of the armies and navies of the United States along a line thousands of miles in extent and with a force of more than 700,000 men.

Such a spectacle the world has never seen before nor such a general either as the one has conducted operations so as to render it possible. I wonder whether England will pooh pooh the victories that have been won in the last month as she has every thing that has happened yet. Mr Russel seems to be a good representative of England does he not. Every thing he says turns out false and every thing he prophecies is evil which we cant help thinking is what he and his nation wish us. How we would like to have a fight with England, at the end of this war and how sympathizing and respectful she will again become when she sees us successful and feels our strength.

Do the rebels mean to fight us anywhere or to keep forever just out of our reach as Price has tried to do in Missouri. I hear that the reason they left their Potomac batteries was that they heard McClellan was going to put two divisions of his army behind them by sending them up the Rappahannock.

I wish every day that I had graduated 3 years ago. It is so hard to be here while such events are progressing. What perfect bores Astronomy and Chemistry are after reading of the attack on Fort Donelson or the evacuation of Manassas or something of that kind.

I don't see how those officers who are detailed to stay here as instructors can stand it at all.

Have you ridden out with the army at all to see the advance. The newspapers say that it presents an entirely diff appearance from that of July, that there is none of that straggling and dirt that was then every where. I should like to ride over the field of Bull Run again, should not you. Love to All

Ever Your Affct Son J. R. Meigs

Autograph letter signed, MCMC, MCMP, Ms. Div., LC.

John Rodgers Meigs to Montgomery C. Meigs

My dear Father, West Point, April 27, 1862

I received a letter yesterday from Mr Parrott of the Cold Spring foundry of which I am quite proud. I wrote to him to support that idea that occurred to me of arming a vessel with a spear having on the end a torpedo and he answered me in a letter of two pages saying that he thought the idea very ingenious and *practicable* the only *drawback* being the danger to the vessel carrying the torpedo. After reading his letter I went over to the library to try and find out as much as I could on this point and I found in the Civil Engineers and Architects Journal for 1852 pages 119 & 160 and also in the Transactions of the Royal Engineers accounts of operations which satisfy me that if the pole was 50 feet long the vessel would be perfectly safe with a charge of *200* pounds of powder. If the charge were *2000* pounds, she would have to be over a hundred feet distant *probably* perhaps not so far. Mr Parrott says he would like to work up a corrected plan if he can possibly spare time and wishes me to come over and see him. I think that if I show the Sup[erintendent] the letter he will allow me to go. When I have done so I will send the letter on to you. He asks me whether I have written to you about it or taken any action in regard to it. I have written to you but heard nothing in reply so I suppose you did nothing about it and said nothing to any one. May I tell him so. A man of Mr Parrots experience would not embrace the idea so favorably without there was something valuable in it would he.

I did write to Capt Ericsson a short letter but was afraid he might think it conceited in a young cadet to make a suggestion and so I left my letter anonymous. I mentioned it to Capt Ed Morgan of the Committee of the Chamber of Commerce on harbor defence and received an answer saying he would give the plan due consideration. Please answer this soon if only in two lines so that I may tell Mr Parrott as he wishes to know whether any one is taking any action on it.

I am very much obliged to you for the book on Landscape Painting which I received the other day. Prof Weir commends it and it is the admiration of every one who has seen it. I have painted but one picture in water colors yet and that was copied from a model in black and white. The Professor has now given me a little gem of a water color picture to copy which will be I suppose my last one. Our class will only have this one year of drawing instead of two as usual on account of the change in the course. I am sorry for it. But most of the men think it the most fortunate thing in the world. Drawing is a great bore unless you like it. I am getting along pretty well with my section in the 3d class. We have finished Calculus and I am now teaching them Surveying, it is very pleasant in this fine weather to take them out into the field. I think they are treating us with injustice in requiring us to do this work while the only equivalent is being excused from one or two roll calls and having a light after taps.

I was delighted to read in a New York paper the other day that Uncle John was to have command of the *Galena.* He will make her useful I am sure. I should think they would be strong enough on our side when she arrives to go into Norfolk Harbor and destroy the *Merrimac* and the other boats they are building there.

Uncle has been remarkably unlucky in this war in having such a position as Com Footes taken away from him and I hope his luck has changed now that he has the *Galena.*

Write to me soon if you can get time.

Ever Your Affct Son J. R. Meigs.

Autograph letter signed, MCMC, MCMP, Ms. Div., LC.

John Rodgers Meigs to Montgomery C. Meigs

My dear Father, West Point, May 5, 1862

I was very glad to receive your letter the other day. Prof Bartlett and Capt Parrott tell me that Fulton did not propose the same plan by any means. He proposed to use a torpedo but the most valuable point of this way of using it was wanting. I mean the perfect control [he] had over it at all times.

I went with Prof Bartlett last Saturday afternoon to see Capt Parrott. The Prof was very kind and took me all over the foundry and I saw all the operations connected with the making of the cheapest and most effective guns in the world and learned some curious and interesting facts. I told Capt Parrott that I had received a letter from you on the subject of that torpedo and what you said. He said that Fulton's proposal was wanting in the most valuable quality of ~~this kind of~~ mine, the perfect control over it all times. He said that he thought it was a valuable idea [illegible] that he had never seen it and it had never occurred to him until he read my first letter. He did not seem to think the objection in regard to steering a very formidable one. He said too that for harbor defence it was especially valuable. That by having a double boat with this spar between them it might be made of almost any length and yet be whirled about in any direction with great rapidity, there would be no difficulty about steering them.

I asked after he had been talking for some time about it whether he thought I had better let it drop, and wait till somebody thought of it who could carry it out. He said that if he were not pressed to the utmost limit of his capacity and time to supply guns and ammunition he should like to perform some experiments on the subject himself, but that as it was he did not have time to perform some that were very necessary to the improvement of his guns. I noticed too that they were doing some work there by hand that might have been much better done by machinery and Prof Bartlett told me that the reason was he had not time to have the machinery made and put in.

But he said that he thought that *you* in your position at Washington might do something to develop it and make it extremely valuable. In conclusion he told me to follow it up work out a plan and if possible get it carried out in some way and *claim* it. I am satisfied myself that it is perfectly easy to carry out and I do not think that Mr Parrott would take so much interest and advise me to work it out unless it was really worth while. I have thought that this would be one of the best forms for the torpedo, and at

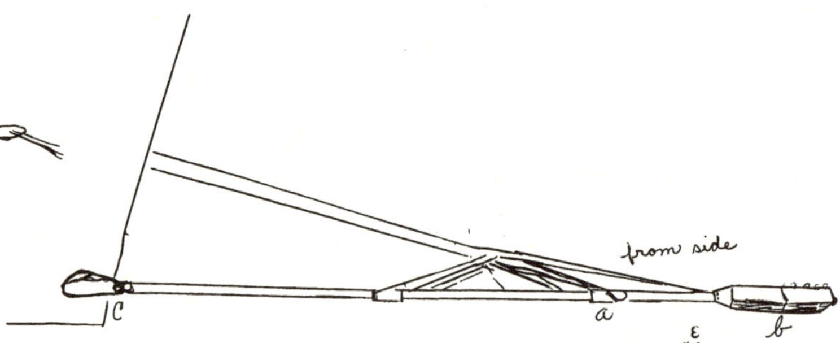

This would be flat and not interfere with the steering a great deal would hold about 200–400 pounds of powder, made of boiler iron rivetted, would have

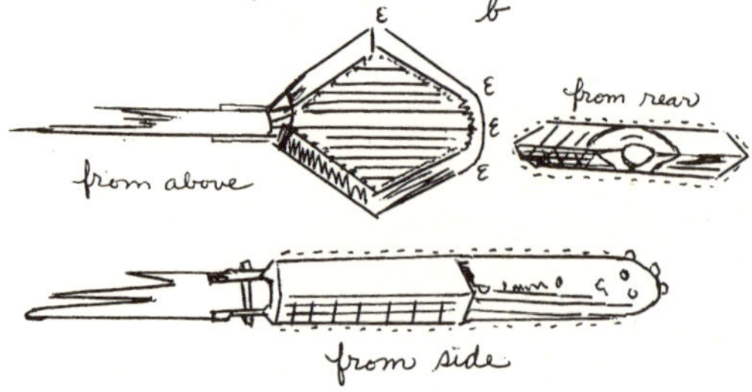

exploders EEE sure to be hit and explode on being struck in any direction against a vessels side, be adjustable by means of pivot at C to any depth, necessary. I was for sometime in trouble to find an exploder which would be sure to go off when struck with a crushing slow and powerful blow but I have one now which Professors Kendrick and Bartlett say would be sure to go off. It is made thus,

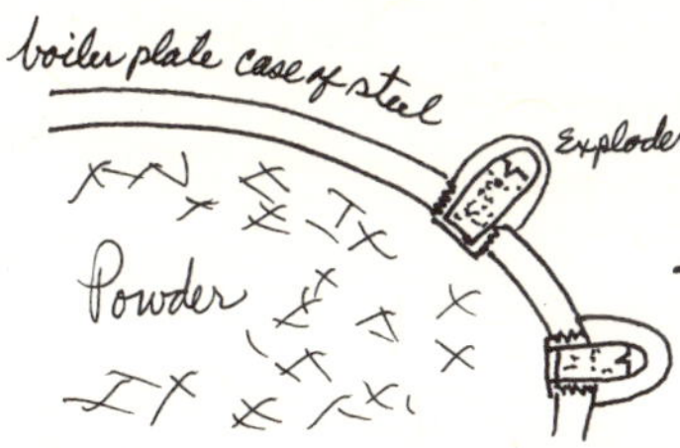

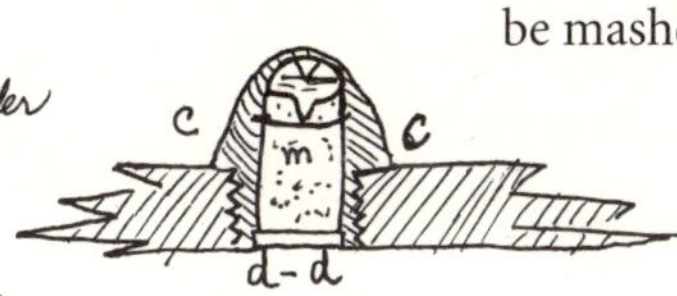

(c) Case of soft white metal to be mashed by a blow on top (V) glass vesicle of sulphuric acid to be broken by the crushing of (c) (M) mixture of chlorate of potassa and sulfa in proportion of 3 [?] parts of former to one of latter.

When the vesicle is crushed the sulphuric acid escapes and instantly explodes the mixture (m) which drives out the little plate (d-d) stamped in and communicates with the powder (p) in the interior of the shell. These would be screwed in only when the shell is to be used and they would be placed on it in such a manner that one of them would be sure to be crushed when the torpedo strikes a vessels side no matter in what direction the blow should come. I think this would be a good kind of fuze for all torpedoes where a slow crushing action can be got. Mr Parrott told me to write to him if I wanted any advice on this subject and to come and see him when I could. In consequence of his advice and Professor Bartletts I am going to make a drawing of this and send it with a modest letter to Mr Stevens of the great battery fame.

Professor Bartlett also advised me to show it to the Sup and get him to let me carry on some few experiments.

My dear Father I suppose you will be wearied out with this letter and as to the department, I know they receive hundreds of them, but perhaps after turning over a great many pebble stones, you may find one diamond.

I saw at the Foundry all the operations incident to the making a gun from the casting to the final finishing. There are 3 or 4 200 pounders in process of construction and one 300 pounder is now cooling in the mould. I saw a 100 pounder fired across the river at a stone apparently about as big as the little centre table in the hall at home, it struck just on the edge each time. I saw many of the steel faced shot now making and

one that had been fired against a wrought iron bar, 8 inches thick, 1 ft wide and 3 or 4 long intended for casemates.

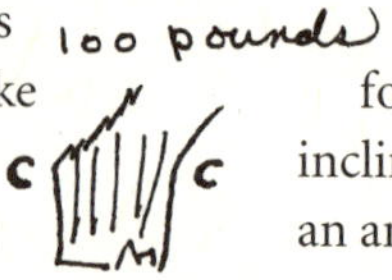

It looked thus completely mashed up and squshed out at cc to a vase like form. It was fired against the surface of wrought iron inclined at an angle of 45 degrees and another one was fired at an angle of 20 degrees neither of them glanced at all.

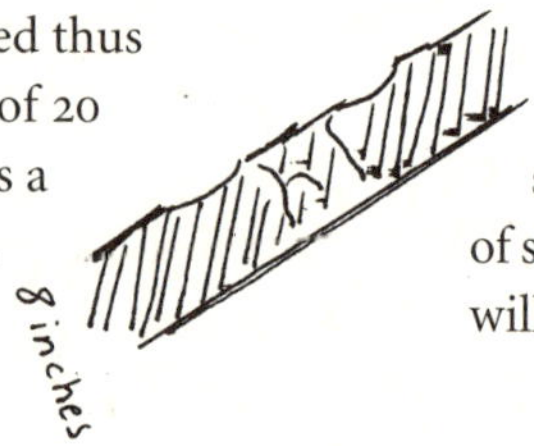

The bar looked thus It had been fired at twice at an angle of 45 and of 20 and was cracked and had two hollows as big as a saucer scooped in it. Mr Parrott thinks this kind of shot destroys the value of *inclined* armor for they will not glance at all.

Capt George Rodgers Commdt of Naval Cadets at Annapolis, and Cousin Raymonds brother was here today and I had quite a talk with him. He was very much pleased with the Point says the bearing of the cadets has improved very much since 2 years ago (all the Southerners are gone) and he has taken some ideas back to Annapolis. Gen'l Saxton was out in front of the Battalion at drill but left before I had any opportunity to speak to him. We had another fire here last week which could have been easily put out if we had had a decent engine. Love to all.

Ever Your Affct Son J. R. Meigs

Autograph letter signed, MCMC, MCMP, Ms. Div., LC.

On May 17, 1862, Cadet Meigs was placed under arrest, charged with "conduct prejudicial to good order and military discipline" for "attacking a Cadet officer for his official acts."[1] The cadet officer was William A. Marye, who was appointed to the academy from California and graduated from West Point in 1862. According to Meigs, Marye's "manner to me has been rather insolent . . . and I have of course hated him pretty well."[2]

In compliance with the rules, Meigs wrote out his explanation of the event. He addressed his explanation to Colonel Kenner Garrard, Commandant of Cadets.

1. Special Order No. 6, in Orders of the Commandant of Cadets, . . . 1861–1910, Orders and Issuances, RG 404: Records of the United States Military Academy, Post Orders, 1838–1904, USMA Archives.
2. JRM to MCM, May 18, 1862, MCMC, MCMP, Ms. Div., LC. This document is included in the present volume.

Explanation of Delinquency

West Point, New York, May 17, 1862

Delinquency. Conduct to the prejudice of good order and military discipline.

Explanation, In compliance with an order I make the following explanation. On the 15th inst. at drill I was watching the guide in front of me to obtain the step, and was perhaps a little out of dress when Cadet (Capt) Marye who was at the time in command of the division, spoke to me and ordered me to dress back in a tone and with a manner which I could not tolerate from any one, especially from him for whom I have not without reason, entertained a most cordial dislike. His manner was so unnecessary and so offensive as to be particularly noticed by those who heard him, several of whom spoke to me about it with indignation. This morning I went up to his room with Cadet (Sergt) Reid and spoke to him as nearly as I can recollect in these words. I told him that he had spoken to me the other day in an exceedingly rough manner and one which I supposed he meant to be intentionally insulting and I wished to know whether it was or not.

He replied "Yes I did speak to you, and I just tell you what it is now, if you are not satisfied with it you can have all the satisfaction you want." I replied that satisfaction was what I wanted.

As he had thus refused to make an explanation of his offensive manner, I was forced to seek redress in the only way practicable in the Corps of Cadets.

I was apprehensive before I went to see Cadet (Capt) Marye that he would refuse me any explanation and I therefore asked Cadet (Sergt) Reid to come with me and any blame attaching to his position must belong to me alone, as he probably did not think he could decline such a request from a friend.

I have on several occasions forbourne having a difficulty with Cadet (Capt) Marye and I thought there could no longer be any merit in so doing but rather the contrary. I submit that it can in no manner be prejudicial to good order and military discipline, for one cadet to require from another in his intercourse with him the language and manners of a gentleman, even though he should be placed for a time in the position of comdt. of a division. Such position can give him no right to insult with impunity those placed under him for a time but who are in all personal matters in no sense his inferiors.

I therefore request to be relieved from arrest under the above charge.

I had hoped to have got through this Academy without having attached to my name the stigma of having been under arrest, and I hope that if the Commdt is pleased to find that any injustice has been done me, through ignorance of the circumstances, he will as far as possible erase the stigma, by giving both Mr Reid and myself an *honorable* release.

I must, if it be allowable, protest against this statement over my own signature, being in any way used as evidence against me, in a Court Martial.

Respectfully Submitted

John R. Meigs

Document signed, JRM Other Papers, MCMC, MCMP, Ms. Div., LC.

On the following day, Meigs wrote to his father in Pitman shorthand.

John Rodgers Meigs to Montgomery C. Meigs (Shorthand Transcription)

My Dear Father: West Point, May 18, 1862

I am sorry to have to tell you that I am afraid I have got into a serious difficulty. There is in the corps a man named Marye, pronounced Moree, whose supercilious airs and overbearing manners and the peculiar stride with which he walks have made him detested in many quarters. His manner to me has been rather insolent ever since I came here, and I have of course hated him pretty well. The other day at drill he spoke to me in a way which I could not very well stand, and which was noticed by everyone.

I spoke to Michie and Reid and told them that I thought I ought to demand satisfaction for it, and they agreed with me. I finally determined that if he reported me, I would fight, and if he did not, I would not. He did not report me, but several men spoke to me about it as if they thought I ought not to stand it, and I did not think so myself.

So yesterday I asked Reid, an orderly sergeant and one of my classmates, to go up to his room. I would have been delighted, of course, to get an explanation or apology for his insolence, for I do not like fighting, and I expected to be whipped in less than 5 minutes, as he was a strong, active fellow, which I am not.

I went up to his room yesterday and found him in bed, and when I asked him whether he meant to insult me or not, he only offered to fight me and commenced putting on his clothes and sent for a second. He chose a rough and tumble fight, and we went at it. Pretty soon, his second said, "Choke him, choke him, Marye," and he did get his fingers on my throat, but I wrenched them off and got mine on his and held them tight, until he commenced to get weaker and weaker every second.

When his second saw that, he said, "There, there, that will do now; that will do for a time." And I, supposing that his second had given up the fight, let go of him and went into an adjoining room and put on my coat to go home. But when he had got his wind again, he came blustering in there, and said he felt fresh yet and had not had satisfaction.

It was not fair at all, for you know in a rough and tumble, there is no cessation until one party is whipped and the one asked for it is considered licked. But as he seemed anxious, I pitched in again, and we had 5 or 6 bloody rounds, at the end of which his floor was all covered with sweat and blood, some of which was even splattered on the walls and ceiling. His nose was broken and both his eyes black as ink, and they became shut in a little while afterwards. My lip was cut, and I had a black eye and skin knocked off my knuckles. I finally got so out of wind that I had to give in or I would have been marked all up, without hurting him anymore. He thought it had been long enough, in all conscience.

Just as my second and I were going down the stone steps of the division into the area, we met Captain Whistler coming in to inspect. It was call to quarters and before breakfast, and we were all, of course, reported. And when he saw all the blood on the floor and Marye in rather a frantic condition, he had all of us put under arrest, on the charge of conduct prejudicial to good order and military discipline.

We had both to go to the hospital after the fight. The commandant told me this morning that he intended to have me tried by a court martial, charge of conduct prejudicial to good order and military discipline; specification: demanding an explanation of and challenging my superior officer for an act performed while on duty. I think he has no business to meddle with matters entirely personal between cadets, and it seems to me absurd to attempt to call Mr. Marye my superior officer because he happened to have had drill, the division in which I was. He is a cadet captain, but the commandant lays no stress on that, for if he did, I might say that he was guilty of striking his superior officer. I am an acting asst professor of the USMA, and the Superintendent says that they are regarded as the first appointments given to cadets.

I enclose an explanation which I sent to the commandant and which I think he will send to the Superintendent.

Colonel Garrard says that he has applied for a court martial to try cases of Cadets Bowen and Lee, who were found off limits the other day, and he will try to bring me before it. He thinks it likely that he will be a member of it. If he is, I believe I shall object to his being a member of it in my case, as he expects to be the prosecutor also; and being the commandant, he must be prejudiced to some extent in favor of *his* captains. I do not see how he can very well prosecute me and at the same time have a vote on my case. The court martial ought to be composed of persons, if possible, unacquainted with the prisoners.

They used to have court martials here almost every week to try the most trivial cases, and they have not been able to do so of late. I hope Mr. Stanton will not grant their request. I have a good character and would have, I think, a strong point in the fact that Marye, as it were, *invited me to challenge him.*

I am not ashamed of what I have done, because I thought it was the only course possible to pursue and that I was only doing my duty to myself. There is a good deal of indignation in the corps about the way in which I have been treated. I am most concerned, however, at the trouble in which I got my friend Reid, who is as gentlemanly and good a fellow as I know. I am afraid that he will not be made an officer next year as he expected to be and ought to be. I can see no justice in keeping us under arrest and releasing the other two men, as was done immediately. Sergeant Lee, Marye's second, who went over there with the certainty of assisting at a fight, is more to blame than mine, who went up with some faint hope of receiving an apology.

I was very much surprised and rejoiced to see how badly I had used him up, as I expected it to be the other way entirely. He will not be able to see any ladies for a month and may carry my mark on his broken nose as a remembrance of his impudence to the day of his death. He is the most absurd looking fellow I ever saw and holds his head back so far, you would think his neck would be dislocated. I could not help, while I was fighting him, from laughing at the way in which he strutted up and down like a peacock and at the scientific manner in which he worked his hands and head and then came at me and got a blow on his nose.

I am afraid that you will be very much displeased with me, but I do not think I am to blame. You could, if you choose, prevent the court martial, I suppose, by telling Secretary Stanton what a trifling thing it is for.

I will write you again soon to tell you what happens.

From your affectionate son, J. R. Meigs

I have written an explanation like this, but with a few differences and some arguments, which I will ask the commandant to forward to the Superintendent tomorrow.

Autograph letter signed, MCMC, MCMP, Ms. Div., LC.

While John and his family awaited the results of the court-martial, by Special Order No. 71 of June 10, 1862, the "suspension of the arrest of Cadet Meigs is removed, and his limits are extended to the usual limits of the other cadets."[1] It is of interest that Montgomery C. Meigs had been tried at a general court-martial on March 31, 1835, charged with neglect of duty and making a false report. Montgomery pled not guilty but was found guilty and sentenced to be reduced in rank to private, to perform six months' extra tours of Sunday guard duty, and to be confined to his room for one month.

1. Special Order No. 71, p. 105, Orders and Issuances, RG 404: Records of the United States Military Academy, Post Orders, 1838–1904, USMA Archives.

John Rodgers Meigs to Montgomery C. Meigs

My dear Father, West Point, New York, June 14, 1862

I am still in arrest though with extended limits waiting for the finding of that Court Martial of which I have heard nothing.

I hoped to hear from you in regard to it some time ago but have not had any letter from home since I wrote to ask you to send it to me. We are going into Court next Monday I guess. There is nothing more to be done now. The board of visitors have got through and the examinations are all finished. My section did very creditably in Calculus and I was complimented upon it by a member of the board after the examination. I do not know his name.

Our whole class passed a very good examination indeed. I myself received the maximum for every thing at every examination this June. I hope you will not be displeased at my standing. I am second in artillery tactics head in infantry tactics head in drawing, head in Chemistry, head in Philosophy astronomy etc and head of the class in general merit. I hear that I am on the total mark about 30 units ahead of Michie, he is about 10 ahead of Rabb and I do not know how the rest goes.

Michie is better pleased at my being head than if he had been himself, for he says he does not think he ought to be above me. I do not believe there ever was a more generous hearted fellow than he is. Rabb is head in Artillery. The standing in tactics was a good deal the result of accident, as we only studied it for a few weeks and one miss in one recitation would throw a man down a few files.

Colonel Bowman gave me a long lecture the other day on the subject of my difficulty and said that he had written you and you had written to him about it. The first class are leaving to day. Just as I am writing I hear one shouting, Good Bye fellows, God bless you all.

It is hard to see so many go off with whom you have been in daily contact for 3 or 4 years. There are some splendid men in that class. Mackenzie and McCrea I think, will not soon find their superiors in manly qualities.

The class is very much put out at the prospect of being ordered to camps of instruction instead of to active service with Maclellan or Halleck.

Goodbye

Ever Your Affct Son J. R. Meigs

Will you give me the address of those who are building gun boats in the US. I am advised to send drawings of my idea to such people.

General Merit

1. Meigs
2. Michie
3. Rabb

4. Townsend
5. King

My varicocele is getting troublesome, and I think of getting a leave to go to New York and have a very good net put on it. This is, I believe, a very successful method, and I have received a letter from a doctor who practices it. These little cotton bags are an indomitable trouble and soon wear out, and I can get no more. Besides, I have to wash them myself. Can you send me 5 dollars, if perfectly able to? The doctor charges that for fitting the thing. Of course, I shall first take the advice of our surgeons here.[1]

Autograph letter signed, MCMC, MCMP, Ms. Div., LC.

1. Shorthand transcription.

The decision of the court-martial was to dismiss Meigs and Reid from the academy. However, their sentences were remitted by Secretary of War Stanton to confinement to the limits of camp. Feeling the sting of injustice, John Rodgers Meigs asked his father for permission to withdraw from the academy that he might join a volunteer unit and then the regular army.

One of the welcomed respites to life at West Point was the summer encampment that began after the arrival of the incoming plebe class and the departure of the graduating seniors in late June or early July and lasted until late August. During this period, cadets lived in tents, drilled, and practiced tactical movements. In effect, they lived as they would in the field, with the exception that they still took their meals in the dining hall. But not all was soldiering. During this period, family and friends visited, hops and balls where held; this was the social season at West Point. In addition to strong feelings of injustice done, Meigs believed his punishment would prevent him from engaging in the social activities associated with the encampment. It was with great disappointment that he wrote to his sister, Mary, who, along with his mother, he had expected to see during the summer.

John Rodgers Meigs to Montgomery C. Meigs

My dear Father, West Point, June 21, 1862

I have just heard that that court has found Reid and myself guilty and sentenced us to be dismissed and that the Sec has *commuted* it to confinement to the limits of this nasty little camp during the whole summer. If it is so I wish that you would send me your permission to resign right off. I can take my class standing I suppose and get a position in the volunteer service and then get into the regulars if I choose. I never did care about the army in time of peace at any rate and intended to resign soon after getting into it. I have now got the principal part of the education of West Point, next years course is nothing but military engineering which will not be necessary to a civilian, and which can easily be learned by reading if necessary.

I do not see how that court without perjuring itself could find Reid guilty and if you should read the evidence in his case you would say so too. There has been nothing criminal at all proven. It is one of the most unjust things altogether that I have ever heard of and such as I never did hear of before I got to this nursery of imposition and tyranny.

That scoundrel and fool Marye was more guilty than I and certainly more despicable in his conduct and yet they never thought of punishing him but have let him graduate as if he done nothing worthy of punishment.

This first class encampment is looked forward to all the time we are here as one of the few things that recompense us for the weary months of study and work and the years of isolation and here I am to have it all taken away. I would rather a thousand times have been dismissed. I have for the last year worked to teach a section, in addition to my other duties and have done *half as much work* as a Lieutenant who receives his salary, has the encampment for furlough and no other duties to perform.

This is all to go for nothing. I have not received any recompense whatever not even a leave of absence in summer and am told by the professor that I will have to do the same thing against my will next year. The country sends us here to study and not to work without pay. I do not want to stay here now at all and often been sorry since I am back that I had not gone into the army when on furlough. I only wanted to come so as to get through Professor Bartletts course of mechanics and astronomy and I have done that and come out head in it.

If I could get a sick leave it would perhaps make it bearable though I should not want to stay even then and I do not want such a thing now as it would seem like shirking the thing while my friend who only went with me and whom they could not justly find guilty would have to stay it out. I hope you will send me on that permission and let me go home and get into Halleck or Maclellans army. All my promised pleasure in Mother and Marys society is taken away. They could only see me twice a day to speak to me and it would be useless for them to attempt that visit that I have looked forward to with such anticipation of happiness for a whole year.

Our kind hearted fool of a Commandant has given the fiat that ladies should not enter the camp except twice a day. He has turned all the regulations and orders of the Secretary of War and the President, that were established for his government as much as ours, topsy turvy. We cannot feel very angry at him because he seems to be kind to us.

Good bye

Ever Your Affct Son J. R. Meigs

Autograph letter signed, MCMC, MCMP, Ms. Div., LC.

John Rodgers Meigs to Mary Montgomery Meigs

My dear Mary, West Point, June 25, 1862

I suppose you know by this time from my letter to Father what a sentence I have got and what I want to do. I could not stay here all summer shut up in this place and tantalyzed by seeing the pleasure of others which I had expected to participate in. Your visit from which I expected to receive the most pleasure would now scarcely afford me any.

The Commdt has said that he is opposed to all leaves for cadets but furlough and wishes that lady friends were not permitted to come here. He has forbidden any citizens to be allowed to enter camp except during parade and guard mounting. And came very near putting a cadet in arrest for asking the Misses Fish to step into one of the guard tents during a shower. That this is all unnecessary is shown by the fact that other commandants did not find it so but pursued rather an opposite policy.

Mrs. Bowman told the cadet the other day that the Colonel [Bowman] had intended to march the cadets through the country and encamp on Boston Common and had asked and obtained permission from W to do so but had forgotten it until it was too late. We might have had a delightful time if this had been done. It is now said that the Commdt in order to get us away from ladies I suppose wants to march us about 8 miles over the Mountains and encamp by the side of a lake there.

The President was reported to be down at Cozzens yesterday and it was even said that he would pass through camp. I asked Colonel Garrard whether I could speak to him and he said that the first class would be presented to him at the Commdt's tent and I could then.

We all went down there and were introduced shook hands and when I was introduced the President: said Ah Mr Meigs they tell me you are a son of the General. Yes sir, said I and President Lincoln you can do me and my friend a great favor if you only choose. I have been sentenced by a court martial to be confined to Camp during the whole summer because I resented an insult given me in ranks by another cadet. The offence consisted I believe in being in ranks. As for my character I hope you will ask the Superintendent there about that.

If you will only sign your name on this card you will relieve me from a great deal of trouble. I had on the card "The punishments of Cadets JRReid and JRMeigs are remitted." The President looked quite kind and said "Well Mr Meigs I should be very glad to do so if it would not be interfering with the authorities here" and he turned towards the Superintendent and officers. ["]I do not know whether I would be doing anything inconsistent with the regulations of the establishment. I must first find out that if not I should be very glad to do so." I very stupidly made two or three attempts to put in a word while he was saying this and when he ended told him that it was the custom to remit punishments when the Pres or Sec visited the Point. "Very well, sir" said he "I will find out that from some of these gentlemen." I was told afterwards that he gave the card to Colonel Bowman and the Colonel said that they had not yet received official information of what the punishment was. The Pres said something about my coming very directly which made them all laugh but I do not know exactly what it was.

An eager crowd gathered about me of course as soon as I came out to know what he said and what I said and whether he had signed the card etc. I am afraid there is no hope for me in that quarter, though the Pres was evidently willing, because the Sup probably was not, he had the Court called and would not like to see me get off, as he probably feels as if what the court did was his act, still I cant see that he has anything to do with the sentence of the court, and I do not see why he should feel a personal interest in my punishment. I am quite sure that if he had chosen to speak a favorable word the card would have been signed. With regard to the sentences of that court, I do not wonder that the Dept thought them too severe. One of the members of it said that a court comprised as it was of officers on the Post could determine impartially from the evidence the guilt or innocence of a cadet (which I do not believe) but would be apt to give too severe a punishment which I do believe. The officers and cadets get combined into antagonistic sets, it is all very well to talk about their impartiality and disinterestedness but it is all bosh and I know absolute facts which prove it.

You do not know what a disappointment it is to be deprived of that visit from yourself and Mother. [illegible][1] You must be having a nice time in W acc to your own account. What has become of West. I saw a map of his published in the *Times* not long ago. Poor fellow he was unlucky in getting into such a division whether its reputation is just or unjust.

They are talking of court marshaling two or three poor fellows who would not turn out a few minutes before breakfast when they were washing and dressing to police camp and pick up dirt. There is a furlough man brought back by telegraph who got drunk in NY. I don't know what they will do to him. It is the first time that such a thing has been done I believe though whole classes used to get drunk. Still I do not feel very sorry for him.

Goodbye. Love to all,

Ever Your Affect Brother J. R. Meigs

Autograph letter signed, JRMC, MCMP, Ms. Div., LC.

1. Montgomery C. Meigs crossed out a sentence written in John's hand.

On June 29, 1862, Meigs wrote to his father that the "request I made of the President has worked me more ill than good." For approaching Lincoln, he received fifteen demerits as he was charged with three separate infractions. His responses to the three charges were deemed unsatisfactory by Kenner Garrard, the commandant of cadets, and Alexander H. Bowman, the superintendent of West Point.

John Rodgers Meigs to Montgomery C. Meigs

My dear Father, West Point, June 29, 1862

I received your long and kind letter on Friday and it made me think of some things more earnestly than I had before. Those ideas were presented with more force than before though I had not asked without thinking for my resignation. I told some of my classmates that I had sent for my resignation in case that report was true about the punishment, I was to receive for doing what anyone would have been compelled to do and they all seemed to think that though I was suffering from injustice it would be foolish and against my own interests to do so and moreover they all seemed sorry that I should go away and not graduate with the class. I think that I might have got on very well in active service but after the war was over I would be in much worse position than if I had stayed here.

There are a good many things to try one though who stays here. I never heard such continual grumbling as there is here and it cannot be entirely without reason.

That request I made of the President has worked me more ill than good for though he said he would remit the punishment if it were not interfering, Colonel Bowman or Garrard said something to him which prevented him from signing the card and the next day the following reports were read out which the Commdt had given me.

Meigs. In arrest addressing the Commander in Chief of the Army on official matters without permission to do so in violation of par 248 Ac[ademy] Regs. Same. In arrest speaking to the Pres of the US uninvited after having been informed by his commanding officer that it would be improper for him to do so. Same. When being brought into the presence of the Pres of the US for the purpose of simply paying his respects to him and retiring introducing matter affecting the discipline of the Military Academy.

June 24, 1862

I think these reports speak for themselves. I have been three times reported by one person for one offence. Many told me that they thought the reports showed the most malicious spite on the part of the Comdt and all were indignant. I do not think so myself because Colonel Garrard's manner has almost always been as polite as possible, still it is pretty hard to understand.

I hope that you will not for one moment think that they are true as they stand.

The Commdt told me quite plainly I thought that I could speak to the President after being introduced though not otherwise. I wrote out this endorsed excuse the next day but afterwards made a much milder one and sent it in.[1]

Love to all.

Ever Your Affct Son JRM

ENCLOSURE

West Point NY June 24 1862

Delinquency, When being brought into the presence of the President of the United States for the purpose of simply paying his respects to him and retiring introducing matter affecting the discipline of the Military Academy.

Explanation, This delinquency if perfectly true is an offence against no rule or regulation prescribed for Cadets—I knew that I had been brought before the notice of the Pres of the U S before and when he spoke to me cordially I felt that I had a right and was expected to make some answer—I did not suppose that any one but the President had the authority to prescribe how many words I should speak or what they should be about—The matter about which I spoke had no ref to the action of the Authorities of the Academy but merely to the sentence of a General Court Martial with which they could not be concerned in any other way than simply to carry it out—

If the Commandant thought it improper for me to speak to the Pres I submit that he might have ordered me to stop that he did not do so shows that he thought I had the virtual permission of the Pres to go on—

My action was not without precedent for Sec Cameron the head of the War Department listened to the similar request of a cadet and granted it showing that he thought it not improper in the cadet to ask him and it would be difficult to find a higher authority than the Sec of War—I am sure from the answer the Pres made me that if the authorities had spoken a favorable word he would have remitted the sentence—

I will remark that I have received 3 distinct reports for the same offence the evident effect and probable intention of which was to give me 15 demerit for the same offence—I think that this will be seen to be in direct violation of what is prescribed on page 37 Academic Regulations and I believe those regulations were made by the Pres as well for the protection as the government of cadets—

Respectfully Submitted, to Col. K. Garrard Comdt Corps Cadets
John R Meigs, Cadet Co D

Autograph letter initialed, with enclosures, JRM Other Papers, MCMC, MCMP, Ms. Div., LC.

1. In the Montgomery C. Meigs Papers at the Library of Congress there are copied documents with a fuller explanation of each of the charges. Whether these are the "milder" explanation is unclear.

By the end of June, John Rodgers had received additional letters from his family. His spirits soared when he was assured that he would be able to see his mother and his sister Mary on their visit to West Point.

John Rodgers Meigs to Louisa Rodgers Meigs

My dear Mother, West Point, June 29, 1862

I was very much delighted on receiving yours and Fathers letters to find that you still intended to come as you did not think my sentence would prevent me from seeing you.

It would have done so if it had been exactly what I heard it was, confinement to camp and not allowed to leave it without the permission of the Superintendent which would not be given for small reasons or for anything in fact that was unnecessary.

I would have written on Sat but I wished to wait until I had an answer from Roe which I enclose. He does not say that the room is large but Mrs Miller says you can get a double room if you choose soon after you come as people are coming and *going* every day. I have seen something of her since she came and have had quite a pleasant talk with her.

I expect that Mrs Judge Hall of Cincinnati will be a companion though I do not know her yet. I hear she is very pleasant, she is a sister of Genl Anderson's.

If my sentence does not interfere I can have a delightful time with you this summer, but I am afraid that if it is confinement to camp I will have but a hard time.

There is a most extraordinary rage for court martials on the part of Colonel Garrard, he is to have another one on soon to try about half a dozen men who have been caught pulling out plebes.

I dont believe there ever was a man with so many insincere motives in his head as he has.

Mary will want, all her dresses here I guess. They wear low necked dresses at these hops but I do not think she had better try that too much.

You know people dress here.

I had a very pleasant time with a new sister in law of Mrs Reales, I did not see Mrs Reale herself, she passed on the boat but did not stop as they did. The last tune is beating for tattoo, good night.

Ever Your Affct JRM

Autograph letter initialed, LRMC, MCMP, Ms. Div., LC.

On Tuesday, July 8, 1862, Louisa Rodgers Meigs and daughter Mary left Washington, D.C., for West Point, for a lengthy stay. Other family members, including Aunt Mollie and Aunt Nannie, visited as well. Regardless of John Rodgers Meigs's renewed spirit and the expected visit from his mother, sister, and other family members, the specter of fifteen demerits for one offense loomed over the family.

On July 2, 1862, Montgomery C. Meigs wrote to Joseph G. Totten, in the latter's capacity as Inspector of the Military Academy. General Meigs found the charges against his son "in direct violation of the regulations and of justice." He called Totten's attention to the fact that he, Meigs, had "known young men of character and talent ruined by just such treatment at West Point." In response, Totten requested a complete review of the case by Superintendent Bowman.[1]

Bowman's seven-page report, received in Washington on July 26, ended, "I do not think the reports in question should be removed, and [I] regard them as a mild punishment for the offense." Based on Bowman's response, Totten wrote to Meigs: "Upon consideration of the facts and views which it presents, I am of opinion that the proceedings . . . should not be disturbed by me."[2]

Rebuffed by the chief of the Engineer Bureau, General Meigs turned to higher authority for redress. On July 31, 1862, he wrote to Secretary of War Stanton, asking that the charges against John Rodgers Meigs be dismissed, pleading "as my excuse for troubling you . . . the interest of a father in seeing justice done to a son." His answer would come the following year.[3]

John Rodgers Meigs's responsibilities as a cadet were never far in the background. On August 30, 1862, in addition to his obligations as a student, he was assigned temporary duty in the mathematics department under Professor Church. Once the August encampment was over, Cadet Meigs returned to the academic regimen.

Beyond West Point, the Civil War continued. On August 29, 1862, the Second Battle of Bull Run began. A Union victory was reported on the first day of the fighting, but the result was a Union debacle. Montgomery wrote to John Rodgers Meigs on the next afternoon.

---

1. MCM to J. G. Totten, July 2, 1862, File 1862/2746, RG 94: Records of the Adjutant General's Office, from Records Relating to the United States Military Academy, 1803–1917, Correspondence Relating to the Military Academy, 1819–66, NA.
2. Alexander H. Bowman to Joseph G. Totten, July 22, 1862, File 1862/2746, RG 94: Records of the Adjutant General's Office, from Records Relating to the United States Military Academy, 1803–1917, Correspondence Relating to the Military Academy, 1819–66, NA; Joseph G. Totten to Montgomery C. Meigs, July 28, 1862, vol. 24, p. 166, Letters Sent Relating to the Military Academy, 1812–67, RG 94: Records of the Adjutant General's Office, from Records Relating to the United States Military Academy, 1803–1917, Letters Sent Relating to the Military Academy, 1812–67, NA.
3. MCM to Secretary of War Edwin M. Stanton, July 31, 1862, File 1862/2746, RG 94: Records of the Adjutant General's Office, from Records Relating to the United States Military Academy, 1803–1917, Correspondence Relating to the Military Academy, 1819–66, NA.

Montgomery C. Meigs to John Rodgers Meigs

My dear John, Washington, August 30, 1862

The Secretary of War has just been here to direct me to send out to the neighborhood of the Bull Run Battlefield all the ambulances in the City & a corps of volunteers from this department to help care for the wounded of whom we have some 8000 patients spread over the field.

A still greater number of rebels are lying there.

Gen Pope sends in a despatch dated last night. He had fought the whole rebel army all day. They on the defensive he attacking & steadily driving them back. The Secretary thinks the whole army of the rebels is concentrated there & is being beaten, fighting again today.

God grant us a decisive victory & put an end to the slaveholding rebellion & treason.

I send this to you to tell your mother if she is still at West Point. I am not informed of her movement. If she is not send it to her.

Macomb breakfasted with me this morning, all night he had not been able to get back to the army. He may have gone out today however as I hear that the communications with Manassas has been restored. I have not seen him since breakfast.

Ever your affec father MCM

My dear Mother,

I received some letters from the Hotel which I enclosed to Mrs. Macomb because you forgot to give me your direction at 8 PM.

We are now safe in our barracks and studying hardly have time to look at a female though we want to just as badly as ever.

I had a delightful day in N.Y. I went to see Miss Legget after leaving you and had quite a pleasant visit. Mrs Overman has gone but the Hyers not yet. Goodbye. I must go to breakfast. I got a table cover in NY myself.

Autograph letter initialed, with JRM note, MCMC, MCMP, Ms. Div., LC.

With the Federal defeat at the Second Battle of Bull Run, General Robert E. Lee moved northward. His army crossed the Potomac River near Leesburg, Virginia, on September 5, 1862. By the seventh, his troops occupied Frederick, Maryland. Attacks on Baltimore, Washington, or Philadelphia were expected. In writing to his grandfather, Charles D. Meigs, John Rodgers Meigs suggested a spirited strategy to defeat the Confederates if they decided to invade Pennsylvania. He believed that Lee's real intent was to attack Washington and Baltimore.

John Rodgers Meigs to Charles D. Meigs

My dear Grandfather, West Point, September 13, 1862

I suppose you are all excitement at the prospect of the invasion of your state. It has certainly been in some respects a fortunate thing it has developed a strength which nothing else could and which when it begins to be felt will bear the rebels back not only to their own lines but far beyond.

What an opportunity an attack on Harrisburgh would present for the capture of the whole rebel army. Everything in the shape of the country would assist this. We could so easily get around in their rear that I cannot believe it is anything more than a feint to distract our attention from Washington & Baltimore which they will of course attack first.

Wherever we have invaded the rebel country it has been either only a short distance from the sea coast where we could have our gun boats and transports or along lines of parallel rivers like the Tennessee and Cumberland where we could have the same advantage.

If the rebels were to invade Penn, they would have to go through a rough country and with a most dangerous river to cut off their rear. If they attempted to leave it for any distance, McClellan by getting in their rear without at all abandoning the defense of Washington or exposing it at all would have them between the upper and nether millstone and ought to crush them to powder.

I hope he will yet turn out a Napoleon though he has had time and opportunity to give more evidence of it than he has.

It is wretched to stay here droning away over studies when such things are happening in the country.

If they would only let us leave in 3 or 4 months we could be satisfied.

If we can only use it right this invasion of Pennsylvania will be the destruction of the Rebels. How I hope we will.

I send you my picture which I should have done sometime ago but that somehow I thought I had. I have had a most delightful summer. The principal pleasure being from the visit of Mother and Mary and Aunt Mollie and Nannie. Remember me to all with love.

Ever Your Affct J. R. Meigs

Autograph letter signed, CDMC, MCMP, Ms. Div., LC.

Lee's campaign plan was effectively thwarted at the Battle of Antietam. On September 18, the Confederate leader recrossed the Potomac. He had shown great courage while his Federal counterpart had displayed a lack of military acumen. For the young cadet who followed these events with great interest, the days at

West Point passed into another winter season. A jubilant John Rodgers Meigs wrote to his mother of an anticipated visit to New York City and of plans to visit his Aunt Nannie over Christmas.

John Rodgers Meigs to Louisa Rodgers Meigs

My dear Mother, West Point, December 12, 1862

I feel in high glee to night at having obtained from Colonel Bowman a leave to go to NY and to see Uncle John. Tomorrow is Saturday and I am to have Sat and Sunday in NY as it will not interfere with my lessons. I received to day a very kind letter from Aunt Nannie containing an invitation to come and see her on Christmas and bring Michie with me. I am going to spend Sat and Sunday nights there. Isn't it glorious to have such a chance. I am very much indebted to you for those articles you sent me by Mrs Elderkin. She is just as kind as ever. They have been just what I wanted. I should have had to get some new socks here, otherwise as those that I had last year are now more holy than righteous.

The handkerchiefs came in too very nicely for I was innocently using those wide bordered ones you got me in NY on all occasions thinking that they must be *à la mode.* You ought to have given me some hint before that they were not quite in the style. But you had no time. Do you recollect how calmly you all started out on that tour of observation just about one minute and a half before the stage had to start. It was lucky I was able to catch you so soon *or* you would have had to spend some more hours at that *very cheap boarding house* the Brevoort. I was made exultant and joyful by the receipt of your letter and Mary's. Mary's was a perfect mammoth. I was interested in learning that Genl Byard proposes to marry Miss Lalie so soon. She denies the engagement positively on her honor. And Colonel Bowman authorized Wharton last summer to deny it in the army. It is very strange but rather mysterious. I did not think that Miss Lalie would tell a downright, well nothing but falsehood will do, for no apparent purpose. She is either perfectly reckless about telling the truth and besides that as deep and cunning as a trained diplomatist or else it is not true that she is engaged to him. She contrives various little expressions and incidents that appear perfectly natural and which unless you believed them to proceed from the most perfect cunning would convince you that there was not a word of truth in the report. She ought to be the wife of the Secretary of State.

The Bowmans are as kind to us cadets as they can be. We have had a fall of snow that afforded a good deal of sleighing. Miss Lalie took four of us out sleigh riding in their big sleigh the other evening, we had a merry ride. I do not think I have been under the robes before since we lived in Rouse's Point. I was sorry to see that you said in one of your letters that you thought Colonel Bowman must be influenced by his daughter. It is not so. He is as kind to all as he is to one. They are no very great privileges after all—Taking a ride in a carriage or getting a leave for a day 2 or 3 times a year when you are a first classman.

There are a good many of us going to NY on Christmas day. If I can go until the 23 of this month without demerit I will have a leave on Christmas in common with Michie and Mcginnes. The Hyatts are going to have a "hop" on Christmas night and want to know how many of us can come. The Fish's are to have a grand ball late on Christmas Eve and though I have not been invited, I suppose I will be if Miss Sue finds out that I am coming to NY.

What glorious news from the army, I wish I was a battery that I might express my feelings on the subject. We have waited so long that we had begun to lose faith in Burnside too. Do you not think that Gen'l McClellan comes out nobly before that Court and shows himself influenced in his actions by the highest and holiest motives.

It was a rare opportunity to defend his policy which he did not use, further than he could help. Love to all.

Ever Your Affct Son J. R. Meigs

P.S. I wish to ask you for a pecuniary favor if you can spare it in order to enable me to go to NY and enjoy the Christmas fun. It costs 3.00 to go to NY and back and about 2.00 or 3.00 more to stay there and to go to Aunt Nannies in the evening. Can you send me that easily. I would not ask you only that I know that you will send it if you can spare it and not if it is, inconvenient. I also remember that you owe me 5 or ten dollars balance from the New York trip. I forget whether you paid me 5 dollars back of my 18 that were used up on that occasion. You sent me some the other day though which reduces it to ten certainly and perhaps five. Have not a good memory for such things.

Autograph letter signed, LRMC, MCMP, Ms. Div., LC.

John Rodgers Meigs was granted a leave of absence on December 24, 1862, from "12 n today, to arrival of first train on the 26th instant" and spent a happy Christmas holiday.[1]

---

1. Special Order No. 178, p. 145, RG 404: Records of the United States Military Academy, Post Orders, 1838–1904, USMA Archives.

CHAPTER 4

Graduation and Military Service at Last

Following the Christmas holidays, John Rodgers Meigs returned to West Point, to his classes, to the routine of a cadet, and to social activities there, including attendance at a minstrel show, an event that allowed participants to act out frustations and to deride authority without danger of punishment.

He continued his education with classroom and practical instruction in civil and military engineering; ethics; constitutional, international, and military law; mineralogy and geology; Spanish; and cavalry tactics. He received punishments including confinement to his quarters for one week for being off limits and for being absent from parade. But generally he did not commit any major offenses. His behavior at this point in his studies was particularly important, since both scholarly proficiency and personal conduct were used to determine class standing, which in turn determined military assignment after graduation. Only the highest-ranked cadets were selected for the U.S. Corps of Engineers.

John Rodgers Meigs to Louisa Rodgers Meigs

My dear Mother, West Point, February 5, 1863

I have been so busy at night lately that I have not been able to write home as I should have done. I received a letter from yourself and one from Mary, and one from Father since I last wrote. I missed Marys usual letter today and thought it a sore deprivation I cannot say I deserved it.

We are now studying the science of Fortification one of the most important and interesting that we have and at the same time the one especially important to officers of the Eng. Corps.

There is a model of a fortification here which Professor Kendrick says is the same one that Father used to study over and another one constructed under Genl Hallecks supervision, when he was an instructor here. It will not be long now only about 4 months before we leave the Academy for service in the field where I hope we may all be lucky.

We are commencing to think of our outfits. You told me when on furlough of a coat that Father had made but which he had not been able to wear. I wish you would show it to him and ask him if he thinks it can be made to suit me and if so send it on by the first opportunity so that I can let our tailor here get it ready. I think you told me also of some other things Father had for me when I graduated. Please let me know what they are.

I suppose Mary has gone to Newark by this time. I hope the change will do her good poor girl. It must be hard for any young person to stay in W. Do you ever hear how grandfather stands the news now adays. They are rather disheartening. It seems as if we could find no one to use to advantage the armies and the means so plentifully gathered. It is strange to hear of such an exploit as that at Charleston with two invulnerable iron clads within a few miles. Were you not all delighted to hear of the perfect success in stories of Uncle John's new vessel.

It is lucky another was able to make such a favorable report. Capt Worden does not seem to have thought much of his.

I wonder whether Com Dupont is entirely free from old fashioned contempt of new innovations. No new project has much chance of success in the hands of those who do not believe in it.

John Macomb [Jr.] was here for a few hours the other day. I showed him what was to be seen. He is I think a very fine fellow, it is a pity he could not have come here in place of some of the hacks that were sent. There was a man named Lewis appointed at large from Illinois who thought he had such wonderful perseverance as to sit up half the night to study for weeks together, was so stupid as to be sent off even before the regular examinations in Jan as there was no hope of his being able to pass it.

Have you received the cards for Miss Clara French's wedding to Lieut Remington of the Artillery. It is to take place from 8 to ten tomorrow evening. Cadets are not invited and I have received no invitation nor the cards.

Miss Clara has not been in good health for the last year or two. I hope it will do her good. Miss Mary French is getting to be quite a pretty blackeyed young lady, I saw her out riding the other day on a horse that Remington sent Miss Clara. She was enjoying it hugely.

I see Professor Bartlett every now and then, he is as kind as can be to me and to all cadets. He wished to be remembered at home and so did Colonel Clitz and Professor Kendrick.

Love to all.

Ever Your Affct Son J. R. Meigs

Autograph letter signed, LRMC, MCMP, Ms. Div., LC.

John Rodgers Meigs to Louisa Rodgers Meigs

My dear Mother, West Point, February 19, 1863

I have been interested in the accounts given by a young engineer officer a classmate of Gillespie of the way things jog along in US army of the Potomac. It was not very encouraging. He is not pleased with his corps or rather with the way in which he is treated in common with the other Engineer officers. It is not very flattering.

He says that their only duty is to survey the roads and make rough maps of them and to oversee the men throw up earthen batteries which have been laid out by some one else. Some little snipe of a beardless aid de camp will come along with an order from Hooker to post some pickets and which says that Mr Suter may accompany him. I guess that his commanders have not yet discovered the high and numerous qualifications which he possesses and have therefore not often asked his opinions. He says however that McKenzie is treated in the same way and McKenzie is really a smart and brave fellow. It does seem a pity to make no use of the education and expense that has been lavished on the graduates of the academy and especially when they have graduated high enough to go into the Engineer Corps and have shown that they are smarter than the common run. If they are to do nothing more than topog roads they might as well not have had that education. On something of the same principle they might give a boot black the education of a Prince.

It seems to be different in the West and in the South. I am afraid though that there may be danger of several of us being ordered to that army where I do not wish to go. Suter says that they have fewer Engineers in proportion to its size than in the others, still as they make no use of those they have got I do not see what they can do with more. I believe I told you that Wharton who is ranked by Suter was ordered to take 1200 men and go and fortify a town in Tennessee. It is rather a singular thing that the chief engineer of that army is a 1st Lieut. In Europe his position would make him a Brigadier or a Major General.

Suter says that there is a great deal of carping between the volunteers and regulars and that the latter being the smallest sometimes get the worst of it. He seems to have some of those stupid prejudices himself and related with some gusto apparently an opportunity he had had of snubbing a volunteer Colonel.

I wish you would ask Father to send me a copy of that bill in regard to the two corps if he can.[1] More of my success and happiness and whole course of life I suppose will depend upon the decision I make in regard to the Corps I am to enter than upon any other one thing I can do and I wish to have all the lights possible. He can easily get it and mail it to me.

Suter says that there is no man in that army who is in any one else['s] opinion a great man. Hooker is better than Burnside a great deal but like everyone else down there he has the Mcclellan fever and of course Hooker cannot compare with Mcclellan.

I am glad now that it is through with that Uncle John has had such an opportunity of testing one of the greatest questions of the time, the practicability of seagoing iron clads. He was lucky in being able to write a letter detailing the success of his vessel instead of one filled with complaints.

I had a short letter from Mary the other day in which she gives me hope of a great pleasure that of seeing her soon at West Point. I do hope she will come. It will be delightful for me and of [the good?] very pleasant for her.

Love to all.

Ever Your Affct Son J. R. Meigs

Autograph letter signed, LRMC, MCMP, Ms. Div., LC.

---

1. House of Representatives Bill H.R. 523 of the 37th Congress, entitled "An act to promote the Efficiency of the Corps of Engineers and of the Ordnance Department, and for other purposes, as amended," combined the Corps of Engineers and the topographical engineers and increased the number and rank of engineer officers. It was signed by Lincoln on March 3, 1863.

John Rodgers Meigs to Louisa Rodgers Meigs

My dear Mother, March 3, 1863

We have had another nigger show and an evening in consequence full of fun and laughter. The fencing academy the room just under the Hop room you know was filled to repletion with the residents and cadets.

Colonel B was induced to go in spite of his foot and the weather and was evidently very much entertained. It was quite amusing to watch the interest exhibited in the way that he received each joke. Sly glances were continually thrown towards him and the Professors by the performers and the rest of the cadets in order to see whether they were laughing. Of course it is no use to attempt to describe such a thing, those jokes have to be caught just fresh from the lips of the performers or all their pungency is lost.

Our Mrs Elderkin is going to leave us for a month and she will be missed a great deal by us all. She is so kind to everyone and so much loved by all. The picture of married life presented at that house is very tempting and must have an effect on all who see it.

Parson's a young graduate was here the other day with his wife. She was perfectly lovely like a beautiful picture and seemed so full of affection for him. We have had another wedding. Captain Chambliss with Colonel Clitz to help him went off to Cincinnati [illegible] married to Miss Shoenburger of that city. He brought his wife here yesterday. Harry Ledyard proudly informed me yesterday that he was the uncle of a daughter of Lizzie's. The telegram said both were doing well.

Your Affct JRM

Autograph letter initialed, LRMC, MCMP, Ms. Div., LC.

John Rodgers Meigs to Louisa Rodgers Meigs

My dear Mother, West Point, March 8, 1863

I have received Fathers letter. I have seen a copy of the bill as it passed the Senate sent to Phipps by some congressman I suppose. I do not think it differs from the former one except in making one general instead of 3 which is an alteration of not much importance to us little folk. I wish I could feel that I was always acting from the motives indicated by Father in his letter. I suppose it can be done by trying. Perhaps he will like to see the speech of an old farmer, that I enclose. It is spoken evidently from the heart and without the graces of rhetoric but it is full of burning eloquence.

How he rears to the idea that he "stands ready to meet the heretics here on this floor." It was delivered in an assembly that sorely needed it.

We are to have still another court martial here for the trial of some cadets who were found absent from their rooms in the night. They have tried to get the class on pledge and let them off but have not succeeded. There is not near so much drunkenness nowadays as there used to be. I think there is less of it every year. There are only one or two members of our class the highest in influence in the Corps who approve of drunkenness. This may seem an odd declaration but it is quite a favorable one.

You say in your last that Father thought it would bother me to have those I loved present at the final examination. I do not think it would. I have always since my first examination felt certain of doing pretty well at least and have sometimes wished when I have done very well that you were present.

Do come on for the two final weeks from the 4th to the 15th of June. You would all enjoy it I think and goodness knows I would.

That period is filled with all the exercises cadets know how to go through with performed as well as it [is] possible for them to perform them. It is decidedly the time to see the Point. Do come on and get Father to spend at least a day if he can be here no longer. He has only been here for two hours since I have been a cadet. When I graduate I could be your escort home.

I have given an order to [illegible] of Phila for a suit of flannels at $30.00. He always gives satisfaction and they are sure of being genteel garments though I might at some risk have got some for a smaller price.

Give my love to Father Aunt Nannie etc.

Ever Your Affect JRM

Autograph letter initialed, LRMC, MCMP, Ms. Div., LC.

John Rodgers Meigs to Louisa Rodgers Meigs

My dear Mother, West Point, April 15, 1863

I received your kind letter and its very acceptable enclosure the other day for which I have to thank you. This will be my last trip to NY until I go there with yourself on my way home and to the army.

I do not expect to go for a week yet and will then have from Saturday noon to Monday morning. Mary speaks in her letter that I got today of the army having started from W and other letters and dispatches say that Hookers main army has started though strange to say the papers have refrained from publishing it. How we do hope he will succeed. He has more of the dashing spirit that we have so much needed than any other general of high position. How much we might have gained if we had taken advantage of the great opportunities that we can now all see were presented to us several times in the Peninsula and in Maryland.

We have now the most interesting of all our studies in hand. "The campaigns of Napoleon." The analysis which is very clear though very concise and short is by Professor Mahan. You say that Father reads a good many such books. He ought to get a copy of the new edition of Mahans *Outposts* which is said to be the best short treatise of the kind in the language. We have been very fortunate in our class in regard to having the professors as instructors in our course through the academy. Prof Mahan to whom we recite every day has taught three quarters or more of the generals whom we have now in the field. I wish I had never been cursed with bashfulness and I might have had the benefit of a personal acquaintance with him and of hearing him talk on the subject of the war. He was very kind when I was a plebe in asking me to his house but I did not go.

It is disgusting to see the anxiety of some officers to get ordered here. There is now at this post an officer of the Engineer Corps named John M Wilson who has after about a years effort obtained the position of Ass Prof of Spanish, his sole occupation here is to teach one hour each day a section of 6 men. I dont see how they can complain of a scarcity of officers in the Corps when they allow them to do such duty as that. He has been married within a year and I suppose that had something to do with it. Almost all of these officers seem to lose their spirit when they get married and become very anxious to arrive here as "Government Nurses" as I understand Mrs Bowman calls them.

Miss Lalie is to all outward appearance recovering from her bereavement though they say it is only in company that she appears merry. Her Mother and Father are trying to make her see a good deal of the society of young people in order to keep her from brooding over her misfortune.

I spent Saturday afternoon and evening over there and took tea and dinner.

There were two young lady friends at the house and we had a very pleasant time.

One does not like to speak of the news from Charleston, it is such a disappointment. Our family though should feel glad that Uncle John came through so gallantly and his vessel uninjured from the terrible ordeal. He had the post of honor and the most responsible and difficult duty to perform of any of the commanders.

It is a pity that they seem to have determined it is no use to try to take the City. It seems that it is the obstructions of the Channel and not the fire that stopped them. They must be able to invent some way yet of tearing up those obstructions. What a fight that was compared with any that had ever before been seen. I hope you will send me Uncle Johns letter about it when it comes and has gone round the family.

Love to all.

Ever Your Affct Son J. R. Meigs

Autograph letter signed, LRMC, MCMP, Ms. Div., LC.

John Rodgers Meigs to Louisa Rodgers Meigs

My dear Mother, West Point, April 30, 1863

To day is the Presidents fast day and feels very much like Sunday as we went to church in the morning and have had no academic duty. Prof French's sermon this morning was I am afraid productive of more impious reflections and observations than pious ones. He preaches in the same style now as ever, his sermons seeming more like extracts from a dictionary than appeals to the feelings. Everyone however has a great respect for his goodness.

I have spent a rather pleasant day. This afternoon I walked out to the Cemetery with Miss Chris Eastman. You recollect that pleasant walk we had there with Mrs Overman one morning.

Miss Chris is not pretty at all when you are in conversation with her, though at a little distance she does look so, but she is very bright and pleasant. She and her mother and in fact the whole family are living now at Bob Eastmans, it must be pleasant for him to have them all with him but I do pity his poor wife. Reid has just been in here and he and Michie and I have been discussing the subject as to whether it is pleasant for a married man to have his former family living with him. They are most decidedly of the opinion that it is not desirable on account of the danger of difficulties. Judging from the way one large family in W used to get along, I should think it was an excellent thing.

It is strange how much we all talk about getting married and what we are going to do before and after, discussing what our future wives are to be. Three of the class are already engaged and I heard to day that poor Townsend was engaged to a young lady who was here last summer and who has had on mourning ever since his death.

Miss Lalie has commenced to lay aside her mourning and now and then wears a blue ribbon or something of that kind. When I was in NY the other day people asked me about it and seemed to think it a little queer. I had a delightful visit. I was there from Saturday afternoon until Monday morning.

I saw all my friends but Miss Leggett who was away from the city. The Hyatts were as kind as ever. I took tea there on Sunday evening in company with Miss Minnie and Miss Mary Vail. Miss Minnie Vail is quite a beauty and a belle in N.Y. She was here last summer perhaps you recollect her. Mrs. Hyatt wanted to be remembered to that sweet little sister Loulie of mine. Miss Phanie is going to Europe in a few weeks for a years tour. She does not seem as much delighted as a young lady would be expected to be because she dont like to leave all her friends. They have mighty good hearts all those Hyatts. I had a walk down 5th Avenue with the pretty Miss Caddie and saw all the people in NY society. I imagine, Miss Caddie must have bowed to two or three hundred people before we got home again, it was just after afternoon church when all the world takes its promenade. Mrs Townsend wanted one of Marys pictures which I wish she would send me to give her if she has it.

I saw Mr. Higbee also and little Tommy. Mr. Higbee informed [me] that he had ridden in the omnibus that morning with a man who was apparently a perfect gentleman except that his gloves were too nice and that he had been able at once to decide that he was a vulgar fellow on seeing him brush a spot of dust from his coat. He showed the Hercules in the action.

He also treated me to a detailed account of Jack's education which had been superintended by himself and concluded it by telling me he would venture to say that there was no young man of Jack's age in this country that could write as well as he could, which I believe of course.

I went through the Central Park and saw its lovely bridges and green lawns. The trees are rather small yet and the leaves were not out. It looks much prettier than when Father and I went through it 3 years ago.

My report last month was not very satisfactory but there is nothing to be judged from one month standing as it depends very much on accident. I was 6 in ethics but have lost only a single tenth the whole six months. My low standing in Engineering was owing to the time that I have to instruct in having to be taken from my hours for drawing so that I got behindhand in it. It has no effect however on the final standing in that study.

Good night. Love to all.

Ever Your Affct JRM

Autograph letter initialed, LRMC, MCMP, Ms. Div., LC.

As graduation approached, concerns about John Rodgers Meigs's standing in class continued. Not only did Montgomery C. Meigs want to correct what he considered an injustice done, but he wanted to make certain that his son graduated first in his class. In May 1863, Montgomery contacted the secretary of war to raise for the second time the issue of the extra demerits related to Lincoln's visit. In his response to the secretary's renewed interest, General Totten changed his position and canceled the second and third charges, reducing the number of demerits given to five from fifteen. The practice of dismissing one-sixth of all demerits given for the last three years of study, coupled with successful examinations, placed John Rodgers Meigs first among his classmates.

John Rodgers Meigs to Louisa Rodgers Meigs

My dear Mother, West Point, May 25, 1863

I am sorry that I have not been able either to write to you to express my joy and gratitude that you are coming on to see me or to congratulate Uncle Em on obtaining the crowning happiness of his life. Come on so as to be here at least on Monday evening. We commence the examination on Tuesday. Come sooner if you can and make my congratulations to Uncle Em and give my love to his little wife if she will accept it.

I am in a great hurry and not well either.

Ever Your Affct Son J. R. Meigs

Autograph letter signed, LRMC, MCMP, Ms. Div., LC.

On May 30, 1863, John Rodgers Meigs, along with his twenty-four classmates, was found fully qualified for military service by the medical board convened at West Point.

John Rodgers Meigs to Montgomery C. Meigs

My dear Father, West Point, June 5, 1863

It is after 12 at night but I must sit down to tell you of my happiness in having Mother here with Mary and Loulie and in coming out pretty well in my studies. I have not come out head in every thing and the grand sum total of my efforts has not yet been published as I do not suppose the *general standing* will be made known for some time. I came out 1st in Ethics 1st in Engineering 1 in Cavalry Tactics 1st in Mineralogy and Geology, 2d in Ordnance, Twining being first. 2d in Spanish, McKee being first.

I have been quite highly complimented by all on the way in which I passed my examinations. Mary is enjoying herself a great deal. She has got among her old cadet friends here and I have hardly a chance to see much of her. I have had several flirtations with Mother and long pleasant walks and talks about you. I feel very happy at having arrived at the end of my four years work and come out well.

I have now been studying for sixteen years preparing to *do* something in what remains of my life and I hope I shall have a chance to do something in the way of using that education instead of only getting more. I do not believe in spending more than half my life in *getting ready* to work, 16 years is quite enough to have learned something that I can make of practical utility, if I have the opportunity; and if I do not have it I will only live on gradually letting what I do know fade out of memory. Whether I shall have this opportunity or not depends on the position I am placed in almost entirely. If I am ordered with twenty others like me who have just graduated within a year or two to the Engineer battalion in the army of the Potomac I will not have that opportunity.

If I am on some rising Generals staff or in a position which might lead to giving me after some months service the Colonelcy of a regiment which is a round in the great and continuous ladder to distinction and usefulness which must be gained if one has any aspirations I will have the opportunity. In any event my reputation as an officer will depend very much on what I can do in the first year and that will evidently depend on my opportunities.

I have a favor to ask you for the Father of my poor dead classmate Townsend who died last December. I send you his letter to me written by his son (which accounts for the spelling). I do not see what better guarantee as to his character could be given than that he was Hub's Father and had taught him the principles of honor and manliness that he showed.

He has been in service now since the war first broke out and the exposure and fatigue of the last campaign on the Rappahannock brought on a severe illness such that it is as likely as not that a renewal of it would kill him. Can you not give him this position which he asks and enable him after having served his country nobly in the field to avoid this risk.

I do not think the family are in over comfortable circumstances and his death would be their destruction. He was at first a chaplain but afterwards was placed in command of a company. He was a clergyman at home and Hubs language and manners show that he must have been a man of education and refinement.

Tell dear Aunt Nannie that I will soon see her to return those kind expressions of love she wrote in her letter to Mother.

Mother enjoys herself a great deal here I think. I was very very much disappointed when I heard that you could not come on to see me leave, it would have been so pleasant to have had my father here at this time when I am about to receive my diploma, the reward of all my study.

I remember when I came here that you told me it would be worth $20,000 in cash to me to graduate head of my class. I suppose at this time in the history of the country it is worth much more.

We may leave here next Thursday with a leave of absence until the first of

July when we will report at the War Department but we do not *know* any thing about it, only guess.

Love to all.

Ever Your Affct Son J. R. Meigs

Just 27 years ago you were first what I am Almost a graduate of the U.S.M.A.

P.S. Will you tell me something favorable to let Mr Townsend know or write it to Mother if you have not time to write it to me. JRM

Autograph letter signed, MCMC, MCMP, Ms. Div., LC.

On June 11, 1863, John Rodgers Meigs graduated first in his class at West Point. His graduation was made even happier by the presence of his mother and his two sisters. Montgomery C. Meigs was gratified by his son's success and asked "God [to] protect him in his career & make it useful to his country." It was now time for John to leave "the Point," travel south to Washington, D.C., and begin his work as a military engineer. On Tuesday, June 16, 1863, the family was reunited in Washington, D.C. During his brief stay at home, John visited the Little Falls of the Potomac and Fort Stevens on the outskirts of the capital.[1]

While young Meigs was traveling south, his future assignment and that of other graduates was decided. On June 15, 1863, along with seven of his fellow classmates, he was assigned to the Corps of Engineers. On the same day, War Department Special Order No. 264 assigned Meigs to duty as an engineer in the Middle Department, Eighth Army Corps, and directed him to report without delay to the departmental commander in Baltimore.

On June 19, 1863, at the age of twenty-one years and four months, Meigs was sworn into his office in the United States Corps of Engineers. The evening of the same day he left Washington to report to Major General Robert C. Schenck, commander at Baltimore, Maryland, where Meigs's first task was the construction of forts and entrenchments to protect the approaches to the city against the new Confederate invasion of Maryland.

Assigned responsibility for fortifying the left division of the city—its southwestern and western sectors—Meigs went to work immediately, providing requisitions for artillery that headquarters dispatched to Washington. General Schenck called on the mayor of Baltimore to furnish Meigs with a thousand laborers and also ordered two companies of infantry to assist him. In less than one week, a number of forts and gun emplacements were under construction in the sector, which, when completed, were to be manned by twenty-three hundred troops servicing at a minimum eighteen artillery pieces of calibers specified by Meigs.

1. MCMD, MCMP, Ms. Div., LC, entries for June 8, 11 (quote), 16, 18, 19, 20, 23, 1863.

John Rodgers Meigs to Colonel Donn Piatt

Chief of Staff
Sir, Baltimore, Maryland, June 26 1863

There will be required to man the Left Division of the Defences of Baltimore extending from tide water around the West front of the city to the Hookstown road the following amounts of men and guns.

Fort A	Minimum amount.	Proper amount. in addition
	2—8 inch howitzers siege	2—24 pdrs siege
	1—30 pdr Parrott gun	1—8 inch columbiad
Fort 1	1—20 pdr Parrott	1—24 pdr barbette
	2—32 pdr brass howitzers	2—8 inch columbiads
	1—barbette 24 pdr	1—30 pdr Parrot.
Battery 1 ½	1—12 pdr gun	} Siege or
	2—32 or 24 pdr howitzers	} field carriage
Fort 2 rifle pit		
Fort 3	20 pdr Parrott	12 pdr guns
	12 pdr gun	12 pdr howitzer.
Fort 4	2—24 pdr barbette guns	30 pdr Parrott or the ordnance gun
Fort 4 ½	2—12 pdrs	20 or 30 pdr Parrott
	2—6 pdrs	

At fort 1 1000 men should be encamped with 2 company of artillery heavy. They should detach a couple of companies and a section of light artillery if possible to the bridge on the Frederick road as a grand guard there and send out their pickets on the other side of the stream.

This force should guard Forts A & 1 and the contiguous rifle pit and be held in readiness to march to support the right of the division by Kirby's Lane or the extreme right of the city by Baltimore Street leaving always some sentinels in the works.

1000 men should be encamped near Fort 4 ½ not far from the Hookstown road and should detail guards for forts 3—4—4 ½ and send out grand guards on the Hookstown and Liberty roads.

There are already 3 or 4 companies encamped in No 2 rifle pit who had better stay there to take care of it and 1 ½ battery.

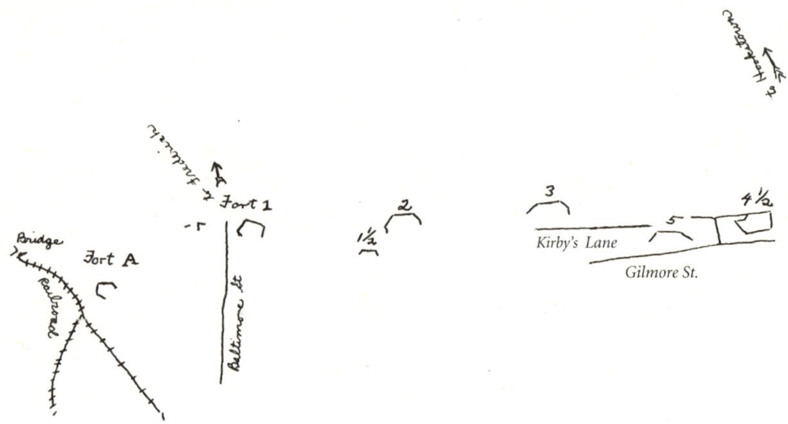

The above is the position of the Entrenchments.
State of the works Fort A available for inf tonight
Fort 1 ready for infantry and rifle pit on left finished
Fort 2 rifle pit finished 1 ½ ready for guns tonight
3 ready for inf 4 available for inf and 4 ½ same.

John R. Meigs

Autograph letter signed, RG 393: Records of U.S. Army Continental Commands, 1821–1920, Middle Department and Eighth Army Corps (P), 1861–66, Letters Received, 1863–66, M-555, NA.

In addition to planning and supervising the construction of the Baltimore forts, John Rodgers Meigs was ordered to defend the Baltimore and Ohio Railroad. After Lee's army had destroyed most of the bridges between Harpers Ferry and Cumberland, Maryland, General Schenck directed the railroad company to build five ironclad "defensive cars" for protection of the road. An order was sent the same day. The company's master of machinery pressed the work forward, employing a "full force of car builders and iron workers, both night and day," and the five cars were completed five days later, on July 3. John Rodgers Meigs made suggestions for additional work to accommodate the "introduction of a field piece" in each car. Meanwhile, Schenck had been communicating with Washington to obtain suitable artillery pieces for armament. With Lee now in retreat from Gettysburg, Schenck, on July 5, placed Meigs in command of the cars and sent him off toward Harpers Ferry as part of the attempt to intercept the Army of Northern Virginia as it moved south.[1]

Meigs was excited to be actively engaged against the Confederates. General Schenck reminded his young subordinate to follow instructions from senior officers and not to initiate an engagement on his own. Although official reports on the activities of Meigs's command during the early days of July apparently do not exist, John wrote a letter to his father on July 12, 1863, that fleshes out the sketchy accounts contained in the telegraphic communications between Meigs and his superiors.

1. John Rodgers Meigs to Colonel Donn Piatt, June 23, 1863, File 1863/M-527, and John W. Garrett, President, B&O Railroad, to Colonel Donn Piatt, 2:35 P.M., June 28, 1863, acknowledging receipt of the order for five defensive cars, File 1863/G-188, both in RG 393: Records of U.S. Army Continental Commands, 1821–1920, Middle Department and Eighth Army Corps (P), 1861–66, Letters Received, 1863–66, NA; Report, Thatcher Perkins, Master of Machinery, B&O Railroad, to John W. Garrett, July 3, 1863, quoted in Edward Hungerford, *The Story of the Baltimore and Ohio Railroad* (New York: Putnam, 1928), vol. 2, p. 29; John W. Garrett to Robert C. Schenck, June 30, 1863, File 1863/G-188, RG 393: Registers of Letters Received, 1863–66, NA; John W. Garrett to Robert C. Schenck, July 3, 1863, Letters Sent by the Baltimore and Ohio Railroad, January 1, 1859–February 9, 1867, MS 1816, Maryland Historical Society; RG 393: Letters Received, June 1862–April 1864, vol. 3, p. 259, #191, NA; Robert C. Schenck to John W. Garrett, July 3, 1863, in *OR*, series 1, vol. 51, part 1, p. 1069; O. Perkins to John W. Garrett, July 3, 1863, Letters Sent by the Baltimore and Ohio Railroad, January 1, 1859–February 9, 1867, Maryland Historical Society; Robert C. Schenck to Brigadier General James Wolfe Ripley, Chief of Ordnance, Telegram, July 3, 1863, RG 393: Dispatches Sent, June–July 1863, vol. 31/34, p. 114, NA; RG 393: Special Orders, June 1862–May 1864, Special Order No. 180 (July 5, 1863), vol. 54, pp. 355–61, paragraph 13, NA.

John Rodgers Meigs to Donn Piatt

Sir, Baltimore, Maryland, June 28, 1863

I have the honor to submit the enclosed order to the General for his signature and immediate issue.

Against any but the largest bodies of troops these defensive cars would be a perfect protection to the points where they might be placed. They could be manned with the railroad guards scattered along the road and could be always placed at the *points threatened.*

John R. Meigs

Autograph letter signed, RG 393: Records of U.S. Army Continental Commands, 1821–1920, Middle Department and Eighth Army Corps (P), 1861–66, Letters Received, 1863–66, M-527, NA.

John Rodgers Meigs to Charles D. Meigs

My dear Grandfather, Baltimore, Maryland, July 1, 1863

I was greatly disappointed when I found I had to go without seeing Grandmother and yourself again. I wanted very much to see Ham too but I had to go.

I was not sorry to get my orders, though I hope after a while to get a leave of absence. I am very busy now fortifying the western front of Baltimore. I hope there will never be any real use for the forts I am building though I should like to see them tried. The forts of Washington I have no doubt have often prevented a campaign being undertaken for its capture. Those of Baltimore must have the same moral effect.

There will probably be another great battle before a great many days when I hope to see Lees Army defeated dispersed and captured. Our new Genl Meade has the confidence of many of his subordinates and may prove to be the coming man for whom we have been waiting so long to finish this war. I do not believe he will be found until Father comes out into the field in active command. The more I see the more I am convinced of that.

I am pretty busy here directing the labor of a thousand or 1500 men.

Goodnight and Love to all.

Ever Your Affct John

Autograph letter signed, CDMC, MCMP, Ms. Div., LC.

John Rodgers Meigs to Montgomery C. Meigs

Baltimore, Maryland, July 5, 1863

Brigadier General Meigs, Received Washington, 12:40 A.M. July 6, 1863

Genl. I go tomorrow morning in command of ten rail clad cars 150 men to escort troop trains & get in Lee's way.

J. R. M.

Telegram, MCMC, MCMP, Ms. Div., LC.

John Rodgers Meigs to Major General William H. French

Relay House, July 6, 1863

General French, Received 7:15 P.M. July 6, 1863

I am coming up the road with five Iron clad cars to report to you. Will you telegraph any orders or information to Marriottsville telegraph station thirty *30* miles from here.

John R. Meigs

Telegram, William Henry French Papers, Special Collections Division, USMA Library.

John Rodgers Meigs to Major General William H. French

Monocacy Junction, Maryland, July 7, 1863

General French, Received 7:10 A.M. July 7, 1863

Iron clads go on to near Point of Rocks to get ahead of trains & get in fighting trim. Orders can be telegraphed there.

John Rodgers Meigs

Telegram, William Henry French Papers, Special Collections Division, USMA Library.

John Rodgers Meigs to Major General William H. French

General French, Point of Rocks, Maryland, July 7, 1863

I am at Point of Rocks. Going to Harpers Ferry to save the bridge, will take the operator with me to Sandy Hook one mile from Harpers Ferry. There are some rebels said to be a cavalry regiment on the other side of the river. Supervisor of the district says fords are impassable. He came down one hour or so ago. Tracks are almost blocked with trains.

John R. Meigs

Telegram, William Henry French Papers, Special Collections Division, USMA Library.

Major General William H. French to John Rodgers Meigs

Lieutenant J. R. Meigs, Frederick City, Maryland, 11:10 A.M. July 7, 1863

A Brigade will be at Knoxville, early to-day with Artillery and Cavalry.

The force across the river is of no account. I destroyed the bridge.

William H. French

Telegram, William Henry French Papers, Special Collections Division, USMA Library.

John Rodgers Meigs to General Robert C. Schenck

General Schenck, Point of Rocks, Maryland, July 7, 1863

All well at Point of Rocks. Forty rebel pickets on other side of the river, opposite here. Track burned off Harper's Ferry bridge, but iron-work unimpaired. Go on immediately, to save what is left. Take the operator belonging to railroad to Sandy Hook, 1 mile this side. Send on for operator.

John R. Meigs

Telegram, OR, series 1, vol. 27, part 3, p. 589.

John Rodgers Meigs to General Robert C. Schenck

General Schenck, Sandy Hook, Maryland, July 7, 1863

I am now opposite Harper's Ferry. Been desultory picket firing across the river, both sides. Iron-work of bridge uninjured. Shelled the rebel bushwhackers quiet. Please have carriages of mountain howitzers sent and carriages of 6-pounders without limbers.

John R. Meigs

Telegram, OR, series 1, vol. 27, part 3, p. 589.

John Rodgers Meigs to General George G. Meade

General Meade, Sandy Hook, Maryland, July 8, 1863

Sir: I am in command at Harper's Ferry of five iron-clad cars, proof against musketry, and, I think, against 12-pounder solid shot. I expect five more to-morrow. I intended to rebuild the track over the Harper's Ferry Bridge, and press on building the road and the telegraph as far as I could go, perhaps even to Martinsburg. I am not afraid of the heavy masses of the enemy, provided they cannot get heavy artillery to work at me. May I build the bridge to-morrow and go on? I will have a corps of railroad men and sufficient material. Will you at least permit me to build a part of the bridge, leaving a gap which can be filled in an hour?

River rising here rapidly; a few pontoons coming down with current.

John R. Meigs

P.S. I have clumsily mounted in my cars a 6-pounder and four mountain howitzers.

Telegram, OR, series 1, vol. 27, part 3, p. 607.

John Rodgers Meigs to General Robert C. Schenck

General Schenck, Sandy Hook, Maryland, 12 M. July 8, 1863

Sir: I am guarding the railroad, but I wish to push over the river with my cars, reconstructing the railroad and telegraph as I go. Have men and materials, and have telegraphed to General Meade for permission to build the bridges, as I understand he has forbidden. Went outside cars while the enemy is at a distance, in order to get room, and they will be fairly protected by cars. Will carry a mountain howitzer up the Heights to-morrow, to keep rebels quiet while building bridge.

John R. Meigs

Telegram, OR, series 1, vol. 27, part 3, p. 608.

General Robert C. Schenck to John Rodgers Meigs

Lieutenant J. R. Meigs Baltimore, Maryland, 4:30 P.M. July 8, 1863

What are your operations and plans? Why do you want gun carriages? You must be cautious as well as active. How many iron-clads have you with you? Have you reported to General French and taken his views? Remember that the duty assigned you is not to seek a fight, but to help keep open and protect the railroad.

Robert C. Schenck

Telegram, OR, series 1, vol. 27, part 3, p. 608.

John Rodgers Meigs to Major General William H. French

General French, Maryland Heights, Maryland, July 9, 1863

General Kenly is at Harpers Ferry & two (2) of my Iron Clads are left there. We have been firing at Bush Whackers. Had to lay our Mountain Howitzers on the ground & shell the town. The Rebels fire has been slow. Wounded 2 men. Leave one iron clad at Sandy Hook & 2 others along the road to Point of Rocks opposite which cavalry pickets were seen this morning.

John R. Meigs

Telegram, William Henry French Papers, Special Collections Division, USMA Library.

John Rodgers Meigs to General Robert C. Schenck

General Schenck, Sandy Hook, Maryland, July 9, 1863

Gen Meade forbids the reconstruction of the RR bridge for the present.

J. R. Meigs

Telegram, RG 393: Records of the U.S. Army Continental Commands, 1821–1920, Records of Middle Department and Eighth Army Corps (P), 1861–66, Press Copies of Telegrams Received, July 1863–January 1865, vol. 1, #63, p. 17, NA.

General Robert C. Schenck to John Rodgers Meigs

Lieutenant J. R. Meigs, Baltimore, Maryland, 9 A.M. July 9, 1863

Your telegram of last evening just received. Do nothing at the Harpers Ferry Bridge without distinct order or permission of Genl Meade or Genl French. You must run no risk of interfering with plans of the Commander of the Army.

Robert C. Schenck

Telegram, RG 393: Records of the U.S. Army Continental Commands, 1821–1920, Records of Middle Department and Eighth Army Corps (P), 1861–66, Press Copies of Telegrams Sent, July 1863–August 1866, vol. 67 (July 9, 1863), NA.

John Rodgers Meigs to General Robert C. Schenck

General Schenck, Sandy Hook, Maryland, July 9, 1863

I climbed on the iron work just now as far as the beginning of the last span, and was not fired at. I do not believe there is an armed rebel nearer than Bolivar Heights, a mile and a half beyond Harper's Ferry, where there seems to be a picket. I have ordered the timbers that can be collected brought to the end of the bridge, so as to be ready the moment General Meade orders its construction. Will you tell Colonel Donaldson it will require 1,200 running feet of 6 by 6 scantling. and 16,500 feet, board measure, of 2-inch plank, to make a good artillery road across it?

John R. Meigs

Telegram, OR, series 1, vol. 27, part 3, p. 618.

John Rodgers Meigs to General Robert C. Schenck

General Schenck, Sandy Hook, Maryland, July 10, 1863

Sir: Had a flag of truce to Harper's Ferry to-day to get a child across. Learned that rebels there are in ignorance of events. They seemed to think we were cut off from Baltimore. Wanted to know the name of commanding general. Found out that the pontoons coming down the river are some that were lost a week and a half ago; that they thought Lee in Frederick, and that he had captured thirty-six regiments. Too hazy to see Williamsport.

John R. Meigs

Telegram, OR, series 1, vol. 27, part 3, p. 632.

John Rodgers Meigs to Montgomery C. Meigs

Headquarters Iron Clads
My dear Father, Opposite Harpers Ferry, West Virginia, July 12, 1863

I have been so busy until now that I have been unable to write home as I ought to have done. I came up here with five of my iron clads the other day in the first train that had been over the road since the evacuation. I brought 60 men with 4 mountain howitzers and one 6 pdr brass. We went into service immediately for when we arrived we found that our men were deployed along the road at the end of the bridge trying to skirmish the rebels out of the town opposite. They did not succeed however and two of our men were wounded one of whom has since died. I ran back the Vicksburg, the car containing the 6 pounder and let fly some spherical case at some of the rascals up on Bolivar

Heights in the road. I had to swing the gun by ropes from the roof of the car and every time it went off it struck against the ceiling and bounced around so that we had to restring it nearly every time on account of the slipping of the ropes. The little mountain howitzers we got out and laid on the ground with their muzzles propped up and fired shell at the town. Every time they went off they turned about half a dozen somersets backwards but they put their shell where we wanted them. After about half an hours firing the rebs concluded to stop. Five more cars came up yesterday of a different construction. They look on the outside exactly like freight cars the port holes being closed with shutters. They are lined on the inside with iron and are musketry proof. I have now about 90 men my full complement which I expect to get from General Naglee being 180. I have my headquarters in the store car of the train where the rations and the pro[ven]d[er] are kept. From the door of the car I am in I can look across the river which is now falling and is said to be fordable already at Shephardstown, and see the rebels lying about some of them in uniform.

I have just been summoned to the end of the bridge there was a rebel officer on the other side who wished to exchange papers but I hallowed back to him that it could not be allowed.

I went over there day before yesterday to bring back a child that they wanted to send over and even talk with them. The crowd increased to about half a dozen in number and some men and women besides. They wanted to know if we had any news. I told them we had pretty good news but I did not feel authorized to tell them what it was. I drew from them that they could hear nothing from Lee or from Meade that they imagined that we at Harpers ferry were cut off from Baltimore and that Genl Lee was in Frederick. That there was a picket guard in H Ferry, that there was a colonel somewhere near and a general in Charlestown.

I thought it best for them to remain in this state of blissful ignorance so that they will not learn anything that will induce them to destroy the long trestle work on the other side. At present they may be under the pleasant delusion that Lee will need the railroad for reinforcements and have even been lately constructing a way over the break in the Shenandoah bridge right under our eyes. I do not know whether it will not be best for us to stop that, however as they can then send their pickets to Loudoun Heights and shoot right down into us.

It seems curious to be in this way, with only a river between us so narrow that we can halloo across. I do not allow this however. Several people have been here to look through my glass at their wives and children in the town opposite.

I have named all my gunboats as they call them, Vicksburg, Port Royal, Gettysburg, Antietam, etc. On those that counterfeit the Military freight cars I have painted in large letters (Govt Stores). When Johnnie Reb comes up to them and sees the little shutters drop out of their ports and the whole thing suddenly transformed into an inapproachable blockhouse I think he will be astonished.

My only sorrow is that Genl Meade will not allow me to build the bridge and get over the river, I could do it in two days. I intended having this force 10 rail clad cars 5 of them proof against light artillery manned by 180 men and with 5 pieces of cannon to push over the bridge and as far as I could go building the railroad and telegraph as I go and leaving a car at the exposed points which would protect them from anything but a heavy force. If I got captured why I would be captured but would have paid for myself and I would not be captured except by a force much greater than my own with artillery.

[Unsigned]

Autograph letter unsigned, MCMC, MCMP, Ms. Div., LC.

John Rodgers Meigs's words of July 12, 1863, regarding his possible capture, ring as a harbinger of events to come. On July 14, 1863, General Schenck ordered the young engineer to turn over the ironclad cars to their tactical commanding officer and return to engineering duties in Baltimore. Before returning to Baltimore, Meigs joined Major Charles Farnsworth of the First Connecticut Cavalry, who with his troops had crossed the Potomac River to reconnoiter Confederate positions on Bolivar Heights, between Harpers Ferry and Charles Town, West Virginia. A sharp engagement ensued; Union troops captured the colonel of the Confederate cavalry force but gave up Farnsworth and twenty-four of his men as prisoners. John Rodgers Meigs escaped on foot after losing a borrowed horse. Official reports of the action do not mention Meigs's participation, but his father recounted the events in a letter to Charles D. Meigs on July 14, 1863.

Following this action, John Rodgers Meigs stopped over briefly with his family in Washington before proceeding to Baltimore, where he resumed work on the fortifications. Earlier, after Lee had crossed into Maryland, the War Department in late June had redesignated the western portion of Schenck's command as the Department of West Virginia, encompassing territory from Hancock, Maryland, to eastern Ohio. In a readjustment of the resources of the two departments, the adjutant general issued new orders on July 23, 1863, to Lieutenant Meigs, transferring him to the new Department of West Virginia under command of Brigadier General Benjamin F. Kelley.

As Kelley's chief engineer, Meigs was responsible for supervising the construction and maintenance of blockhouses to protect the Baltimore and Ohio Railroad, for undertaking topographical surveys, and for procuring and preparing maps. In addition, from time to time Kelley ordered his young engineer to accompany troops commanded by Brigadier General William W. Averell in a series of raids against Confederate forces and lines of communication in the mountains and valleys of West Virginia and southwestern Virginia. In his official reports, Averell cited Meigs, among other aides, for "untiring energy and hearty cooperation."[1] The following correspondence provides a record of Meigs's activities at this time.

1. Report of Brigadier General William W. Averell, August 30, 1863, *OR*, series 1, vol. 29, part 1, pp. 32–38.

John Rodgers Meigs to Lousia Rodgers Meigs

Dear Mother, Baltimore, Maryland, July 23, 1863

I got back here safe and no not quite sound though I am getting better. The Misses Schenck have come back from Princeton and informed me to day that they had not seen me for three days. Gov Bradford is to give a party on next Tuesday to which I am invited. I have nothing to wear but think I shall go anyway. I went to see Mrs. Ledyard Hodge this evening and found her not well. Still she came down to see me. She has to mix with a great many secessionists or go without seeing anyone, she says, and she finds it pretty hard. I wish you would find my watch, the old one Uncle John gave me years ago. I should like to have it made into a pedometer and my razor. If you can not, I shall have to get another which will be an expense. My salary has not commenced to accumulate yet.

Tell Father I sold the two bonds for $216.44. I have now got my choice for the Dr's Horse down to two. I expect to pay 150 for him. Love to Mary & Monty & Father.

Ever Your Affct Son J. R. Meigs

Autograph letter signed, LRMC, MCMP, Ms. Div., LC.

Montgomery C. Meigs to Charles D. Meigs

My dear Father, Washington, July 24, 1863

I send you one of Johns late letters. He is now in Baltimore making up his accounts of expenditures & works upon the fortifications. He spent Sunday with us. He came very near being captured the other day having accompanied a reconnaissance [illegible] at Harpers Ferry. Maj Farnsworth & some 28 or 29 of his Connecticut Cavalry were captured. John escaped by concealing himself in the woods which was also tried by two privates who with him & Major Farnsworth had become separated from the main body of the Cavalry detachment. They had taken a Col a Capt & five privates & in pursuing through the woods another rebel separated from their companions & were cut off by 200 rebels.

Of the four John alone escaped afoot. The others & the horse he rode were discovered & captured. I had to lend him money to replace the borrowed horse. He went through the charge in which they captured the Colonel & Capt unarmed having run across the river afoot to carry an order without his arms or horse & then joined the event, to see this fun as he expressed it.

It came near being serious fun to him as it was to our major & the rebel Colonel & Captain. . . .

Love to all. Ever your affc MCMeigs. . . .

Autograph letter signed, MCMC, MCMP, Ms. Div., LC.

John Rodgers Meigs to Brigadier General Lorenzo Thomas

Sir, Pittsburgh, Pennsylvania, August 4, 1863

I have the honor to acknowledge the receipt on the 29th ultimo of Special Orders No 327, Adjutant General's office July 23d and to report that yesterday I received corresponding orders from the Headquarters of the Middle Dept and am now here on my way to join Gen Kelley.

John R. Meigs

Autograph letter signed, RG 94: Records of the Adjutant General's Office, Records of Appointment, Commission, and Personal Branch, 1783–1917, Letters Received, 1863–94, CB 1863, M-1033, NA.

John Rodgers Meigs to Brigadier General Joseph G. Totten

Clarksburg, West Virginia, August 6, 1863

Sir,

I respectfully request that the sum of $3.500 three thousand five hundred dollars be placed to my credit at the subtreasury in New York to be expended in erecting military defences in W.V with the object of checking in some measure the rebel raids which issue from the gaps in the Alleghany Ridge and wind through the state. Now that the main army of the Rebels appears about to remain inactive these raids, in order to give employment to their cavalry, will probably become more frequent.

I further request that $2.500 be placed to my credit to enable me to make such surveys of the strategic roads and important points as will supply the knowledge necessary to enable the Genl Commanding to carry on military operations.

I find on inquiry in Wheeling that the reason such information is so difficult to obtain is that there never have been any detailed maps made of the western part of Va.

I also request authority to purchase such materials and employ such assistants at the usual rates and with the approval of the General Commanding as maybe necessary to the furtherance of the above objects.

Your Obdt Servt, John R. Meigs

Autograph letter signed, RG 77: Records of the Office of the Chief of Engineers, Letters Received, 1826–66, M-4388, NA.

John Rodgers Meigs to Louisa Rodgers Meigs

Dear Mother, Cumberland, Wva. [Maryland], August 8, 1863

I arrived here yesterday evening from Wheeling, Va.

I rode nearly all the way upon the Engine. The view in coming over the mountains was perfectly grand. In riding on the front of the Engine just over the cow catcher you avoid all the unpleasant dust and smoke and have a most extended view. The road climbs along the valley of Cheat River and as you rush along up the grade you can see hundreds of feet below you the black waters of the river foaming over its rocky bed while towering farther still above you are the steep precipitous sides of the mountain. The valley of the river runs on before far in among the blue masses of the Alleghany Range with its sides still clothed with the wild uncut forest that flourished there ages ago. It gives one a peculiar sensation of awe to plunge into one of the long tunnels, one of them nearly a mile long, with which the route abounds. Sometime before you reach it you can feel its icy breath and then suddenly with a deafening roar you dash into the blackness ahead. Far away in the front is a bright round spot with two long streaks of light leading up to it that are reflected from the wet and slippery tracks one single break or irregularity in which would send you to destruction. The bright spot grows constantly larger & larger until finally you flash out again into the open air and daylight. It was quite fatiguing holding on to the engine it jumps and pitches about so. The wind too blows very strongly and makes one drowsy. I started once at feeling myself giving way and found that I was on the point of going to sleep. It made me more careful to keep myself awake afterwards.

I fell in at Wheeling with Dr Stewart of Alexandria one of the Vol surgeons who finding that I like himself was going to Gen Kelley struck up quite a friendship. He is a curious genius a little pedantic I think. Instead of telling you that a certain man is a skilful physician he will say that he found him eminently well versed in physiology, pathology and, theraputics. He is good humored however and likes me which is pleasant. I found Gen Kelly a pleasant person and a man apparently of some force of character. His son and aid de camp is I think a very fine fellow. I took with him quite a ride in an ambulance to day up to the town of Frostburg, one of the highest villages in the US being 2300 feet above tide. It was cool and pleasant there to day. I am going to Clarksburg tomorrow where I think it probable I shall open my office and where letters should be directed. Please write.

Ever Your affct Son John R Meigs

Autograph letter signed, LRMC, MCMP, Ms. Div., LC.

On August 12, 1863, John Rodgers Meigs was directed to report to Brigadier General William W. Averell at Moorefield, West Virginia, for temporary duty on the Huntersville expedition to oppose Colonel William L. Jackson and to destroy saltpeter and gunpowder works.

John Rodgers Meigs to Montgomery C. Meigs

My dear Father, Huntersville, [West] Virginia August 23, 1863

We arrived in this old town yesterday at 3 PM.

We had a hard but pleasant march through the mountains and narrow valleys of this part of Va. We first encountered the enemys pickets on the day before yesterday and drove them before us with continual skirmishing from 9 AM until we encamped at Gibsons store 12 miles N of Huntersville.

The bullets sometimes came pretty close but it was rather better fun being in the advance than with the main body and so I was generally there. We had on that day Friday one horse shot one captured and one mans hand hurt. We think we killed one of the rebs and we hit one in the head one in the shoulder and shot a finger off, so that we got rather the best of them. We captured several prisoners. I had the pleasure of taking one with a squad of 4 men I was leading through the woods. He jumped up suddenly in front of us and hollered "don't shoot me."

They bushwacked the trains in our rear and shot one poor fellow, a teamster in the small of the back. I saw him today, he cannot expect to live. Another was wounded in the arm. It was rather a small kind of fighting however. It is very strange that more men were not killed on our side. We had to go at one place through a gap in the Alleghany you will see it marked on Lloyds Map at Warwicks Creek SW of Monterey and just at the SW corner of Highland County.

The road was through a dense brush and made a bend back on itself thus *a* was about 60 feet higher than b. Our advance guard of about 10 men were moving cautiously along the lower part of the road when suddenly a volley was poured into them from right above, a horse was killed but no one was hurt.

We have passed through many such places but with little loss.

Gen W L Jackson was encamped on the Warm Sprs and Huntersville pike he had about 1500 men scattered around his headquarters were three miles E of this place just on the other side of a very rough gap in Browns Mt.

He was so much afraid of being cut off from W Springs that he left precipitously. We have lost a good many horses from exhaustion and have not found many for a remount as they are driven off when news of our advance arrives. We have received reenforcements and rest here to day Sunday. We have by moving down from Petersburg to Franklin & Monterey and Huntersville occasioned a general skedaddle of all detachments in the Western Mts for fear of being cut off. There are no troops but guerrillas infesting the Alleghanys & W of this place. Our information leads us to believe that all rebel troops intend to go East but that many of the soldiers will desert the army and take up bushwacking rather than leave the section to fight. We will we hope either capture or chase out of the Country all the rebel Force and the whole of W Va would then be in our possession if we choose to hold it.

There is now occupying W Va or so much of it as the rebels have about 6000 men. Genl Kelly has you know how many to oppose or drive them with.

He has forty thousand men, he thinks. If we succeed as we hope in clearing the force out by a kind of raking movement along their line of outposts, it would be—I will write on a new piece of paper because[1]

We have reached here with no other opposition than a few skirmishes in which we have lost not so many as the enemy, one mortally wounded one leg lost one arm injured and some scratches. Our horses are a good deal pulled down and we are out of nails. Having rested to day we are in good feeling for a days march.

On our march which we hope will have been a sort of raking movement along the rebel line of outposts passing successfully in their rear we were a complete surprise as far as Monterey. The whole country is now however alarmed and up to bushwhacking.

Jackson who was in his *Camp Northwest,* east of a gap 3 miles long whose Western entrance is at this place, hurried over the Alleghany Mts for fear that we were advancing along the Back Creek road and would thus cut him off from Warm Springs. He has about 1500 & has sent for reinforcements probably to Staunton. He has 6 pieces [of artillery], many of his men are newly organized.

There is a force of about the same size at Lewisburg anxious about Scammon who is in front of them rumored to be at Coal Knob. We are mostly Cav have 300 dismounted infantry. In case we either capture or clear the country west of our line of march of *rebels forces* it will be a pity to abandon it again to be occupied by a force of 8 or 10 thousand as it has been while we watch along the B&O.R.R with 30,000. To hold it we would have to occupy each gap with some block houses and field works. The block houses only need be garrisoned except on approach of the enemy. We could thus with a very small force hold the line of the Alleghanys as far as Lewisburg. Then we would strike westward to the Big Sandy. This line would not be as easy to defend as the Alleghany but then we would have no large force in front from which detachments can be sent in quiet times to worry us. Lees army is too far off from this line. On the other line a small force fights to great advantage. This new line of defense could be supplied by 3 routes. One is up the Kanawha as far as Charleston by water from hence by wagons. The other two routes would be from Clarksburg on the railroad by wagons and from New Creek also by wagons. These lines of supply are very long and through a bushwhacking country but the difficulty is diminished by the fact that there is plenty of forage in the country which should be bought not with *Greenbacks* but with *Groceries* of which the people are entirely destitute. This would make the supply trains smaller and the continual travel through the country would free it of bushwhackers. The raids against the B&O.R.R would be too dangerous for the Enemy and besides he could not get in unless he went round by the Kanawha and dodged through. Perhaps most of the people in W Va would know that there is such a state. Many of its inhabitants at present do not know that they live in it. Tomorrow we go over the Alleghany to fight Jackson about 12 miles from here.

Of course a dispatch will be sent when it is over.

I send this back by our train and sick which we wish to get rid of as far as possible. Good night.

Ever your affct Son JRM

Gen Averills Staff and the Genl are very kind and very pleasant. We have a jolly Dutchman with us who sends his regards to Meickes' father. He is a very fine soldier and a friend and admirer of Sec Chase. Give my respect & warmest love to dear Mother & Mary & Monty.

Autograph letter initialed, MCMC, MCMP, Ms. Div., LC.

1. This paragraph in shorthand. Missing text in original not found.

John Rodgers Meigs to Julius E. Hilgard

Dear Sir, Clarksburg, West Virginia, September 25, 1863

I am very much obliged to you for the maps which I have received and your offers of assistance. I should have answered your letter of the eleventh inst at once but that I thought I should be in W as soon almost as my answer. I shall be in Washington in a day or two when I intend to get at Baltimore some supplies of paper etc. which I wished to have Mr. Lindenkoh select or I would have sent for him at once. The topographer whom you proposed to send to Williamsport, I should like you to send here Clarksburg from which place when supplied with a horse I shall send him to Beverly to make a topographical sketch. There will be plenty of work in the field for all the topographers you can send me as long as they can stand the weather.

The map which you send me of Va is not accurate in the part of Va through which I have travelled, I am now having a copy made of a diagram of some of the roads between Lewisburg & Petersburg which I will send you.

John R. Meigs

Autograph letter signed, Chicago Historical Society.

John Rodgers Meigs to Brigadier General Joseph G. Totten

Sir, Clarksburg, West Virginia, September 25, 1863

I have been charged by the General comdg this Department with the erection of blockhouses for the protection of the bridges on the B&O R.R. between Cumberland and Harpers Ferry. Blockhouses were erected but were burned during the invasion of Pennsylvania by Lee's Army.

They proved effectual in saving the bridge over the Monocacy river on that occasion and are much needed to insure the safety of communication against the roving bands of guerrillas in that neighborhood.

There will be required for the erection of these blockhouses $1.000.00 during October.

For 40000 feet of lumber for 8 blockhouses	$800.00
For iron work for same	$200.00
	$1000.00

The labor expended will be that of soldiers so that the cost will be barely that of the flooring roofing and ironwork. I have the honor to be Sir John R. Meigs

Autograph letter signed, RG 77: Records of the Office of the Chief of Engineers, Letters Received, 1826–66, M-4449, NA.

John Rodgers Meigs to Brigadier General Joseph G. Totten

Sir, Clarksburg, West Virginia, September 30, 1863

I have the honor to request authority to employ such assistance as I may deem necessary for carrying on the surveys for military defences in the Department of W Va.

John R. Meigs

Autograph letter signed, RG 77: Records of the Office of the Chief of Engineers, Letters Received, 1832–65, M-850, NA.

John Rodgers Meigs to Brigadier General Joseph G. Totten

Sir, Clarksburg, West Virginia, September 30, 1863

There will be required for surveys in W Virginia.

1 Large Sextant.
1 Art Horizon
2 Surveying compasses
2 Smalchalder compasses
4 Triangular Scales
8 Copies Lloyds Map Va backed
Profile paper
1 Notebook
2 100 ft tapes
2 100 ft chains & pins.
1 Odometer
1 Copy Nautical Almanac
1 [illegible] Tables
1 Logarithms
4 Wooden Rulers
4 Large W Triangles
4 Small do
4 hand compasses
1 box instruments

John R. Meigs

Autograph letter signed, RG 77: Records of the Office of the Chief of Engineers, Letters Received, 1832–65, M-852, NA.

John Rodgers Meigs to Major Israel C. Woodruff

Sir, Clarksburg, West Virginia, October 11, 1863

Your two letters of the 7th inst have been received and the invoices desired are enclosed. I respectfully request that the instruments for which I made requisition may be sent me as soon as possible. Also that all of the maps for which I asked that can be immediately obtained should be sent now and the others as fast as they are prepared. I require about a dozen copies of the Map of the Upper Potomac from McCoys ferry to Conrads ferry by Lieut Col Macomb with additions by Lt Col Strother. Lithograph Engineer Dept 1863. I wish also 4 copies of the large photograph map of Jefferson Cy.

John R. Meigs

Autograph letter signed, RG 77: Records of the Office of the Chief of Engineers, Letters Received, 1832–65, M-893, NA.

John Rodgers Meigs to Major Israel C. Woodruff

Sir, Clarksburg, West Virginia, October 17, 1863

I should like to have for use in this office a copy of the last edition whatever its date of the *American* Nautical Almanac. It contains a number of useful astronomical positions.

I wish to know particularly the lat & lon of the SW corner of Penn and of some point in Wheeling WVa. Can you discover whether any good maps of the Ohio River have ever been made.

John R. Meigs

Autograph letter signed, RG 77: Records of the Office of the Chief of Engineers, Letters Received, 1832–65, M-940, NA.

Report to Brigadier General Joseph G. Totten from John Rodgers Meigs on Operations in the Department of West Virginia August to October 25, 1863

Sir, Clarksburg, West Virginia, October 25, 1863

I have the honor to submit the following report of the Engineering Operations in this dept. since my arrival in August last.

I found myself entirely unacquainted with the department and the only regular Engineer Officer in it.

I could find no organization which could be called a department of Engineering and had some trouble in finding persons competent to make surveys. The field works, which had been thrown up had generally been designed, by the officers commanding the different posts, and the repairs of roads, erections of bridges, and houses, were generally executed by the Quarter Masters Dept.

I have been ordered by the general commanding the dept, to make a military survey of the dept. as far as possible. For this purpose, I obtained from the Engr. Dept. 2.500$ and this with 1.400$ that I had unexpended, in my possession, when I was relieved from duty in the Middle Dept, made about 3.900$, available for this purpose. I have employed several assistants, and have collected a quantity of material, and instruments, with which to prosecute the work. I have been particularly fortunate in obtaining the cordial assistance, of the Coast Survey, in this undertaking. Their aid will enable me to make the maps, much more cheaply, than it could be done under ordinary circumstances.

I have furnished no maps to the Engineer Dept. yet because all the copies which I could make, have been demanded for immediate use here. I hope soon however to be able to send some.

Blockhouses were erected all along the line of the B.&.O.R.R. but during the invasion of Pennsylvania, last summer all those between Harpers Ferry, and Cumberland, were destroyed by the Rebels under Genl. Lee. I have received a 1.000$ with which to rebuild them, and the order has been issued, for their erection. There are in addition to the blockhouses, ten iron clad cars, upon this road, which can be moved from point to point thus having an advantage over blockhouses, besides being much stronger. Five of them are thought strong enough to resist small artillery, and are mounted with mountain howitzers, one of them has a six pounder brass piece aboard.

Maryland Heights, have been strongly fortified by various officers, the last of whom was Capt. Michler of the Engineer Corps who I understand has made a detailed report of them. Petersburg [West]Va. one of the outposts of the army occupying this dept. has been fortified lately by an Officer who was formerly in the Company of Sappers & Miners. I have not been able yet however to inspect the works. I laid out a line of defences at Beverly for about 3.000 men which have been executed and should make it a very strong place. At Bulltown another of the outposts some light entrenchments in connection with a naturally strong position and the bravery of our men enabled 144 Men to defeat a force of between 7 & 800 sent against them with the purpose of capturing them.

Not having been as yet among the troops along the Kanawha I can not speak of the state of the defences there. I have taken charge of the erection of a bridge over the Buchannan River in place of one of 150 feet span which was destroyed. This I expect to be paid for by the Quarter Masters Dept.

The work of repairing the roads is now performed though very insufficiently by that department, and I have not attempted to relieve them of it.

John R. Meigs

Document signed, RG 77: Records of the Office of the Chief of Engineers, Reports Relating to Topography, Bridges, Roads, and Geology of Virginia, M-4466, NA.

John Rodgers Meigs to Major Israel C. Woodruff

Sir, Clarksburg, West Virginia, October 25, 1863

Your letter of the 17th inst and the maps therein enumerated have been received.

I have bought from publishers and private individuals some few maps and some stationery. I wish to know whether I shall have to obtain the approval of the General Comdg this department for all such expenditures or whether a simple order from him to execute a military survey of his department in connection with the letter from the E Department dated Oct 2d. 1863 will cover them. That letter apparently only gives me authority to employ competent assistants, and not to purchase any article whatever.

Should such authority be necessary I respectfully request that it may be furnished.

John R. Meigs

Autograph letter signed, RG 77: Records of the Office of the Chief of Engineers, Letters Received, 1832–65, M-943, NA.

Montgomery C. Meigs wrote to his wife, Louisa, on October 25, 1863: "General Grant has asked that John may be ordered here. This will be a great opportunity for John. He can I think be very useful here."[1]

A week earlier, on October 18, 1863, Ulysses S. Grant had assumed command of the Military Division of the Mississippi, controlling Union operations from the Mississippi River to the Appalachian Mountains, and was engaged in raising the Confederate siege of Union forces at Chattanooga, Tennessee. Montgomery was in Chattanooga and most likely lobbied in behalf of his son. On October 21, 1863, Grant wrote to Secretary of War Edwin M. Stanton to request the transfer of young Meigs to his command.

Once again, on October 28, 1863, Grant requested the transfer—this time in a letter to General-in-Chief Henry W. Halleck. On the following day, Halleck responded: "Lieutenant Meigs is the only engineer officer in General Kelley's department, and cannot possibly be spared. Captain Comstock will soon be able to report to you." In a letter to J. E. Hilgard of the United States Coast Survey Office, Montgomery C. Meigs expressed his disappointment: "I am sorry to learn that Lt. Meigs cannot be spared from Western Va. so that he will not be able to take part in the interesting and important operations of this army." John Rodgers Meigs would remain with Kelley.[2]

---

1. MCM to Louisa Rodgers Meigs, October 25, 1863 (extract), MCM Letterbook, "Letters Sent No. 2," pp. 5–6, part 1, RG 92: Records of the Office of the Quartermaster General, Consolidated Correspondence File, 1794–1890, NA; Ulysses S. Grant to Edwin M. Stanton, October 21, 1863, File M-1003 (1863), RG 94: Records of the Adjutant General's Office, Letters Received, 1863–94, NA; Ulysses S. Grant to Henry W. Halleck, October 28, 1863, RG 107: Records of the Secretary of War, Telegrams Received by Major General Halleck, September 21, 1863–October 30, 1863, NA; *OR,* series 1, vol. 31, part 1, p. 774.

2. MCM to J. E. Hilgard, November 2, 1863 (extract), MCM Letterbook, "Letters Sent No. 2," pp. 5–6 and 55, part 1, RG 92: Records of the Office of the Quartermaster General, Consolidated Correspondence File, 1794–1890, NA.

John Rodgers Meigs to Louisa Rodgers Meigs

My dear Mother, November 10, 1863

I am safe though still in Rebeldom after long marches in every direction. We had a severe battle 31 miles north of Lewisburg at Droop Mt in which we routed the Rebels completely and have kicked up a great fuss among all their armies in this region having even aroused their Potomac army. They are now marching and counter marching in every direction while we promenade their country in confidence and with all the comforts needed at hand when we camp. In the battle we fought we drove an equal force from the top of a high mountain on which they were forted in a position overlooking the country for miles around. One of their [illegible] regiments which entered with 500 men, had but 100 left to answer to roll call. This is from the statement of their own officers.

Hoping to see you again. Love to all.

Ever Your Afft Son JRMeigs

Autograph letter signed, LRMC, MCMP, Ms. Div., LC.

John Rodgers Meigs to Major Israel C. Woodruff

Sir, Clarksburg, West Virginia, November 19, 1863

I have the honor to acknowledge the receipt of your letter of the 30th Ult and the three maps accompanying it. I have just returned from an expedition to Central Va under the command of Brig Gen Averell. I was fortunate enough to capture a copy of the Edition of 28 of the nine sheet map of the state. With the exception of a break of a few miles I have the notes of a horse back and Schmalchalder compass survey of the whole route 300 miles long, which was taken by the expedition. I discovered that Lloyds map and those like it were entirely untrustworthy with regard to the roads in Pocahontas and Pendlleton Counties. The topography of the mountains is also very falsely represented.

John R. Meigs

Autograph letter signed, RG 77: Records of the Office of the Chief of Engineers, Letters Received, 1832–65, M-1003, NA.

John Rodgers Meigs to Major Israel C. Woodruff

Sir, Clarksburg, West Virginia, November 30, 1863

There are needed at this office 12 more copies of the map of the Upper Potomac from McCoys ferry Lithograph, and ten of the small compasses 2 inch exterior diameter. There is a great deal of difference in their quality. They should be selected from those which vibrate the longest when disturbed by a magnet and in which the needle does not strike the arc when turned in any direction.

John R. Meigs

Autograph letter signed, RG 77: Records of the Office of the Chief of Engineers, Letters Received, 1832–65, M-1086, NA.

John Rodgers Meigs to Major Israel C. Woodruff

Sir, Clarksburg, West Virginia, December 3, 1863

I have learned that a map of several of the neighboring counties of Western Va was made by Lt Col Nicholas Bowen of the Corps for General Rosecrans. If you can tell me where it is, or how to get it I can make good use of it.

John R. Meigs

Autograph letter signed, RG 77: Records of the Office of the Chief of Engineers, Letters Received, 1832–65, M-1085, NA.

Report to Brigadier General Joseph G. Totten from John Rodgers Meigs on Operations in the Department of West Virginia, October 25 to November 30, 1863

Chief of the Corps of Engineers

Sir, Clarksburg, West Virginia, December 6, 1863

I have the honor to submit the following report of the operations of my department from the 25th of Oct to the 30th November 1863.

The $1,000 which was received for the purpose of constructing blockhouses between Harpers Ferry and Cumberland is yet unexpended. The order was issued by General Kelley for their construction, and minute directions were sent from this office to the officer commanding the troops along the line, but on my return from an expedition to Lewisburg, I found them not yet commenced as the officer to whom the orders were sent had been arrested.

In accordance with orders from Genl Kelley on the 29th of October I reported to Brig Gen Averell and accompanied a second expedition into the enemy's country. Our forces were victorious in a well fought action at Droop Mt and afterwards proceeded as far into the interior as their supply of rations would permit them occasioning a great commotion among all of the enemys forces between East Tennessee and Northern Va, the enemy having been led into the belief that we were 7000 strong. While with Genl Averell I was occupied in the usual duties of an engineer officer selecting positions for troops gaining information about roads and movements of the enemy etc.

A horseback survey of the whole route of the expedition with the exception of about ten miles was obtained and much information collected useful in correcting the existing maps.

A copy of the "nine sheet map" was captured at Lewisburg and was found generally correct except with regard to the roads and towns.

The Coast Survey are about to send me another of their officers who will be employed in fixing astronomical positions in various parts of the State, which will be a valuable aid in constructing a reliable map. I have been promised by Genl Kelley that an order shall be issued detailing some officer in each brigade within the department as an Acting Asst Engineer who will report to this office for instructions. When this is done I shall be able to accomplish much more than I can now single handed.

I have been directed to lay out a work at this place on the top of a high hill north of the town which will render it always a secure spot and give employment to some new troops quartered in the vicinity.

I found a work in process of construction at Parkersburg, which showed the necessity of having engineer officers with each command.

The profile adopted for the work was this.

and in some places this.

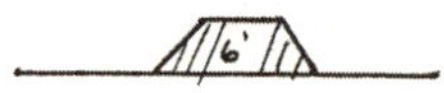

I had it altered.

Upon inspecting the works at Beverly I did not find that they had been completed as I had directed and do not regard them as so strong as I stated in my last report.

I have the honor to enclose a copy of an order from Dept Headquarters.

John R. Meigs

Autograph letter signed, RG 77: Records of the Office of the Chief of Engineers, Reports Relating to Topography, Bridges, Roads, and Geology of Virginia, M-4508, NA.

John Rodgers Meigs to Louisa Rodgers Meigs

My dear Mother, New Creek, West Virginia, December 8, 1863

I am "off to the wars again." Start tomorrow.

I received your letters and Mary's. I cannot tell whether I shall be able to come home on Christmas. How I should like to you can guess. It is going to be very cold campaigning but I guess we can stand it. I had a letter from Mont the other day with which I was much pleased.

You have never yet told me how I can send a letter to Father.

Good Bye with greatest love to yourself and dear Mary.

Ever Your Affct Son JRM

Autograph letter initialed, LRMC, MCMP, Ms. Div., LC.

Troops under the command of William W. Averell left New Creek, West Virginia, on December 8, 1863, with orders to travel south to destroy the Virginia and Tennessee Railroad at Bonsack's Station in Botetourt County or at Salem in Roanoke County, Virginia. Traveling over little-used trails and ridges, advance units of Union cavalry entered Salem on December 16, 1863. The troops were ordered to destroy bridges and track and to burn other railroad equipment and buildings. Later the same day, the Union forces began their return trip north. Hampered by severe weather and rain-swollen rivers and creeks, and engaged by Confederate troops along the way, the last of Averell's men reached Beverly, West Virginia, on the twenty-sixth of December.

As a result of his participation in this raid, John Rodgers Meigs, along with many other soldiers, suffered from the effects of the weather and the fighting. On December 28, 1863, he requested and was granted a leave of absence for two weeks to recover. After spending several days in bed at his parents' home in Washington, D.C., by January 4, 1864, he was able to visit some neighbors for a "candy pulling" and to learn "a very stupid game of cards called 'bezique.'" By January 8, he was well enough to visit the War Department, to deliver "some US property for Mrs. Kirby Smith," and to see Major Israel C. Woodruff. While in the nation's capital, he purchased a number of articles for his own use, including clothing, spurs, and a pistol with ammunition. On January 12, 1864, he left to return to Cumberland "to go on duty again Frozen feet cured."[1]

1. JRMD, January 1–3, 1864 (regarding days in bed), January 4–9, 1864 (activities in Washington, "candy pulling," "stupid game"), MCMP, Ms. Div., LC; see also MCMD, January 12, 1864, MCMP, Ms. Div., LC.

CHAPTER 5

Battles and Bridges

By 1864, the Union's determination to achieve victory, coupled with the material superiority of the North, was beginning to tell on the battlefields. However, there was still much fighting and devastation to come before the end. In the Shenandoah Valley, the conflict became a bitter and long-remembered struggle as Union commanders attempted to drive out Confederate troops and suppress any opposition to Federal victory. The burning of homes and property begun under Brigadier General David Hunter, but undertaken more systematically by Major General Philip H. Sheridan, was carried out in part as Union retaliation for the activities of "bushwhackers" and irregular troops, the Confederate guerrilla fighters or raiders.

The Shenandoah Valley Campaign of 1864 commenced when Union forces marched toward Richmond from the north and west as part of the overall strategy to defeat the Army of Northern Virginia. While Ulysses S. Grant pursued Robert E. Lee, Brigadier General George Crook of the Union moved through the Kanawha Valley in West Virginia southeasterly toward the Virginia and Tennessee Railroad at New River Bridge; at the same time, Major General William W. Averell pushed more directly south toward Saltville and Wytheville, Virginia. These troops were to link up with the army of General Franz Sigel, who was ordered to advance up the valley, that is, in a southern direction—"up the Shenandoah River"—towards Staunton from Beverly, West Virginia.

Following his recuperation at home, John Rodgers Meigs returned to the field and became actively engaged in the fighting and the logistics of warfare. As part of his duties, Meigs was ordered to build blockhouses at the bridges on the B&O Railroad between Cumberland and Harpers Ferry. (The original blockhouses were destroyed by Lee's army during the invasion of Pennsylvania.) As the work progressed, it was found necessary to increase the number of blockhouses and to modify their character in accordance with instructions from army headquarters.

Captain Robert S. Gardner to John Rodgers Meigs

Lieutenant, Clarksburg, West Virginia, January 13, 1864

Will you please accept a pair of Turkies with my compliments and best wishes.

I hope you have entirely recovered from the effects of *Jack Frost* on your feet, and that we may have you *here before shortly*. My kindest regards to your Father and Family.

R S Gardner

Autograph letter signed, JRMC, MCMP, Ms. Div., LC.

John Rodgers Meigs to Major Israel C. Woodruff

Sir, Clarksburg, West Virginia, January 14, 1864

I have employed in the Engineer Dept here several noncommissioned officers and soldiers detailed on special duty. In almost all the surveys undertaken it is impracticable for them to carry rations with them and some of them have been subsisting at their own expense. No one here seems to know whether their rations can be commuted for them or not. Will you be kind enough to tell me whether they can or not and by whose order. Also whether I can recompense them for the expense they have already been to in obeying orders.

John R. Meigs

Autograph letter signed, RG 77: Records of the Office of the Chief of Engineers, Letters Received, 1832–65, M-57, NA.

John Rodgers Meigs to Lieutenant Charles A. Freeman

Sir, Clarksburg, West Virginia, January 15, 1864

I have the honor to request an order directing me to proceed from Cumberland to Clarksburg and move my office to Cumberland in order that I may obtain transportation.

John R. Meigs

Autograph letter signed, RG 393: Records of the U.S. Army Continental Commands, 1821–1920, Department of West Virginia (P), 1863–65, Letters Received, 1863–65, 1864-M, NA.

John Rodgers Meigs to Major Israel C. Woodruff

Sir, Clarksburg, West Virginia, January 15, 1864

In my letter of yesterday I forgot to enquire the means of obtaining for my employees who are noncommissioned or privates the extra pay which I believe is their due when employed as surveyors.

John R. Meigs

Autograph letter signed, RG 77: Records of the Office of the Chief of Engineers, Letters Received, 1832–65, M-59, NA.

John Rodgers Meigs to Major Israel C. Woodruff

Sir, Cumberland, Maryland, February 3, 1864

I transmit herewith my Engineer accounts for the 4th quarter of 1863 with one set of my money accounts for Dec and January last. I also enclose my accounts of QM property, in accordance with advice received from you, and will be obliged to you, if you will forward them to the QM General.

John R. Meigs

Autograph letter signed, RG 77: Records of the Office of the Chief of Engineers, Letters Received, 1832–65, M-113, NA.

At the end of February 20, 1864, John Rodgers Meigs received an order by General Henry W. Halleck to rebuild blockhouses for the protection of the Baltimore and Ohio Railroad bridges and to build new ones to protect the line.

John Rodgers Meigs Diary Entries of February 25 and 26, 1864

February 25, 1864

At Harpers Ferry with Gen Kelley. Got Lieutenant Joseph E Conway as Asst Engineer for Sullivans division. Came to Martinsburg and saw Genl Averell and Col Rodgers, and left at 2 ½ at night for Cumberland.

February 26, 1864

Arrived at Cumberland. Received Jan pay and cancelled some debts.

JRMD, MCMP, Ms. Div., LC.

John Rodgers Meigs to Brigadier General Joseph G. Totten

General, Cumberland, Maryland, March 5, 1864

Having been ordered by Genl Kelley to build blockhouses at the bridges on the Balt & Ohio R.R. I applied for $1,000 to expend in their construction which was duly furnished me from the appropriation for Fortifications on the Northern Frontier. I find that I shall require in addition to this the sum of 1600 dollars to enable me to complete them. Timber has risen very much in price since I made my first estimate and the number and character of the blockhouses has been increased in accordance with instructions from the HeadQuarters of the Army which are enclosed.

John R. Meigs

Autograph letter signed, RG 77: Records of the Office of the Chief of Engineers, Letters Received, 1826–66, M-4611, NA.

Montgomery C. Meigs wrote in his diary on March 20, 1864, "My son John writes that he is pleased with Sigel." Sigel found young Meigs "well acquainted with the Topography of the Country and the position of our troops in this Department" and wished him to continue in his position as chief engineer.[1]

Sigel arrived in Cumberland, Maryland on March 11, 1864, and by midmonth began to reorganize the department to get ready for the spring campaign. One of his first activities was to tour the garrisons. He strengthened existing fortifications, built new ones, and created a better-prepared army, which consisted of 23,000 men and 118 cannon. His army's primary responsibility was the protection of the B&O Railroad, the shortest line of communication between Washington and Cincinnati. His troops repeatedly clashed with Confederate raiders who attacked Union troops and supplies along the line.

Sigel's responsibilities changed when Ulysess S. Grant, newly appointed general-in-chief of the armies, ordered him to gather eight thousand soldiers, three batteries, and fifteen thousand cavalry at Beverly, West Virginia.

According to Grant's plan, Sigel's army was to march through the Shenandoah Valley in a southerly direction, "to lure the Confederates further away from Richmond and destroy Lee's communications to that region." Although "doubting great results" on the part of Sigel, Grant wanted the army to keep "Confederate Generals John C. Breckenridge and John Imboden occupied in the Valley." As a member of Sigel's staff, John Rodgers Meigs was included in these plans and took an active role in the ensuing battles. While Sigel's army advanced up the valley, Union troops under the command of Crook and Averell moved east and south through the Kanawha Valley and the Allegheny Mountains.[2]

In addition to the responsibilities of combat, as chief engineer Meigs had the continued responsibility for the engineer work required of the army. This included erecting and reinforcing fortifications; building and repairing bridges; securing materials and tools, including supplies for producing drawings and maps; employing surveyors and surveying the area; and preparing maps. These duties necessitated an extensive correspondence with military offices in Washington and Baltimore, with civilian purveyors of materials and supplies, as well as with military units in the area. Accounts of much of these responsibilities can be read in the administrative correspondence saved by Montgomery Meigs, some of it on flimsy letterpress sheets, or in the records of the Office of the Chief of Engineers at the National Archives. Many of these letters reveal the extent of the young lieutenant's duties. From them, for example, we learn that April 1864 was a particularly active month for Meigs, especially with regard to pontoon bridges. These bridges were used extensively during the Civil War and Meigs was in charge of purchasing materials for them, constructing them, and moving them on specially designed pontoon wagons in his area.

Personal matters as well as the business of war occupied John Rodgers Meigs. In March 1864, he applied for and received a three-day leave of absence to travel to Washington to attend the wedding of his sister, Mary Montgomery Meigs, to Lieutenant Colonel Joseph Hancock Taylor. John also had an opportunity to accompany his father on a visit to General Halleck.

1. MCMD, March 20, 1864, MCMP, Ms. Div., LC.

2. Stephen D. Engle, *Yankee Dutchman: The Life of Franz Sigel* (Fayetteville, Ark.: University of Arkansas Press, 1993), pp. 170–74.

Major John D. Kurtz to John Rodgers Meigs

Sir, Washington, D.C., March 31, 1864

Your letter of the 18th inst, is received. To afford you the information you desire in getting up the proposed light ponton train, there will be mailed for you today, a copy of Genl Cullum's "Systems of Military Bridges." Further information on the subject has been applied for, and will be sent to you as soon as received.

Funds for that object will be supplied you on your estimates approved by the Commanding General on the Department in which you are serving.

For information in regard to contractors for ponton trains, you should address Mr. Wm P Trowbridge, Agent Engineer Department, No 7 Bowling Green, New York City, who has had a number of trains prepared.

J D Kurtz

Autograph letter signed, RG 77: Records of the Office of the Chief of Engineers, Letters Sent to Engineer Officers, 1812–69, NA.

John Rodgers Meigs to Brigadier General Joseph G. Totten

General, Cumberland, Maryland, April 2, 1864

In order to avoid the necessity for frequent and small estimates and the pernicious delay consequent thereupon, I have the honor to submit the following estimate for Contingencies during the months of April and May next in addition to the estimates already sent in, and to request that the money be placed to my credit with the US Depositary at New York.

For repairing roads and bridges and purchasing of entrenching tools and carrying on other operations ordered by the Genl Comdg this Dept $20,000.00 Twenty thousand dollars.

The exigencies of the service are so pressing that I have to request that I may be notified by telegraph as soon as the draft has been issued in my favor.

John R. Meigs

Autograph letter signed, RG 77: Records of the Office of the Chief of Engineers, Letters Received, 1826–66, M-1647, NA.

John Rodgers Meigs to Major General Franz Sigel

General, Cumberland, Maryland, April 2, 1864

You expressed a desire to have three strong fortified posts between the Ohio and the Monocacy which might serve as secure depots of supplies and enable you with the few troops which could be rapidly concentrated at these points successfully to oppose any movement of the enemy which did not involve the cooperation of their main army.

Harpers Ferry is at present the only such point in the Dept of W Va, it was to be one of the three.

Cumberland next to Wheeling the most important city between Washington and Cincinnati was to be another and the third was to be somewhere West of the Alleghanies and near the junction of the railroad to Wheeling with that to Parkersburg.

The fortifications of Cumberland now in progress are costing nothing except the price of the tools used and a little lumber which has been expended in their construction. To arm them effectively will require at least 20 siege guns. Twelve of these if possible should be 30 pdr Parrotts and the remainder smooth bore guns.

Half of these guns can be mounted immediately upon their arrival here and the rest as soon as the fortifications can be made ready to receive them.

John R. Meigs

Letterpress copy, MCM Letterbook, April 1864, pp. 4–5, Ms. Div., LC.

John Rodgers Meigs to Captain Patrick J. McDermott

Sir, Cumberland, Maryland, April 2, 1864

I have the honor to acknowledge the receipt of diagram sent me in your letter of the 30th.

Make as much progress as possible with your blockhouse as the Genl is very anxious to see them all completed.

John R. Meigs

Letterpress copy, MCM Letterbook, April 1864, p. 3, Ms. Div., LC.

John Rodgers Meigs to William P. Trowbridge

Sir, Cumberland, Maryland, April 2, 1864

I wish to know the prices, weights, and number of teams required for (200') Two hundred feet of Pontoon bridges of the different kinds now in use and found most useful. In this department when the distances to be overcome are so great and the means of transportation limited the very lightest kind which can be relied upon must be used. Before this reaches you, you will have received two telegraphs from me in regard to pontoon bridges.

John R. Meigs

Letterpress copy, MCM Letterbook, April 1864, p. 6, Ms. Div., LC.

John Rodgers Meigs to Colonel Hiram Leonard

Sir, Cumberland, Maryland, April 2, 1864

I have about a dozen men on detached duty in my office some of whom have not drawn their pay for 5 months. I wish to know how they can draw their pay without returning to their regiments.

John R. Meigs

Letterpress copy, MCM Letterbook, April 1864, p. 7, Ms. Div., LC.

John Rodgers Meigs to Captain Charles W. Thomas

Sir, Cumberland, Maryland, April 3, 1864

Please purchase for me and ship as soon as possible 40 gallons of camphene put up in tin cans from which it can be easily poured. They should be hermetically sealed and contain ½ gallon each. Send me also 400 inextinguishable matches such as are used for lighting cigars in high winds. Also 200 lbs of blasting powder in tins containing 25 lbs with 100' of blasting fuzes. Also two sets of tools for drilling for blasts. Send me the bills.

John R. Meigs

Letterpress copy, MCM Letterbook, April 1864, p. 22, Ms. Div., LC.

John Rodgers Meigs to Major Israel C. Woodruff

Sir, Cumberland, Maryland, April 3, 1864

Please send me the following books for distribution to the Officers Actg as Engineers in this Dept.

15 Copies Engineer Dept Regs
15 Copies Mahans Field Fortifications
10 Copies Duanes Manual
10 " Cullums Bridges
10 " Pamphlet in regard to Military Suspension Bridge.

Will I have to take receipts from all the officers to whom I issue these books, or can I expend them.

John R. Meigs

Autograph letter signed, RG 77: Records of the Office of the Chief of Engineers, Letters Received, 1826–66, M-4649, NA.

John Rodgers Meigs to Montgomery C. Meigs (Shorthand Transcription)

My dear Father: Cumberland, Maryland, April 4, 1864

I enclose some recommendations for *canvas pontoons,* of which I spoke to you when I was at home.

General Sigel keeps everyone about him busy and seems anxious to have even more assistants than he has now. He has transferred one of my engineers to his office and seems to keep him pretty busy, though I do not think it is at engineering work. Still, I have no doubt he is as useful as he would be if at my office.

I have applied for $20,000 for timbers and am afraid I will have to ask for $20,000 more in a few weeks to enable me to buy a pontoon bridge. . . .

John R. Meigs

Letterpress copy, MCM Letterbook, April 1864, Ms. Div., LC.

John Rodgers Meigs to Major John D. Kurtz

Sir, Cumberland, Maryland, April 4, 1864

Genl Sigel wishes me to obtain the drawings and specifications of a pontoon bridge used by the 11th Corps when under his command at Fredericksburg.[1] These papers were referred to the Engineer Dept by the Q.M. Dept on the 27th Jan 1863.

John R. Meigs

Autograph letter signed, RG 77: Records of the Office of the Chief of Engineers, Letters Received, 1826–66, M-4648, NA.

1. Drawings not found.

John Rodgers Meigs to Colonel C. H. Schapp

Sir, Cumberland, Maryland, April 5, 1864

I am desirous of obtaining the services of a good construction Engineer who can take charge of the erection of bridges and fortifications and the construction of roads.

I have been informed by General Sigel that you were such a person that he would like you to be employed in this Department. My engineers are getting from 3 to 4.00 per day according to ability.

John R. Meigs

Letterpress copy, MCM Letterbook, April 1864, p. 43, Ms. Div., LC.

John Rodgers Meigs to Brigadier General Joseph G. Totten

General, Cumberland, Maryland, April 6, 1864

After correspondence with the US Engineer Agency at New York, I have the honor to submit the following estimate and to request that the money be placed to my credit with the US Depositary at New York.

For construction of a canvass
Pontoon train 400' long
complete with wagons and
harness, Twenty thousand dollars. $20,000.00

John R. Meigs

Autograph letter signed, RG 77: Records of the Office of the Chief of Engineers, Letters Received, 1826–66, M-4657, NA.

John Rodgers Meigs to Captain Charles W. Thomas

Sir, Cumberland, Maryland, April 6, 1864

I have to acknowledge your prompt attention to the request contained in my letter of the 3d inst.

I wish to have the following articles sent me in addition.

400 yds of quick match
300 lbs of blasting powder in
25 lb Kegs
100 yds of cotton rope or waste
for salinating and inflaming
12 Waterproof bags containing
about 25 lbs of Powder with
10' of tube attached to protect fuze.

40 [illegible] Will send tomorrow drawing for construction of 6 canvass pontoons which I wish in four days if possible or 6 at [illegible]. The work can be distributed to different shops. They will require from 150 to 200 sq yds of the very best waterproof cotton canvass which will hold water like a bucket.

Please make arrangements for the prompt construction of these boats. The bills for all these to be sent to me.

John R. Meigs

Send me also 4000' of ½ inch manilla rope and 1000' of ⅝ rope or what is called 2 inch rope, around the circumference.

Letterpress copy, MCM Letterbook, April 1864, p. 57, Ms. Div., LC.

John Rodgers Meigs to Captain Charles W. Thomas

Sir, Cumberland, Maryland, April 7, 1864

I enclose a drawing from which the canvass covers for pontoon frame can be commenced at once. The work should be distributed to two or more parties if one cannot perform it within the prescribed time.

I have been unable to finish the drawing to day but will send it by the next mail. If the bottom boards have not yet been made I would rather have the slits which were to be cut in their ends left unmade.

The following materials will be required in the construction of the 6 frames.

120 pieces fine white pine 5' × 6" × 1" when finished
82 " " " " 2'½ × 4" × 1" " "

48 " " " " 5' × 4" × 1" " "
72 butts 4" × 2" ½ or near this size with screws 1" long.

Besides these the bottom boards and hasps described in my last letter and a few other articles which will appear on the drawings I am to send.

John R. Meigs

Perspective drawing showing shape of canvass cover for ponton frame. The outside dimensions of the frame are given. The canvass laps over on the inside 4 inches all round and is stretched by a cord as represented. The eylet holes should be large enough to admit ½ inch rope and 9 inches apart. 6 of these are wanted. The canvass must be of the best waterproof cotton canvass. The angles and corners should be double except those at the upper edge of the sides and ends.

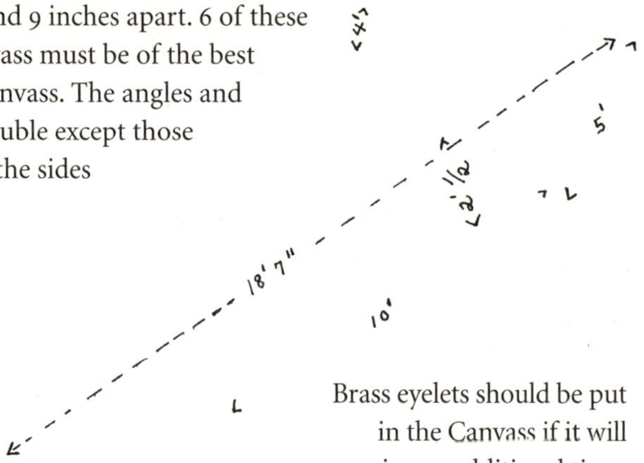

Brass eyelets should be put in the Canvass if it will require no additional time.

John R. Meigs

Letterpress copy, MCM Letterbook, April 1864, pp. 64–66, Ms. Div., LC.

John Rodgers Meigs to Julius Bien

Sir, Cumberland, Maryland, April 7, 1864

I wish to know the probable time, and cost, of lithographing a map, of the style of the enclosed specimen, and 36. x 30.

I also wish to know, whether it would take longer, and cost more, to print the roads red, and the streams blue and what percentage would thus be added.

I will have the map ready in a few days, for the lithographer.

John R. Meigs

Letterpress copy, MCM Letterbook, April 1864, p. 62, Ms. Div., LC.

John Rodgers Meigs to Captain Charles W. Thomas

Capt Thomas Cumberland, Maryland, April 8, 1864

12 additional boats have been ordered this afternoon making 24 in all will you have them underway.

John R. Meigs

Letterpress copy, MCM Letterbook, April 1864, p. 69, Ms. Div., LC.

John Rodgers Meigs to Colonel Jacob M. Campbell

Sir, April 8, 1864

Please send one of the iron clad cars now here to Green Spring Run. It would be a great protection to that port which is very much exposed.

John R. Meigs

Letterpress copy, MCM Letterbook, April 1864, p. 70, Ms. Div., LC.

John Rodgers Meigs to Brigadier General Joseph G. Totten

General, Cumberland, Maryland, April 11, 1864

I have the honor to request that 6 of the improved trucks for wooden bateaus be sent to me at Cumberland. I am informed by Genl Benham that he has some to spare.

John R. Meigs

Autograph letter signed, RG 77: Records of the Office of the Chief of Engineers, Letters Received, 1826–66, M-4661, NA.

John Rodgers Meigs to Captain Thayer Melvin

Sir, Cumberland, Maryland, April 13, 1864

I respectfully request that a detail of from 50 to 100 infantry be ordered from Martinsburg to report to Lieut S C Smith Comdg Detachment Indpt Co PV & Acg Engineers at Falling Waters to assist in laying the bridge of pontoons at that place and to do general duty there.

The force of the Engineer Co is insufficient for that purpose.

John R. Meigs

Autograph letter signed, RG 77: Records of the Office of the Chief of Engineers, Letters Received, 1832–65, M-91, NA.

John Rodgers Meigs to Captain William P. Gaskell

Sir, Cumberland, Maryland, April 13, 1864

I have requested that a detail of from 50 to 100 infantry be sent to Lieut Smith at Falling Waters. 6 of your boats are now at Clarksburg and wagons have been prepared for 3 of them. They will probably all be returned to you in a few days.

I have ordered from Washington 6 regular 6-team wagons for use with your ponton bridge in case necessity requires it. I have taken steps for putting up the trestle bridge at Harpers Ferry which will relieve the ponton bridge now at that place. I will have the necessary repairing tools and materials sent to Lieut Smith.

John R. Meigs

Letterpress copy, MCM Letterbook, April 1864, p. 85, Ms. Div., LC.

John Rodgers Meigs to Captain Charles W. Thomas

Sir, Cumberland, Maryland, April 15, 1864

Will you please order the bridge of which I enclose drawings[1] constructed and sent to Capt Wm Penn Gaskell Indpt Co PV & Actg Engineers Harpers Ferry or if you can do so please send it to Lieut S C Smith same Co at Falling Waters Chesapeake and Ohio Canal.

The bridge is to cross the canal at Falling Waters. Notify me of the probable time of the completion of the bridge and send me the bills. I will pay the bills I have already contracted as soon as my money arrives.

John R. Meigs

Letterpress copy, MCM Letterbook, April 1864, p. 108, Ms. Div., LC.

1. Copies of drawings not identified. See MCM Letterbook, April 1864, p. 55, Ms. Div., LC, for drawing of pontoon construction.

John Rodgers Meigs to Julius Bien

Sir, Cumberland, Maryland, April 15, 1864

I think I shall send you the map to lithograph in a few days of which I spoke. I wish it done in the same style as the Clarksburg Map except that the shading on the mts has a lined character so as to show the contour instead of being a smooth shade. The maps which you wish to give to General Sigel you had better send to me and I will deliver them to him from you. I have charge of all the maps used at Headquarters. If you prefer to send them to him his address is at Cumberland Md.

John R. Meigs

Letterpress copy, MCM Letterbook, April 1864, p. 104, Ms. Div., LC.

John Rodgers Meigs to Colonel E. Jay Allen

Sir, Cumberland, Maryland, April 15, 1864

Please inform [Alexander] Allen that I wish him to make 12 more pontoons on very much the same plan as the others but with some modifications. I send you a list of the pieces of wood which will be required for the 12 boats which you can show him. I will send in a day or two the drawings. Grafflin had better be getting the canvass for them.

I telegraphed you at the Fountain but the boy could not find you how was it? Have three sets of stamps made with the letters USED [United States Engineer Department] upon them for stamping iron and steel and 6 for burning on handles etc.

John R. Meigs

Letterpress copy, MCM Letterbook, April 1864, p. 102, Ms. Div., LC.

John Rodgers Meigs to Henry Pollock

Sir, Cumberland, Maryland, April 15, 1864

You will please make for me two copies on albumen paper and two copies "positive" on glass 1/20 the size of the original map which I have sent you by to days mail when finished send them to me by Express with your bill enclosing the freight.

Send me both the negative and positive and if they are successfully and well done, and everything turns out perfectly plain, I will probably have some work for you to do [illegible].

Other maps which I shall likely have done will be on a larger scale some of them being 6 or 8 [illegible] long but not requiring a [illegible] larger dimension than 6 × 8 inches.

John R. Meigs

Letterpress copy, MCM Letterbook, April 1864, p. 100, Ms. Div., LC.

John Rodgers Meigs to Captain William P. Gaskell

Sir, Cumberland, Maryland, April 16, 1864

I have sent you by todays mail a copy in pamphlet form, of instructions for Transport and Erection of Military Suspension Bridges, which I would like for you to examine and make yourself acquainted with the details, as I expect to send materials for the purpose of building the two spans on the Virginia side, and wish you to take charge of the construction.

John R. Meigs

Letterpress copy, MCM Letterbook, April 1864, p. 105, Ms. Div., LC.

John Rodgers Meigs to Captain William P. Gaskell

Sir, Cumberland, Maryland, April 17, 1864

I enclose drawings which you will send Lieut Smith of a bridge which I have ordered to be framed in Baltimore and sent to you or to him whichever the QM finds most convenient. It will arrive complete. It is as you see the one that he submitted a plan for. I have ordered 10,000 feet of 2 inch plank to be sent you at Harpers Ferry as part of the lumber for building the bridge there. The two spans on the Va side which are now [illegible] I will put a suspension bridge over.

John R. Meigs

P.S. I will send you receipts in a few days to sign for the ponton wagons forwarded you.

Letterpress copy, MCM Letterbook, April 1864, p. 111, Ms. Div., LC.

John Rodgers Meigs to Brigadier General William W. Averell

Sir, Cumberland, Maryland, April 19, 1864

I send you a sketch of Lynchburg from information and a copy of Lloyds map with the main roads marked in red so far as they could be traced by all the *reliable* information in my office. They are all good enough to serve unless in very bad weather as lines of supply. I also send a diagram with which you can increase their number by information from the scouts and deserters.

John R. Meigs

Letterpress copy, MCM Letterbook, April 1864, p. 128, Ms. Div., LC.

John Rodgers Meigs to Corporals Henry Elfers and William Steilberg

Sirs, Cumberland, Maryland, April 19, 1864

I have received your note which will prove useful. Those you sent before we had difficulty in using because we did not know the number of paces in a mile.

I enclose requisitions and receipts for a horse for each one of you with complete equipment.

You will present these to the Quartermaster and on receiving the horses and equipment you will leave them with him and tell him to send me invoices corresponding to the receipts.

Each of you will sign the enclosed memorandum receipt for your horse and send them back to me. Should anything happen to your horse you will at once notify me of the fact and if he dies you will send me an affidavit to that effect.

Should you be obliged to turn him in to a quartermaster you will be careful to give him only to a *Capt* and AQM and will tell him to send me receipts for him. Let me know his name.

You can probably get the commander of the Post to draw forage for him as a public horse in service of Engineer Dept. If not tell the QM to send me each month the forage returns for signatures.

Remember that these horses are only to be used in the execution of your duty as surveyors. So on with your work and send in as many notes as you can.

John R. Meigs

Keep this letter for reference.

Letterpress copy, MCM Letterbook, April 1864, pp. 123–24, Ms. Div., LC.

John Rodgers Meigs to Brigadier General Henry W. Benham

General, Cumberland, Maryland, April 19, 1864

Will you be kind enough to have a sketch made of the scouting ponton devised by you by which I can have one constructed. I should like to have it made of No 1 canvass instead of india rubber. With many thanks for your kindness.

John R. Meigs

Letterpress copy, MCM Letterbook, April 1864, p. 120, Ms. Div., LC.

John Rodgers Meigs to Lieutenant Colonel William C. Starr

Sir, Cumberland, Maryland, April 20, 1864

The General Comdg directs that all prisoners from the enemy refugees and deserters who are likely to have any topographical information in regard to the enemys country or knowledge of the positions of their troops be sent to my office upon their arrival here for examination. I will be at my office at 8 o'clock this evening to see those who are now in Cumberland.

John R. Meigs

Letterpress copy, MCM Letterbook, April 1864, p. 134, Ms. Div., LC.

John Rodgers Meigs to E. Jay Allen

Sir, Cumberland, Maryland, April 25, 1864

The canvass boats came today. I have not yet received bills of lading. They should have been sent. A Forge has come with them, and a box of tools. I have about concluded to abandon the idea of a suspension bridge and build a trestle bridge. I cannot find any carpenter [at] $5.00 a day. I do not pay my Engineers more than $4.00 a day without rations. Push on the pontons. You must keep an account of the labor spent on them for my guidance.

I want you to send me 300 blank address tags for tying to goods.

John R. Meigs

Letterpress copy, MCM Letterbook, April 1864, p. 176, Ms. Div., LC.

John Rodgers Meigs to Brigadier General Joseph P. Taylor

General Cumberland, Maryland, April 25 1864

I have a number of soldiers on detached duty under my orders employed as surveyors. In the execution of their duties these men are frequently obliged to proceed far from military posts where it is impossible for them to carry rations or to draw them. As they generally proceed alone or with but 3 or 4 men it is impracticable to send a team with them. They are thus obliged to purchase their subsistence from the inhabitants and it costs them sometimes more and sometimes less than 75 cents a day depending upon various circumstances. As these men receive no extra pay they are entirely unable to afford to pay this out of their own pockets as they have frequently done and several have asked to be returned to their regiments on the ground that they could not afford this expense.

The commutation of the cost of the ration is entirely insufficient to cover the expense and I therefore respectfully request that a letter may be sent me authorizing the officers of the commissary department to pay them commutation at 75 cents a day on form 21.

I have been much impeded in my operations and my soldiers put to considerable expense for the want of such authority.

John R. Meigs

Letterpress copy, MCM Letterbook, April 1864, pp. 170–71, Ms. Div., LC.

John Rodgers Meigs to Captain Sidney F. Shaw

Sir, Cumberland, Maryland, April 28, 1864

Press the teams I will pay just bills for them. I wish you to cut out immediately 16 sticks 12" × 12" and from 32 to 35 feet long. Also 2 sticks 15 × 15 × 35' long.

2 sticks	15" × 15" & 20' long
16 Sticks	12" × 12" & from 32 to 35 feet long
2 "	15" × 15" & 35' long
4 "	15 × 15 & 20' long
2 "	15 × 15 & 30' long
5 "	12 × 12 & 22' long

Send them all to Capt Penn Gaskell H Ferry for the bridge there and honor any requisitions he may make on you for timber. Use the first boards you come upon for great and little Cacapon bridges.

John R. Meigs

Letterpress copy, MCM Letterbook, April 1864, p. 193, Ms. Div., LC.

John Rodgers to William P. Trowbridge

Sir Cumberland, Maryland, April 28, 1864

I enclose a drawing and letter from General Benham. I wish you would have two of these pontons made and sent to me at Cumberland. Address letters to me hereafter at Martinsburg.

John R. Meigs

Letterpress copy, MCM Letterbook, April 1864, p. 192, Ms. Div., LC.

On May 1, 1864, John Rodgers Meigs reported to Israel C. Woodruff that the Engineer Department had been moved to Martinsburg, West Virginia.

Major Israel C. Woodruff to John Rodgers Meigs

Sir, Washington, D.C., May 10, 1864

There will be mailed for you today the following photographs sent this morning, to be forwarded from this office, by Col D. C. McCallum Superintendent U.S. Military Rail Roads, office No 250 G. Street Washington DC. viz.

Photographs illustrative of operations, in construction & transportation as used to facilitate the movements of the Armies of the Rappahannock of Virginia & of the Potomac. Including Experiments made to determine the most practical & expeditious modes to be resorted to in the construction, destruction & reconstruction of roads & bridges. By Herman Haupt Brig Genl in charge of U.S. Mil. R. Roads, with a pamphlet describing the photographs.

Please acknowledge the receipt of them to Col McCallum.

J. C. Woodruff

Letter signed, JRM Other Papers, MCMC, MCMP, Ms. Div., LC.

E. Jay Allen to John Rodgers Meigs

Sir, Baltimore, Maryland, May 11, 1864

Letter of 4th recd this morning. I shipped as I before advised you Ten boats complete to Cumberland. I shall ship ten more day after tomorrow and will telegraph you one day before shipment and come up with them as you desire. I called to see Mr Quincy who has been ill and therefore did not answer your telegram.

He will connect the Roadway bridge with his Railroad Lines and will endeavor to have it completed before the June Freshet which will be between the 10th and 15th. It would assist him materially if you could get him out some 12/12 inch oak sticks 40 feet long. Larger than twelve by twelve would be better. If you notify him if you can get them out and at what point in the road he will bring them up. He claims that the Govt can get them on private lands which he is debarred from doing. He will communicate with you now he is better. The price of two 35' trusses cannot be ascertained without presenting a plan showing dimensions of timber, amount of framing & I will endeavor to ascertain before leaving Baltimore who can build it.

The anchors for Pontons I am obliged to get in New York. I can only find them in this market and cannot get any made in time. If this letter is detained as was yours I shall see you before you receive it. If you do not instruct differently I shall push on with the pontons to the front.

Rumored to day that Sherman has been whipped. Gold now up since yesterday 9 cts.

Ed Jay Allen

Letter signed, JRM Other Papers, MCMC, MCMP, Ms. Div., LC.

Major General Franz Sigel's army struggled to clear the Valley of Virginia of rebel forces. He was defeated on May 15, 1864, by Major General John C. Breckinridge at New Market, Virginia. The diarist David Hunter Strother recorded the rout of Union troops at New Market. In response to the soldiers running to the rear, Strother and other officers, including John Rodgers Meigs, attempted to rally them. The diarist described Meigs as "especially excited" and "saw him cut down a straggler with his sabre." Following the defeat, Union troops were demoralized and even though "usually so zealous," John Rodgers Meigs was "adverse to the advance" planned against Confederate troops by Sigel's replacement, Major General David Hunter.[1]

Hunter's orders were the same as Sigel's. Under his command Federal troops drove further up the Valley, defeated the Confederates at Piedmont on June 5, and arrived at Staunton on June 6. Unable to reach Charlottesville because Southern troops blocked the mountain passes, Hunter turned his attention to Lynchburg, Virginia, a major source of supplies and an important railroad center.

1. Cecil B. Eby Jr., ed., *A Virginia Yankee in the Civil War: The Diaries of David Strother* (Chapel Hill: University of North Carolina Press, 1961), pp. 227, 232.

John Rodgers Meigs to Montgomery C. Meigs

My dear Father, In the Field near Cedar Creek, Virginia, May 25, 1864

I have just recd your letter dated off Fortress Monroe and suppose you must have seen some of Butler's doings.

We had quite a severe action the other day, losing ten percent of our whole command in killed and wounded. The defeat was caused by the hurried manner in which the engagement was entered into and the bad way in which the troops were handled.

General Sigel had sent one of his brigades 10 or 15 miles from the main body and in the face of the enemy. They were attacked at night and but for the soldiership of the Colonel Commanding would probably have been captured. It was Colonel Moor who was dismissed but reinstated last winter. He had anticipated their attack, which was most determined (as they charged within 20 yds of his line) and changed his position after dark thus throwing them out in their calculations. The attack was repulsed.

Next day Sigel marched to his relief but the command was a good deal straggled out upon the pike and did not arrive upon the field in time to take position until the action was in progress and before the rear got up it was decided.

The position was as I have sketched on the opposite side. The enemy did not take advantage of one of its most dangerous features to us, either from inability or fear.

General Sigel behaved with a great deal of personal gallantry but did not manage his troops well. I think he was entirely surprised in finding in such force and with such fine troops and still more so in finding that they made the attack upon him instead of awaiting his onset.

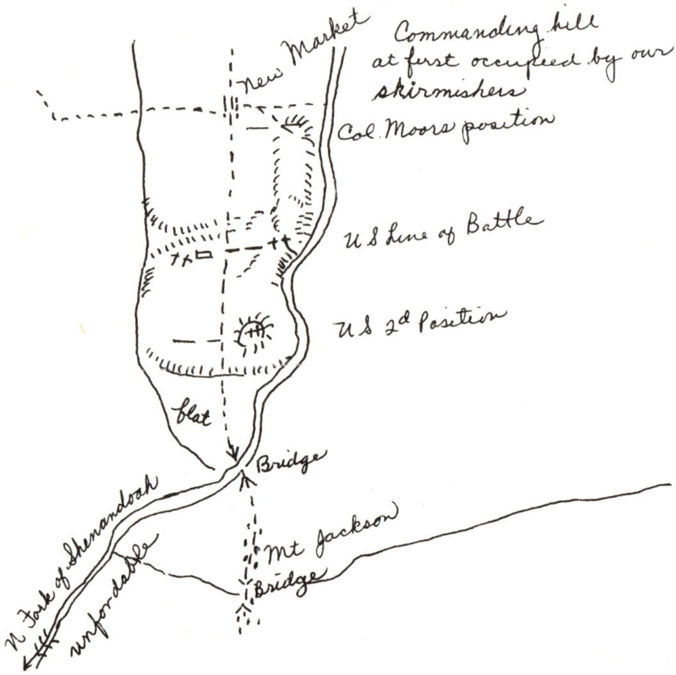

We found Col Moor occupying a tolerable position with his men well disposed and our men or those of them that arrived first were brought up to his support. Soon afterwards the enemy was discovered advancing and we fell back about a mile to a much finer position in the rear leaving some cavalry and skirmishers in front. The enemy came on in beautiful order of battle of three lines with reserves, entirely infantry with a few batteries in his rear. Our skirmishers ran in and the enemy arriving at our first position and finding it deserted came on rapidly and struck us before General Sigel had got done arranging his infantry. Our artillery received him nobly and by its stubborn resistance and the boggy ground over which they had to retreat lost 4 guns. Some of the infantry though far outnumbered and overlapped by the doubled lines of the enemy fought stubbornly. Other regiments which were lying in reserve got up and ran almost before a shot had been fired. The cavalry was scattered all over the field in small bodies some of which remained inactively looking on while others decamped, by their hasty retreat encouraging the enemy. Gen Stahel repeated a charge three times with a small party which he had near him & which behaved very well but could not drive the enemy. In about half an hour after the first onslaught our line was driven back in disorder to a second position. This was finally forced and a third taken up under cover of which our force crossed the single narrow bridge which with the road leading to it over the bottom would have been completely enfiladed for ¾ mile if the enemy had shown more nerve and pushed us from our third position and planted a battery there before we could have crossed the ~~single~~ bridge.

As it was, a couple of squadrons and a section of artillery held them in check long enough for us to cross in safety.

We burnt the bridge and after putting the river between us and the foe hurried on past some dangerous points which he could have occupied in security and closed the road to us by a fire from the opposite side of the river which runs nearly parallel with the pike.

A few days afterwards General Hunter appeared and took command.[1] Indications were immediately given that we would move in a day upon the enemy again, and the idea seemed very doubtful to a major portion of our officers. We had just been whipped and, from the information we had received, would have been almost certain of being whipped again if we made a [illegible] advance upon the enemy.

The Washington theories seem to have now [bubbled?] out, and General Hunter is going at the reb in a cool sort of way which will, I think, insure success. Had [he] made an attack, I think he would have been whipped.

Love to all, from your affectionate son J. R. Meigs

Autograph letter signed, MCMC, MCMP, Ms. Div., LC.

1. The remainder of the body of this letter is a shorthand transcription.

David Hunter Strother recorded some relaxing moments for the young Meigs in late May 1864 when the latter was in Woodstock, Virginia. Strother saw Meigs and Horace V. Field, "two greenhorns," talking to "two sharp girls at a window." The girls "twitted them politely about New Market and said that General Hunter's burning order was uncivilized and unmilitary, as the attacking of trains and cutting off of escorts were legitimate acts of war." There were more sober moments for the young lieutenant. While in Woodstock, he went scouting and burned the house of a man accused of assisting in "killing and capturing stragglers during Sigel's retreat."[1]

By early June, Hunter's troops were at Harrisonburg, Virginia. On June 4, 1864, John Rodgers Meigs accompanied a cavalry unit "to amuse" Confederate forces at Mount Crawford, Virginia.[2] The following day, the Battle of Piedmont was fought. This Union victory erased some of the disgrace of the defeat at New Market; however, some eight hundred Union troops were killed, wounded, or missing. Confederate losses included six hundred casualties. This victory exposed Staunton to Federal occupation and secured the valley for the Union for a brief period. The victory also cut off valuable supplies for the Confederate troops.

Part of John Rodgers Meigs's responsibilities involved the destruction of buildings and railroads that had strategic value to the Confederate forces. Strother provided a picture of Meigs at this task. On June 7, 1864, Strother recorded that Meigs allowed local citizens in Staunton to burn a carriage factory rather than have the army do it and increase the chance of danger to private property. On the following day, orders were given for troops to assist Meigs in destroying the railroad in front of the American House in Staunton.[3]

1. Cecil B. Eby Jr., ed., *A Virginia Yankee in the Civil War: The Diaries of David Strother* (Chapel Hill: University of North Carolina Press, 1961), p. 238. Woodstock, county seat of Shenandoah County, Virginia, was approximately eighteen miles down the valley, north of the site of the Battle of New Market.

2. Report of David Hunter, June 8, 1864, *OR*, Series 1, vol. 37, part 1, p. 94.

3. The American House was some forty yards from the railroad tracks, making it convenient for travelers staying there.

John Rodgers Meigs to Montgomery C. Meigs

My dear Father, Staunton, Virginia, June 8, 1864

After much tribulation and hard marching and one severe fight we have arrived at Stanton. We started under Sigel for this place but did not get here as you know. We rested at Cedar Creek found a new commander and some additional troops and started again having made arrangements before starting for meeting Crook and Averell in this place.

Pushing slowly up the valley in order to time with Crook & Averell who were delayed by the slowness with which their supplies reached them. We found that Breckinridge had joined Lee immediately after the battle at New Market and there was nothing but Imboden in our front. He was collecting *all* the inhabitants justices ministers doctors old & young and arming them as reserves.

He retired to a fortified position at Mt Crawford filling the fords of the river there with upturned harrows teeth, sharpened telegraph wires etc. and having his guns commanding our side completely. We laid at Harrisonburg a day just in his front. 4 days ago Genl Hunter sent me with a regiment of Cav to make a false advance against the enemy, and the main column struck off to Port Republic. I drove in their pickets fought their skirmish line and by my movements and keeping my men in the woods magnified my numbers and indicated an intention of turning their left. After fooling with them all the morning about 12 o'clock I retired. They seemed completely mystified at the direction I took in retreating. We have since learned that they did think until two o'clock that it was our real advance and that we killed 5 and wounded 3 of their men without having one of our men scratched.

I joined the column at Port Republic where it was waiting for the completion of the pontoon bridge. We camped there that night and next morning started for New Hope intending to take ~~Wan~~ Stanton or Wa[y]nesboro acc to circumstances. After marching about an hour we had a lively skirmish with the 18 Va Reb Cav and took some 30 prisoners among them Capt Imboden the Genl's brother. This was done in a gallant charge by the 21 NY Col Tibbetts.

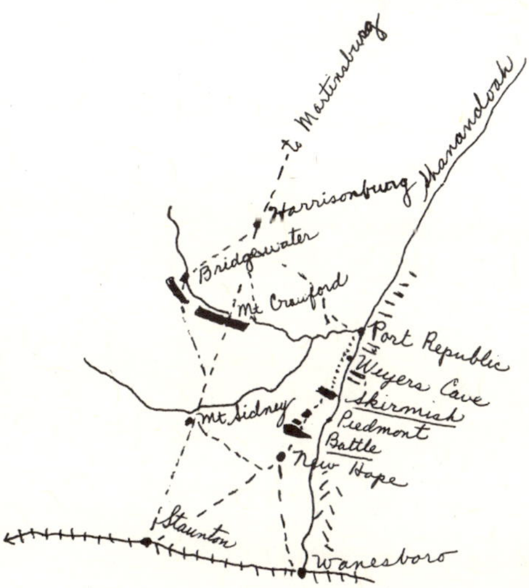

We then formed line of battle and pushed on cautiously till we came upon the main body in a very strong position such as this. Our Cav drove them back upon the main line. We advanced in line of battle and came in heavy contest with them in a piece of woods at D from which the first brigade drove them to a second piece at C. The 2d brigade advanced to E and met them, all the way; in dense woods it disappeared from sight entirely.

The first brigade fought for several hours at D but could not drive the enemy from their fortifications at C. Towards one or two o'clock our case became critical we expected every moment a charge from C which it was much feared we could not resist. We steadied the men however by encouraging speeches and quietly withdrew the 2d brigade from E changed its front to the right and hurled it against the flank of the enemys advanced brigade.

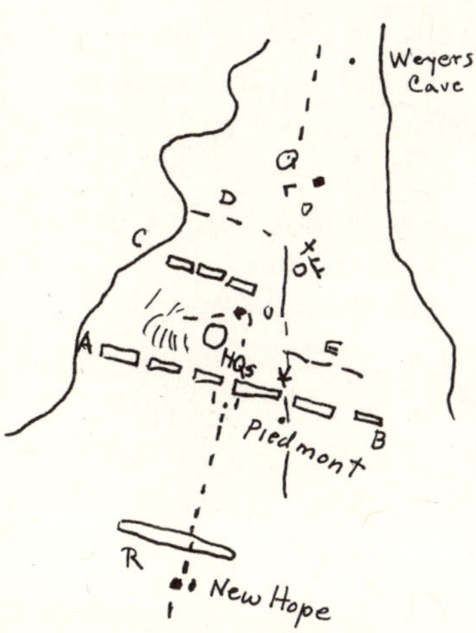

The first brigade under a splendid soldier Col Moor who was dismissed but reinstated held fast while the 2d came up the hill on the flank and opened a raking fire through the enemys line. 3 Corps of Cav passed at the same time into the rear of the Rebel force and all rushed in together with a yell. In a moment more we had the delight of securing 888 eight-hundred and eighty-eight Rebels disarmed and prisoners come trotting out of the woods at the double quick guarded by our Cavalry. The rest of their brigade threw away their arms and rushed wildly down a steep bank into the river in crossing which many of them were drowned. Others were shot after they had got across. The rest of their army rushed pell mell from the field with the exception of one or two regiments. A single regiment of cav charged them and got as far as their wagon train but shut in by high fences and crowded in the road they lost so many men that they had to retire.

If they had been properly supported we could have captured the whole of them. The enemy now occupy a strong position in Rockfish Gap having been reinforced by all the SW Va force and expect more from Lee. Crook and Averell have joined us having torn the track completely from the RR and burnt all the bridges up to this point. There are a few unimportant stretches of track unimpaired but many miles have been utterly destroyed.

A regiment of infantry in line just picks up one side of the track and throws it down the embankment with the greatest ease. They took the rails while cold and twisted them by main force around the trees winding them up like a corkscrew. The enemy had in the fight with us according to their own statement 8 or 9000 men. We had as many thousand as Mother has had children.

We learn from Rebel sources that Grants right wing was driven yesterday or day before by Lee. It *must* not be so.

The Rebels make strong breastworks in five minutes of fence rails, laid sloping against a fence and with chinks open everywhere. *Our Napoleon shell glanced from these inclined shields in some instances.*

Give great love to Mother & Mary.

Ever Your Affct Son JRM

Stanton has been quite an important depot. We destroyed large ~~deposits~~ store houses and wagon works and [illegible] etc. And captured a great many valuable supplies. We do not find the country teeming with corn as there has been a great [illegible].

The enemy in the fight had Imbodens & Jones commands and about 3000 reserves. Genl W E Jones their commander at the battle was killed. We found on his person a letter from Imboden containing a plan for baiting Hunter to Mt Crawford whipping him well thrashing Crook & Averell and then "paying their respects together to Mr Hunter." The plan has not worked. Hunter swallowed bait fishing rod fisherman and all. Uncle Roberts regiment was left in the woods to cover the movement of the 2d brigade when it retired to aid the 1st. 4 of Uncle Rs men were wounded he is well and unhurt. Col Moors regiment lost in killed and wounded 25 per cent.

[illegible] command has lost in its [illegible] last battles over a thousand men.[1]

We send back tomorrow over 1000 prisoners including about 50 officers. We hear the heavy guns at Richmond quite distinctly, at Harrisonburg we heard the guns of the 3d of June.

Autograph letter initialed, MCMC, MCMP, Ms. Div., LC.

1. The preceding sentence is a shorthand transcription

Following the battle, Hunter's troops moved toward Lynchburg, via Lexington, Virginia. While at Lexington, the troops set fire to the Virginia Military Institute and plundered the school of various articles. John Rodgers Meigs took mathematical instruments and directed that a brass bust of George Washington be sent to West Point. Before leaving, he recorded an entry in the record book of the Dialectic Literary Society.

John Rodgers Meigs's Entry into the Dialectic Literary Society Record Book

Lexington, Virginia, June 12, 1864

The next meeting of the Di was probably interfered with by the threatened advance of the Yankees under Sigel, who marched up the Valley of Virginia against them. The cadets and other Southern forces defeated Sigel at New Market, but three weeks afterwards, in a battle at Piedmont, near Weyer's Cave, Gen. Hunter, Sigel's successor, utterly routed their force, and took 1,000 prisoners. At this moment, the Virginia Military Institute is a mass of flames.

John R. Meigs

William Couper, "Claudius Crozet," Southern Sketches, First Series, No. 8 (Charlottesville, Va.: Historical Publishing Co., 1936), p. 183.

After burning the Virginia Military Institute at Lexington, Major General David Hunter turned his attention toward Lynchburg. In reaction, the commander of the Army of Northern Virginia, Robert E. Lee, sent Lieutenant General Jubal Early to help Brigadier General John C. Breckinridge, whose troops had returned to the area. Skirmishes occurred near Lynchburg on Otter Creek, near Liberty, and at New London. After light attacks against Early in mid-June, in the face of the combined Confederate force, Hunter decided to move out of the Shenandoah

Valley. With Early in pursuit, Hunter moved his troops into the Kanawha Valley of West Virginia. With the Shenandoah Valley undefended, Early moved northward and his army arrived at Winchester with little opposition on July 2, 1864. The following day, Confederate troops were in the Harpers Ferry area fighting Sigel's troops. Unable to secure Harpers Ferry, Early's army crossed the Potomac River at Shepherdstown into Maryland and soon captured Hagerstown. After severe fighting, Early reached the outskirts of the nation's capital. Seeing the Federal defensive measures, Early began to move west with plans to cross the Potomac at Leesburg, Virginia.

While Early was heading north, in southwest Virginia Hunter's army passed within three miles of Natural Bridge. John Rodgers Meigs and others were able to see the rock formation and climbed South Peak.

Engineering responsibilities were never far away. Returning to the northern end of the Shenandoah Valley, John Rodgers Meigs traveled to Pittsburgh, Pennsylvania, to confer about pontoon bridges and other supplies; and he ordered maps from Israel C. Woodruff for Harpers Ferry and the upper Potomac. Once more Meigs was placed on the field of battle when on July 20, 1864, Hunter ordered him to accompany Colonel Rutherford B. Hayes to Charles Town to open communications with William W. Averell.

The first part of August saw another change in command. Philip H. Sheridan was appointed commander of the Middle Military Division on August 7, 1864. By August 15, 1864, young Meigs received a commission as brevet major and, in addition to his duties as chief engineer, was appointed aide-de-camp by Sheridan. The commander was often referred to as "Little Phil" because of his physical size. He in turn referred to his new aide-de-camp as "Little Meigs."

John Rodgers Meigs to Brigadier General Richard Delafield

General, Loup Creek, West Virginia, June 28, 1864

I have the honor to request that the sum of $7,000 Seven thousand dollars be placed to my credit with the Asst Treasurer at New York to enable [me] to reconstruct the Military Wire Suspension Bridge over the Gauley River at its mouth.

This bridge was burned by General Lightburn when he abandoned the Kanawha Valley two years ago. The want of it has been frequently felt since. The whole of the transportation for General Hunters Army had to be crossed on a single flatboat.

John R. Meigs

Autograph letter signed, RG 77: Records of the Office of the Chief of Engineers, Letters Received, 1826–66, M-4760, NA.

Brigadier General Richard Delafield to John Rodgers Meigs

Sir, Washington, D.C., July 6, 1864

Your requisition for funds to build a permanent mil bridge over the Gauley River, approved by General Hunter, has been received. The requisition for the amount will be made on the Treasury Department for Engineer Service and duties confided to you.

On the subject of this permanent structure, I ask your attention to some reflections made upon information obtained here, and hence not so reliable as that you can command on the ground. The propriety of building this expensive iron structure cannot be understood. If left without a permanent guard, liable at all times to call for re-inforcements, it will most probably be again destroyed, not only as a hindrance to our operations but for the value of the material to an enemy. If built of wood it will in like manner be destroyed and for the same reasons excepting for the material. If such be the case, should we not resort to some other expedient than a costly iron structure demanding a permanent force for its protection.

I am informed of the following facts:

1st—The piers and abutments of the old bridge are still standing and from 20 to 30 feet above water

2nd—The maximum rise of the freshet is from 8 to 10 feet, with about 6 to 8 feet at low water

3d—The position at the intersection of the Gauley and New River is favorable for a bridge on piers, a ponton bridge or a Swing bridge, or flat boat bridge

4th—There is no difficulty in keeping a floating bridge structure at this locality through the year, neither drifts nor freshets offering any serious difficulty

5th—Guerrillas would avail themselves of every opportunity to destroy anything we may make

6th—The Banks of the river slope gradually to the water presenting favorable circumstances for ponton-swing, or flat boat raft bridges

7th—Three stone piers are believed to be standing in the river, besides the two permanent abutments

8th—No craft comes down the Gauley River

9th—A foot bridge was put up over the Gauley as a temporary structure that lasted two months. And a swing ferry bridge is supposed well suited for the locality as an expedient

10th—The swing ferry bridge at Chattanooga conveys on its deck four six-mule wagon teams and four companies of infantry at a time, the flat boat being about 75' x 25'

11th—Charleston and Galleopols are said to be suitable and good places for storing and keeping a moveable bridge for use when required over the Gauley.

12th—The Kanawha is always navigable to one or the other of these places, from whence a good wagon road of about 30 miles or 1 ½ days journey exists between Charleston and the contemplated site of the bridge

13th—The river at Gauley to be bridged is about 250 feet wide requiring about 12 pontons, either wood or canvas.

Now with these views the expediency and propriety of constructing an iron bridge cannot be understood. Placing 15 or twenty canvass pontons or the regular ponton boat and wagons in depot at Charleston, or even at Parkersburg, would seem to be more advantageous, to be used and taken up at pleasure. Or building a good flat boat and maneuvering it, as a swing bridge to be left there hauled up, or taken to Charleston and towed up when wanted, may be another expedient, or building a rough wooden bridge on the existing stone structures. And again you may devise even some other and better system. The subject is thus brought to your notice, that you may adopt the best and most advantageous management to subserve the public welfare.

Richard Delafield

Autograph letter signed, RG 393: Records of U.S. Army Continental Commands, 1821–1920, Department of West Virginia (P), 1863–66, Special Orders, July 1863–December 1864, vol. 20, W.Va., p. 353, NA.

John Rodgers Meigs to Montgomery C. Meigs (Shorthand Transcription)

My dear Father: Harper's Ferry, West Virginia, 9 P.M. July 25, 1864

I had the pleasure of seeing Mrs. [Hunter?] the other day and learning from her something about home. I had no idea until I saw her whether you were all living in tents or not, and she told that Mother had gone north.

I suppose I can interest you most by telling you of what has been going on here for a day or two.

General Crook the other day crossed the Shenandoah River with General Wright in his rear at Snikers Ferry [Virginia]. He supposed, I believe, that the enemy's cavalry only were in front. After he had got across, he was attacked by a superior force of the enemy, who he twice repulsed. Finding that they were bringing up still more troops, he galloped up to General Ricketts and asked him to send over one of his divisions to the support of his men, and Ricketts declined to do it, although in plain sight of the fighting, saying that he did not "see the use of holding that point of land." General Crook then ordered his division back, and they came back in pretty good order, not closely followed by the enemy, to enter a pretty severe artillery fire, wading across the river.

The next day, the two parties lay and looked at each other, and the next the rebels went off and we followed them. On this day, Averell, with one brigade of infantry and one of cavalry, had attacked part of them, [one?] Winchester, defeated them, capturing 4 Napoleon guns.

The next we heard was that Averell was following them and had reached Middletown, halfway between Winchester and Strasburg. This morning we learned from runaway soldiers that Averell had been pushed back and that Crook had gone out to help him and had driven the enemy and had reached a position with his left resting on the Front Royal Road. The enemy attacked him here in force and flanked him and drove him. I feel that his troops stampeded, though I do not feel he says so himself.

Cannonading opened this morning at Bunker Hill and soon moved to Martinsburg. Telegraphic communication ceased with the west about 11 o'clock. We have enough militia here, I think, to feel secure, but I should like to hear that Crook had got away from the river safely.

According to his account, the enemy have shown a continual disposition to flank him, as if they want perhaps to get ahead of him. There is nothing that makes troops and generals travel so fast and wear themselves out so much as this fear of being flanked. Crook is supposed to be crossing at Williamsport but would rather have had at Shepherdstown because the enemy may get between him and us by taking that route. It seems to have been impossible to find out what the enemy's force is. Crook, I believe, thinks it quite large. It may be. We do not know what.

Give my love to dear Mary.

From your affectionate son, J R M

Autograph letter initialed, MCMC, MCMP, Ms. Div., LC.

Through the month of July, Early's army fought several engagements and portions of it reentered Maryland and Pennsylvania. Because of Hunter's destruction in the Shenandoah Valley, Brigadier General John McClausland demanded payment from the citizens of Chambersburg, Pennsylvania. Without funds to meet the demand for $500,000 in currency or $100,000 in gold, the townspeople watched as the Confederate commander burned the town.

By the end of July 1864, John Rodgers Meigs was at Frederick, Maryland, with Union troops fighting Early's army. By August 8, 1864, he was once again at Harpers Ferry.

John Rodgers Meigs to Major General Henry W. Halleck

Major-General Halleck: Frederick, Maryland, 12 Midnight, July 30, 1864

General Hunter sent me here, just after the wires were broken, to telegraph you that the head of his column would start in the direction ordered toward

E ____ [Emmitsburg, Maryland] about 3 P.M. The men are very tired; some of the Sixth Corps stragglers were half-way between Frederick and Harper's Ferry this afternoon. General Hunter sent out some cavalry to the gap near Boonsborough. Communications will reach him by courier from Frederick.

John R. Meigs

Printed telegram, OR, series 1, vol. 37, part 2, p. 512.

John Rodgers Meigs to Montgomery C. Meigs

My dear Father, Harpers Ferry, West Virginia, August 8, 1864

You have not written to me since last April though I have sent several letters to you.

One that I received from Mary the other day is the first I have had for some time from any one at home.

I thought a couple of days ago that we were to make one of the most brilliant and important moves of the war, resulting in the destruction of the Rebel forces which has been so insolently operating in Maryland and Pennsylvania. It seemed to me and to every one to be perfectly feasable but as usual *delay* has taken the chance away from us and I can see nothing ahead now but arduous and tedious movements full of danger and barren of results.

It is disheartening indeed to see the Rebels almost stultify themselves by their rash moves as they have been doing lately, presenting us opportunities for their destruction ~~almost~~ time and again and then to see those opportunities neglected and the Rebels escape not through their merit but our fault.

Just about a week ago the situation was as follows.

Crook having been whipped had brought his forces to Harpers Ferry. Averell was at Green castle. Gen Kelley at Cumberland. The victorious rebels apparently thinking that the 6th Corps was on transports on its way to Genl Grant were disporting themselves making raids and manouvering with the idea that there was no necessity for concentration of their forces because we had nothing big enough to attack them.

A large detachment was over in Maryland and Penn, that their whole force was there was not believed here. It should have been positively known. There is no difficulty in finding men brave enough to go into the enemys camps as spies or in sending out cavalry enough to catch some picket or get some information. However it was not believed that their whole force was in Penn but that a large portion had been left at Bunker Hill whose defeat would have been fatal to them. In their divided condition a defeat would have been almost certain if we had pushed vigorously at them. We could have safely attacked them from the S East and a victory would have ensured the destruction of their stores and plunder and have driven them into the Mts. Just as this plan was about to be executed orders came to move N.E. in an exactly opposite direction to the one intended and next day our footsore and weary soldiers were making a fruitless march to Frederick.

Here we staid while the Enemy operated in Penn and towards Cumberland, where he got thrashed. Grown bold through our inactivity he commenced another raid towards Penn. Day before yesterday we learned at Monocacy that he had crossed a force at Sharpsburg. The troops were immediately marched to Harpers Ferry apparently with the idea of cutting off this force and fighting the enemy from the South. When we got there though instead of pushing on with an ample force we waited for some cavalry to come up meantime the enemy heard of our movement and hurried out back out of danger. I can see nothing now but a long stern chase in which we cant catch them unless they *want us to,* in which case I doubt whether we ought to.

We have not the supplies and the transportation for a move up the Valley but we did have plenty to take a short trip of 25 miles to Bunker Hill and cut off the return of those rascals. I have been ordered to report to Genl Sheridan for duty. Whether I am on his staff or not I do not know.

[Unsigned]

Autograph letter unsigned, MCMC, MCMP, Ms. Div., LC.

John Rodgers Meigs to Louisa Rodgers Meigs

My dear Mother, Harpers Ferry, West Virginia, August 8, 1864

I have just read your welcome letter of the 3d describing your life among the New Englanders. I am glad to see that during the tumults, stampedes, and excitements daily occurring in these regions you are far away from them in a secluded place where you can enjoy a telescopic view of the whole ground without the violent emotions that those poor ladies have to feel who live here. Father I am sure must feel more comfortable now that you are away than if you were near him even and he was enjoying your society. Aunt Sarah I understand was prepared the other day for a skedaddle as the Rebels came within 14 miles of Sion Hill. Jennie writes to Uncle Robert that Calbraith first scared them all off and had the things packed up and then went out on picket. In some parts of this country the farmers regularly put their ~~Rebels~~ pickets out before they go to ploughing or reaping so as to get timely notice to run off and conceal their horses. I have often noticed in moving through the South that the people there used their children against us as pickets.

Often on arriving within ¼ mile of a house which it was desirable to surprise I have seen a little curly head going trotting away from a fence corner and in at the door to alarm the inhabitants by whom if possible it would be carried to the next house before we could arrive.

I share with you your feeling towards the contemptible puppy who dawdles about the sea beach and takes no interest in the war.

I sometimes almost wish that the calamity of a raid might be visited on those secluded people to waken them up. I fancy that McCauslands villains would excite an interest even in the fellow you write of should they approach him within a mile or two. I stand at the house [Araby] of Mr Thomas the other day where our headquarters were camped on the Monocacy battlefield. I have rarely seen a house more scarred by battle than was his. His daughter Miss Alice a lovely and accomplished girl was driven for safety with her Mother and the rest of the family to the cellar. She declares that she did not feel very badly frightened though the muskets were popping out of the windows and the balls rattling against the house until a shell crashed through the wall of the dining room and burst just over their heads with only a thin flooring between. 7 shells struck the house and I counted the marks of 26 musket balls on *one side* of the house and discovered more afterwards. Her father she thinks caused them a great deal of unnecessary anxiety by continually going upstairs to see how the fight came on. Our men and the Rebels fought hand to hand around the house and the marks of the bloody contest were everywhere visible.

You know that Uncle Robert is here with his regiment or the remains of it. He looks as the saying is a good deal "demoralized" that is as if he had not shaken hands with the paymaster for some time. I asked him today if he wanted any money but he said no he had borrowed ten dollars and was all right for a while. He came very near being captured the other day being separated for a long time from his regiment in the darkness.

When you write send your letters addressed to Father.

Ever Your Affct Son JRMeigs.

I have finished my letter but as I still feel like writing on and talking a little while longer I have taken another sheet. I have read over the first one and feel a good deal tempted to tear it up as I see several of those inelegancies and negligencies which I know you will notice as indeed I can myself on reading the letter.

I have often told you that writing and speaking are gifts which do not fall to the lot of everyone though a moderate share of them is not uncommon.

I have just been talking to a poor Mother who has come all the way from Boston to get her child 13 years old who joined a regiment leaving for the war although too young to enlist. His mother had the officer who persuaded him off arrested and made arrangements for getting him home again. So he slipped off to another regiment and in his last letter informed her that he was having a splendid time. He was with the company blacksmith and they were going to take Richmond in two weeks.

Genl Sheridan has taken command here. I have been ordered to report to him. I do not know what I am to do yet, having received no further orders.

My most earnest hopes are at Atlanta now. I am sometimes afraid Sherman will find the enemy too tough for him.

You speak as if you supposed I had been saving some of my pay. I am sorry to say that I have not.

Things have ~~cost so~~ risen so much in price and campaigning uses up an outfit so rapidly that I have now in my pocket only my last months pay. I have no debts however except those which I owe Father amounting to 250 dollars.

I have not been able to buy myself a horse and hope I shall not get into trouble by using govt horses as at present though it is against orders.

Love to all Ever your affct son John R Meigs

The address of poor [illegible] mother is Mrs. Field, Milton near Boston, Mass. I promised to write to her but have not yet done so.

Autograph letter signed, LRMC, MCMP, Ms. Div., LC.

In August 1864, Montgomery C. Meigs recorded in his diary that "John is known as the Lieut General. Hand full of maps & head full of plans always." As part of his duties as chief engineer, John Rodgers Meigs continued to prepare and correct maps and to repair and build bridges. But his status was about to change. On August 9, 1864, he wrote to Richard Delafield that he had been assigned by Major General Philip H. Sheridan to report to George Crook, who commanded the Department of Western Virginia, near Harpers Ferry. By August 15, 1864, young Meigs received a commission as brevet major and, in addition to his duties as chief engineer of the Department of West Virginia, was appointed aide-de-camp by Sheridan.[1]

Upon his arrival at Harpers Ferry, Sheridan met immediately with Meigs to study maps of the area. He found Meigs "most intelligent in his profession" and "familiar with every important road and stream" and important geographical locations west of the Blue Ridge Mountains, particularly in the Shenandoah Valley "even down to the farm-houses."[2]

1. MCMD, August 10, 1864, MCMP, Ms. Div., LC.
2. Philip H. Sheridan, *Personal Memoirs of Philip H. Sheridan, General United States Army* (New York: DaCapo, 1992), p. 256.

John Rodgers Meigs to Montgomery C. Meigs

My dear Father, Winchester Aug 17, 1864

I have been ordered to perform the duties of an aid de camp at these headquarters of Genl Sheridans in addition to my duties as Chief Engr in Dept WVa at Genl Crooks Hd Qrs thus serving on two staffs at the same time.

I received your looked for letter the other day. I intended to pay my debt to you with the money sent you as far as it will go—I owe you $250.00.

We have been to Cedar Creek chasing the Rebs when we got there they stopped and we stopped too. The Cavalry had a fight near New Town on the way down and last night Merritt who was at Front Royal handsomely repulsed an attempt on the part of Kershaws Div Longstreet and F Lees Cavalry to force a passage. He killed and wounded a number, captured 260 prisoners losing himself 60 men.

Prisoners say Longstreet is there with his Corps and will cross today. Early laid in front of us at Cedar Creek which we left during last night apparently without being discovered. Smoke is rising from the wheat stacks which our Cav are burning. I am afraid they cannot do it effectively however. There are now two good bridges at Harpers Ferry. One I had built alongside the RR bridge the other is a wide pontoon. The river can be crossed almost every 5 miles above Muddy Branch the fords are rough most of them.

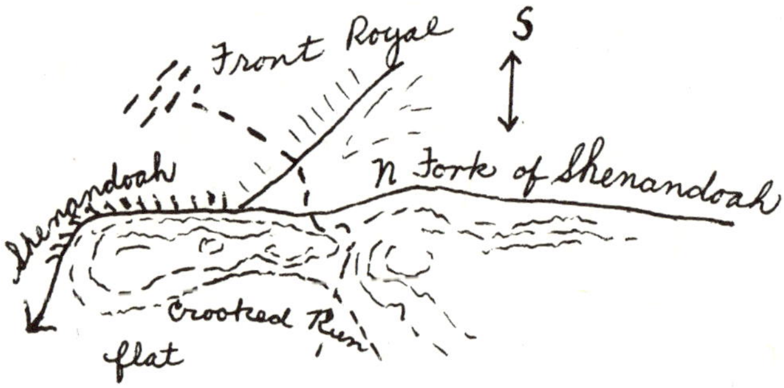

[Unsigned]

Autograph letter unsigned, MCMC, MCMP, Ms. Div., LC.

John Rodgers Meigs to Montgomery C. Meigs

My dear Father, August 24, 1864

We are unchanged yet in front. Skirmishing is going on all the time at the salient point of our lines but nothing of the nature of a battle has taken place since we started from Frederick some time ago. The Rebs do not seem disposed to attack our position and are apparently rather afraid of a sortie.

Love to all.

Ever Your Affct Son JRM

Position in which entrenched Aug 22d & 23d. All positions strong. Crooks especially well fortified and concealed by woods.

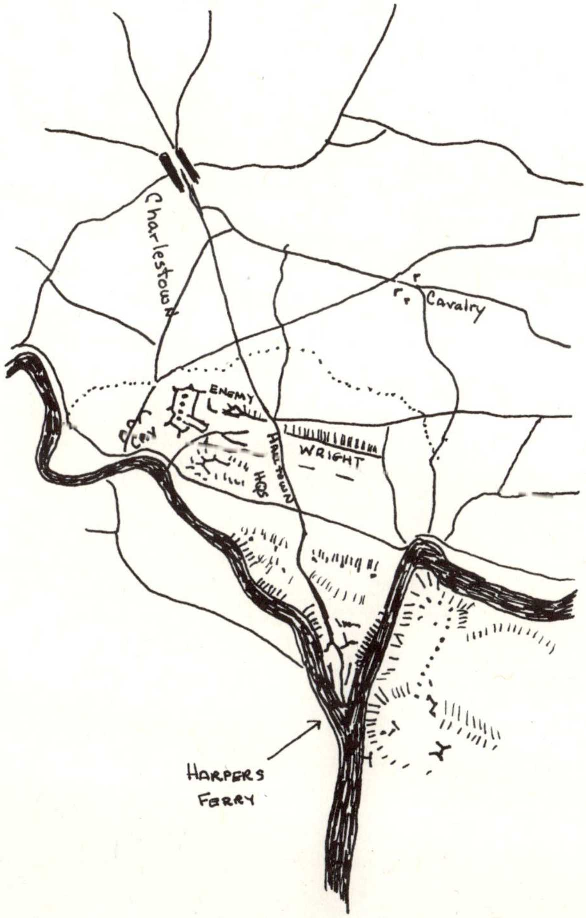

Autograph letter initialed, MCMC, MCMP, Ms., Div., LC.

John Rodgers Meigs to Montgomery C. Meigs

My dear Father, In the field, September 1, 1864

We are again in our old position near Charlestown and have been for a day or two. I wrote you some time ago enclosing a diagram of the situation. The Rebs have retired across the Opequon and our operations have been confined to Cav. reconnaissances. A day or two ago Merritts Division drove the enemy out of Smithfield and across the Opequon to within 3 Miles of Earlys HdQrs at Bunker Hill. Finding the enemy's line much to long for them overlapping on the right and left they retired again closely followed by the enemy. Genl Sheridan & Staff rode out to Genl Merritts to see how the thing was going.

After a while on reaching a ridge between Middleway and Charlestown the division was ordered to hold its position which was done without any difficulty as the enemy did not press us.

After a while the order to advance was given and it was discovered that the enemy had hastily retired across the Opequon leaving only a few pickets in our front. They had employed two divisions of infantry against one of Cavalry so that it was not to be wondered at that we had found it prudent to retire.

I noticed one thing in the fighting which I have never seen before and that was that our cavalry never approached a piece of woods without first firing a volley or two into it apparently to clean it out or develope what was there.

Genl Sheridan and General Torbert expose themselves I think a good deal more than they should. Genl Sheridan kept hanging around our batteries placing himself under quite a hot fire from the enemys guns for some time. He and General Torbert persisted in manoeuvering about first in rear of the skirmish line until finally Dr Rulison of Genl Torberts staff who was talking to Genl Torbert at the time was struck in the bowels. He put his hand to his side and said quickly. "I'm shot through. I shall die. My God My God I'm dying" and in less than an hour he was dead.

I saw the Rebel that shot him. He walked out all by himself into the middle of a green field in our front way in advance of the Rebel line so cooly that several of us thought he must be one of our men who was wounded. He finally stopped however raised his gun and fired. I heard the whistle of the ball as it came heard it strike somewhere in the group but as no one fell I thought at first that it must have struck one of the horses, but then I saw the Dr put his hand to his side and some one cried out that he was wounded.

Some of our dead men were left on the ground and when we recovered it a little while afterwards they were found stripped naked, one who had been mortally wounded but was still groaning when we found him had nothing left on him but his shirt. We have had one fine opportunity if we had only known it in time. The Rebs marched away from us and left one division in our front to hold us and watch us for a whole day. We did not know it till afterwards.

I should think that it would be getting rather cool at Rye Beach now. We in camp here sleep with two blankets over us at night and feel chilly in the morning even then. The weather we are having now is the finest that can be imagined for active campaigning.

I got a kind letter from dear Grandfather a few days ago which I have not answered yet. We are so uncomfortable sitting on our blankets on the ground without any of the conveniences that it is a great trouble to write.

Love to all.

Ever Your Affct Son JRM

I enclose the draft for $192. I sent Sergt McIlvane to you with a note to get it before I received your letter containing it.

Autograph letter initialed, MCMC, MCMP, Ms. Div., LC.

John Rodgers Meigs to Brigadier General Richard Delafield

General, In the field, Charlestown, West Virginia, September 2, 1864

General Crook Comdg the Department, thinks, that as there are no troops of any consequence serving in the Kanawha Valley, and it does not seem likely to be the theatre of operations again, for a considerable army, that the public interest would be subserved by diverting the funds originally obtained for the construction of a bridge at the mouth of the Gauley to the reerection of the wire bridge over the Shenandoah at Harpers Ferry. This bridge was burned by the Rebels last July & the suspenders and cables are lying on the ground only slightly injured I think by the fire. On receiving permission to use the funds for this purpose as much of them as may be necessary I shall issue advertisements for proposals for its reconstruction.

I have the honor to request that a list be sent me of the papers in which Engineer Officers are allowed to advertise.

John R. Meigs

Autograph letter signed, RG 77: Records of the Office of the Chief of Engineers, Letters Received, 1826–66, M-4849, NA.

John Rodgers Meigs to Louisa Rodgers Meigs

My dear Mother, Army of the Valley, September 8, 1864

I received your welcome letter the other day and read it with the pleasure which all absentees must feel on receiving assurances that they are not forgotten there at home.

The song you have often heard commencing "do they think of me at home" expresses a feeling which we often experience. It sometimes seems not improbable that the constant danger to which we are exposed may make our friends as it does ourselves forget that we are running such terrible risks.

At the beginning of the war when such things were new soldiers wrote letters to their friends and made little arrangements about their watches etc before going into action. Now they march upon an expected battlefield with little more emotion than an ordinary parade ground. The other night I went to bed in my tent here with the full expectation of participating next morning in a pitched battle.

The enemy had brought up his whole army and put it in position in the edge of timber about half or three quarters of a mile distant and had engaged in a fight with a large part of our force the previous evening which continued until long after dark. His troops in this contest were driven back and broken but the supposed presence of the main army prevented any advantage being taken of it by us.

We listened anxiously next morning for the roll of musketry or the roar of cannon but the gray mist of early dawn brightened into full daylight without our hearing it. The enemy had come to fight us but had been frightened at our resolute demeanor and the entrenchments which he saw and quietly drew off under the shadow of the woods and returned to Winchester.

I am still at Genl S headquarters. Capt O Keefe a friend of Colonel Taylors and others who talk about him with affection are here. Colonel Taylor came up here a few days ago but I most unluckily missed seeing him.

I enjoyed a treat yesterday in the perusal of a bundle of letters from you and Montie which he had received. I saw none from Mary though your letters compensated for it by speaking of her constantly. I cannot yet rid myself of the feeling that I have somehow partially lost a darling sister in her marriage. Give my love to her and to dear Aunt Nannie.

Ever Your Affct Son J. R. Meigs

Autograph letter signed, LRMC, MCMP, Ms. Div., LC.

John Rodgers Meigs to Louisa Rodgers Meigs

My dear Mother, September 18th, Evening. 1864

Before I go to bed I must acknowledge the receipt of your and Mary's Kind letters. It makes me quite happy to read these letters from home and see that though widely separated from you all I am still thought of.

We are going to have a terrible fight tomorrow and will have to be up at two o'clock in the morning. I feel that the chances are in our favor and if our troops behave well we must win the day and a glorious victory it will be.

Still God only knows what the result will be and if it is not for us I am afraid it will be fearfully against us.

Give my love to all.[1]

Ever Your Affct Son

JRMeigs

Autograph letter signed, courtesy of the Honorable John Rodgers Meigs Alger and Mrs. Alger, Rumney, New Hampshire.

1. At the top of JRM's letter, Louisa wrote: "The last letter received by me from My dear Son."

On October 4, 1864, Montgomery C. Meigs sent the following telegram to his brother Emlen in Philadelphia, with whom his younger son was staying: "Send Montgomery Home, His Brother is Dead."[1]

John Rodgers Meigs had been shot to death on the evening of Monday, October 3, 1864, on Swift Run Gap Road near Dayton, Virginia, while returning to camp with two orderlies. According to his father, "One bullet had entered his left cheek just below the eye another his left breast over the heart. He must have died instantly & could hardly have spoken after either shot struck him."[2]

1. Telegram, MCM to Samuel Emlen Meigs, October 4, 1864, MCMC, MCMP, Ms. Div., LC.
2. MCM to John Navarre Macomb, October 10, 1864, MCMC, MCMP, Ms. Div., LC.

Montgomery C. Meigs Diary Entries of Monday, October 3, 1864, and Tuesday, October 4, 1864

By order of Sec'y of War left at 7:30 P.M. to inspect Q.M. Dept. New York City and return.

About this hour my son, 1st Lt John Rodgers Meigs Corps of U.S. Engineers, while in discharge of his duty reconnoitering near Gen Sheridans Head Quarters and a little South of Harrisonburg Va. was killed by guerillas after surrender one orderly with him has not been heard of the other escaped and reached Gen Custers Head Qrs the next morning, a party under Col. Forsyth, Chief of Staff to Gen Sheridan,[1] went out and found the body of my son where he had fallen unrifled, the left arm raised, the right extended at his side. He lay upon his short cape or cloak, a bullet through the head and another through the heart. The orderly reports that he distinctly heard him say when seized by horsemen whom he was passing, supposing them to be our cavalry men, twice: "I surrender I surrender." They had waited, seized the bridle reins of the three horses as they passed slowly and commenced firing their revolvers & they continued after his surrender. Horses and orderly have not been heard of since.

MCMD, MCMP, Ms. Div., LC.

1. MCM is incorrect in identifying Forsyth. JRM's body was recovered by Major George A. "Sandy" Forsyth, not George W. Forsyth, Sheridan's chief of staff.

George A. Forsyth to Colonel Joseph H. Taylor

Dear Colonel, Harrisonburg, Virginia, October 4, 1864

Upon reaching Gen Custer's Headquarters early this morning, I was very much shocked to learn that one of Lt Meigs' orderlies had just come in bringing information that while Lt Meigs was returning [from] Bridgewater to our headquarters accompanied by his two orderlies he was attacked by guerillas and he and one of his orderlies either captured or killed. Fearing the worst I immediately started back with a party sent out by Gen Custer and when near the scene of his attack the night before, I met a sergeant of the Provost Guard belonging to the army of West Va, who informed me that one of our officers was lying dead in the road a few hundred yards beyond where we then stood.

Upon reaching the spot my worst fears were realized. Poor Meigs lay dead upon the ground, his body extended at full length; one arm partially raised the other extended by his side, his face looking upward, while his hat and cloak lay under his body. He was shot just under the right eye and in the left breast and must have died instantly as there were no indications of his having struggled or in fact stirred from the moment he fell to the earth.

His revolver and all his papers had just been removed by the sergeant of the guard who turned them over to me. I had his body placed in a wagon and conveyed to these headquarters at once. He was killed about four hundred yards from the Dayton pike, on what is known as the Swift Run gap road, which leads in to the Valley pike & thence to our headquarters. I found him just at the edge of the wood which was where the orderly left him. The orderly makes in substance the following statement. "Lt Meigs was riding first ahead of myself and the other orderly. About thirty yards ahead of us in the road were three men with rubber overcoats on. I thought they were our men. They kept ahead of us for a little while, but allowed us to gain upon them and at the edge of the wood they stopped entirely. We turned out to the side of the road to go by when they suddenly wheeled, grasped our horses by the bridles and without saying a word commenced firing upon us with their revolvers. I distinctly heard Lt Meigs say *twice* 'I *surrender,*' but they kept on firing. I tried to get away but finding I could not I sprang from my saddle jumped over the fence into the woods and escaped. I did not see anything more of Lt Meigs or the other orderly. It was between 7 and 8 P.M. that this occurred. I reached Gen Custers' at daylight the next morning."

Thus ends the sad story. Poor Meigs was *murdered* by the members of either Mosby's or White's gang. Lt Meigs was highly respected and esteemed upon the staff and a universal favorite with all, and I think with none more than myself.

Though not personally acquainted with his family allow me to assure you that they have my warmest sympathies in this their hour of sad trial.

George A. Forsyth

Autograph letter signed, Copies of Correspondence Concerning Death of John Rodgers Meigs, 1864, Rodgers Family Papers, Ms. Div., LC.

Montgomery C. Meigs Diary Entry of Wednesday, October 5, 1864

For this act of murder all the houses within 5 miles area have been burned.[1]

And so has perished my first born a noble boy, gallant generous gifted who had already made himself a name in the land. At the age of 19 he had fought with distinction at the first battle of war Bull Run, 21 July 61. He had been in most active service since he graduated in July 63, and had passed through many battles unscathed a martyr in the cause of Liberty.

MCMD, MCMP, Ms. Div., LC.

1. Before the order was rescinded by Sheridan, an area of approximately three square miles was burned (see map of the "Burnt District").

John Rodgers Meigs Diary Entry in the Hand of Louisa Rodgers Meigs

Friday, October 7, 1864

Was brought home on this day the sacred remains of my noble precious son.

JRMD, MCMP, Ms. Div., LC.

Montgomery C. Meigs Diary Entry of Saturday, October 8, 1864

Funeral of my son whose body was transferred with military honors to the Oak Hill Cemetery Georgetown.

The President and Secretary of War honored his remains by their presence.

The time did not permit preparing a grave, and he lies for a time in the Chapel.

MCMD, MCMP, Ms. Div., LC.

Montgomery C. Meigs Diary Entry of Monday, October 10, 1864

Transferred his remains to their final resting place alongside of his brothers Charles & Vincent and the little nameless child who lie together in our family plot in the Oak Hill Cemetery.[1] His mother and brother (my only remaining son) & his uncle William were with us.

We planted an ivy at the foot of the oak under which he lies & left him alone in his glory.

MCMD, MCMP, Ms. Div., LC.

1. Montgomery C. Meigs moved the body of John Rodgers Meigs to Arlington National Cemetery, November 23, 1880.

Louisa Rodgers Meigs to Ann Minerva "Nannie" Rodgers Macomb

My dear Sister, After October 4, 1864

Monty has written or copied at my request a letter which I received a short time since. It has been more comfort to me than any other in giving me an insight into the spiritual condition of our dear boy. There came no parting words from him. With his bright mind full of plans of private usefulness, his active young frame all glowing with health and vigor, he was summoned in a moment of time before the great Tribunal which awaits us all. My dear dear Nannie, how vain seem all the Earthly Asperations I had for that dear boy, how full of solemnty the hopes of his immortal Condition.

When I look back over the years that have passed when God first gave him to me, how short & how precious they seem, how little I anticipated the aging in store for me when I bade you adieu so short a time since. I have been crushed to the Earth. I begin to look up and feel that my duties to the living must be done. Ah what an effort it is, how [illegible] and how vain, all life seems. My darling precious John, I seem to hear his step in the hall, & I see his bright happy face as I last saw him. I crave to be alone, to sit & think of the past, those sweet happy days at West Point how they return and what bright memories they bring of young forms & faces that will never meet there again. How happy our darling boy was there when we were with him. Is it not a pleasure to you now to remember the satisfaction he showed at seeing you, at knowing that you had come especially to see him. Letters from every quarter come to me and Mont to assure us how much he was loved, & what a reputation he had already achieved. It seems an increase of agony to know what a brillant future was before him. All that he was, and all that we have lost. I have at the foot of my bed, his trunk, his poncho and his camp stool, the hat & jacket and short cloak he wore, that fatal night. The hat & cloak are splattered with mud, slightly. You know how it rained a little that evening. The cloak shows where he lay upon the ground, and the jacket is pierced through with a bullet just over the heart. These cast off garments seem to tell me the whole story. I take them out & look at them again & again and kiss and hold in my embrace the dear hat which still retains the perfume of his hair. I seem to feel that I have him again with me. My darling precious boy. My dear Sister, there is no one who can sympathize with me more truly than yourself, for no one knew him better than you did, and next to his parents & Mary, I think his tenderest love was given to you. Brother Robert says he was so affectionate to him, & during all the past summer he has had frequent visits from him. He says he used to spend an hour sometimes with him, would ride up & jumping off his horse, run in & throw himself down on the ground upon the buffalo robe, he having no seat to offer him. The two would lie beside each other on the ground & he says John most frequent talk was of the dear ones at home. He would seem to recall every one of the home circle with so much tenderness. He would take his Uncle Roberts hand and often in his boyish affectionate way put his arm around his neck as he talked. Poor brother Robert seemed overcome with grief as he talked to me. What a tender heart he has. He seemed to love John as a son and no doubt it was a comfort to our dear boy to have his uncle near him. We have nothing but tender memories of our dear John. Mary & I both talk of him incessantly, and the sweet memory of our visits to West Point during those two happy summers are blessed to us now. Dear Mary was much over come at first but you know she is never violent in the expression of her feeling. She loved our darling boy tenderly & he will ever be to her as to all who loved him a source of the happiest recollection. I do not know how I should have sustained myself in this dreadful grief without My dear Husband to lean upon. He has been a source of strength to me. His cheerful submission to Gods will is beautiful & wonderful, seeing how proud he was of his son and how much he loved him. He feels that he died a martyr in the great cause of human liberty, and could not have perished more nobly. He does not [illegible] though he has had moments of dreadful grief. I must not write more dear Nannie. I have longed much to see you. You would have pleasure with me in recalling the past, & those sweet memories of dear John's lovly [illegible]. How fleeting it makes all life seem. I wonder sometimes that with the knowledge that we are never to see the dear boy again, that I can attempt to resume my daily duty. The power & mercy of God alone sustains us, and the knowledge that we are going nearer to him every hour. Give much love to dear Macomb. I have received his letter & Mont also had a most kind & affectionate one from him. I know his kind heart is deeply touched by our sorrow and also that he loved and appreciated our noble boy.

Ever Your loving sister L

Autograph letter initialed, Correspondence of
Ann Minerva "Nannie" Rodgers Macomb,
Rodgers Family Papers, Ms. Div., LC.

Louisa Rodgers Meigs to Ann Minerva "Nannie" Rodgers Macomb

My own dear Nannie, Sunday Night, November 27, 1864

I have not written to you for a long time, not since I received your last beautiful & most touching letter. I do not seem able to write. I have written several letters in answer to the very many kind ones I have received, but it is a painful effort.

My eyes too are worse than ever so that I am often physically unable to write. My thoughts my words, have but one tone, the loss of my precious noble boy seems to overshadow every joy & Comfort that I possess. I feel deeply grateful to the Almighty for all his blessings and I know that I am surrounded by them, yet when I remember that I have seen the grass growing green already over his young head my heart overflows with sorrow and I am only another Rachael weeping for her child & refusing to be Comforted because *he is not.* I have returned to the daily routine of household duty. It is well perhaps for me that I have been obliged to do so, since it compels me sometimes to think of other things than those sad thoughts which force themselves upon me when I am alone. My room is filled with memories of the dear boy. His beautiful drawings have been framed, and hang many of them on the wall, his Cap & sash, the dear sash which we used to see him wear at W Point and which he has worn ever since hang over my bed at whose side stands his camp stool. His trunk & poncho are still at the foot of the bed, and that trunk filled with his cast off earthly garments is still the altar where I most love to pray. I seem to meet him sometimes in prayer and hold him close to my embrace. I love much to sit alone for I find even the cheerful conversation of the family often jars upon me. It cannot be expected that any one can mourn as I do, I, *the Mother* who bore him and have lost him. Dear Mont grieves for him most deeply and tenderly, but has so much pride in remembering what he *was* and so much patriotism which encourages him to remember the holy cause, in which our happiness was sacrificed that he does not give way to despondency. He has a courageous & noble spirit, & since our loss is *irremediable* he feels that we must look forward to that Eternity where we hope to meet our dear son with those who have gone before him, not upon the past which can never return to us. He consoles himself with [illegible] how many virtues were concentrated in the noble boy & feels like the Duke of Ormand when he said "I had rather be the father of my dead son, than of any living one." Our dear Mary is cheerful though she feels that she has lost what can never be replaced, & though she lingers with me in tender memories over the dear scenes where we have so many mutual & happy associations of our boy, she does not grieve with agony for his loss. Hers is a tender mournful grief, made of sweet memories and sad regrets. She has not one painful thought associated with the dear boy's life, nothing but love & tenderness. She has determined to name her child for her dear brother if it proves to be a boy. It will be a pleasure to have his name perpetuated through her whom he so fondly loved. . . .

from yr ever loving L

Autograph letter initialed, Correspondence of Ann Minerva "Nannie" Rodgers Macomb, Rodgers Family Papers, Ms. Div., LC.

A year later John Rodgers Meigs's parents continued to grieve for their son.

Louisa Rodgers Meigs to
Ann Minerva "Nannie" Rodgers Macomb

My dear Nannie, December 8, 1865

. . . In all the past sad year I have longed incessantly to see you for there is no one who has better understood or appreciated than yourself the grief which in all this time has weighed on my heart, "a grief which trembles at my heart alway," & that will terminate only with my own life. I have very many blessings to be grateful for, and I trust that I am not unthankful. I find in household cares & occupations, employment which compels me to exertion & which forces me to put aside the sad thoughts which at other times fill my mind. Yet whenever I am alone or not occupied the sense of the dreadful loss we have sustained, & the same longing yearning for the dear precious boy we have lost sight of, forever in this world, fills my soul. He seems to have left his footprints every where in this house, traces of his hand in books or work of some kind I encounter every day. He has left such a void, such an aching void in Mont's heart & mine that we must go down to our graves sorrowing. It seems to me that the sense of his loss grows heavier day by day, as time passes and he comes no more to cheer us with his voice & presence. Now that the War is over & things are settling down to the usual routine, when the armies are disbanded & every thing is assuming the aspect of Peace, when we see the long absent returning and *he* does not come with them our sorrow is pressed upon our hearts with a heavier weight. As we grow older we cling more & more fondly to our children, and understand how dependent we are upon them for most of our happiness in this world. Dear John was so full of life, of enthusiasm and intelligence & his heart was so filled with affection for all of us, he had too just reached that period of life when it is so full of interest to a parent that we cannot find in this world any consolation for the grief we have suffered. Mont never dwells on this sorrow, he seldom speaks of our dear boy. I know it pains him to do so, but he could not find indulgence for his grief as I do, but it has entered his inmost soul. I found him alone in the parlor the other day looking at a *bust* which he has lately had made of Monty & as I came in he said how much he wished he had one of our dear John. "As the war is over, and as time wears away I seem to miss our boys more & more. I *want* them," he said in such a tone, that I could scarely refrain from bursting into tears.

Ever your loving L

Autograph letter initialed, Correspondence of
Ann Minerva "Nannie" Rodgers Macomb,
Rodgers Family Papers, Ms. Div., LC.

EPILOGUE:

THE SEARCH FOR TRUTH

Montgomery C. Meigs and Union officers serving with John Rodgers Meigs believed that he was bushwhacked on the evening of October 3, 1864, by either Confederate citizens or Confederate guerrillas. Therefore it was to be expected that when Sheridan heard the news of his engineer's death, he ordered the burning of houses and barns within five miles of where Meigs was shot. The order included the town of Dayton, Virginia, the closest settlement. This was an immediate response in accordance with Union practice. However, in this case, after the burning of a number of homes and barns close by, the intervention of the Union's Lieutenant Colonel Thomas F. Wildes saved Dayton from the torch.

There was another response to Meigs's death. Following young Meigs's funeral, Montgomery C. Meigs set out to identify those whom he considered the "murderers" of his son. He worked with Lafayette C. Baker, an intelligence agent, and others to gather information. He read newspaper accounts, including one in the *Richmond Sentinel* printed on November 10, 1864, that gave a likely account of the encounter. He also met with William Cutterson, Meigs's orderly, who escaped, but he described Cutterson's account as "confused . . . & in fact he seems to have run at the first moment and did not see much."[1]

Meigs focused his attention on Confederate guerrillas, particularly John H. McNeill and John S. Mosby. He learned that McNeill had been shot on October 3, 1864, the same day as John Rodgers Meigs, but that this action had taken place outside the Dayton area. With McNeill eliminated from Montgomery C. Meigs's list, he turned his attention to Mosby and was angered to find him "practicing law!" Apparently sure that Mosby was not involved, he continued his search. By September 1865, Meigs learned that an individual by the name of Martin "living near Warrenton . . . states that he killed my son."[2]

Three Confederate soldiers—George W. Martin and F. M. Campbell, both of Company H (the Black Horse Troop), Fourth Virginia Cavalry, and Benjamin Franklin Shaver, of Company I, First Virginia Cavalry—were involved in the death of the young soldier. At the time of the incident the three were serving as Confederate scouts charged with gathering information within Union lines about Union troops in the Dayton area. Shaver was part of the scouting party because of his knowledge of the area, which was where his home was located.

Following the death of Meigs, it was George W. Martin who claimed responsibility. On October 6, 1865, Martin presented the facts as he knew them. His sworn statement was most likely in response to the elder Meigs's search for him.

1. MCM to James A. Garrington, March 25, 1871, MCMC, MCMP, Ms. Div., LC.
2. MCM to Thomas T. Eckert, September 21, 1865, MCMC, MCMP, Ms. Div., LC.

Affidavit of George W. Martin

Warrenton, October 6, 1865

On the 3d of October, 1864, I was scouting, in company with F. M. Campbell of the same company and regiment as myself (Black-Horse Troop, Fourth Virginia cavalry), and ____ Shaver, of the First Virginia cavalry, inside the lines of the Federal army in the county of Rockingham, near the village of Dayton. It was near dark, the sun having about gone down, and the evening cloudy and rainy. We were wearing oil-cloths over our uniforms, so that it was difficult to ascertain to which army we belonged. We discovered riding in the same direction, but behind us, three soldiers, whom we supposed belonged to the Federal army. We were in such a position—so near the camp of the enemy—and they on the only road by which we could escape, and between us and our own troops, that it was a matter of necessity that we should either elude them by passing ourselves as Federal soldiers, or capture or kill them. Holding a hasty consultation with each other we determined to make the attempt to capture them. The three Federal soldiers were riding by file and we abreast. Riding slowly along until the foremost man came up by my side I immediately presented my pistol, which I had drawn under my oil-cloth; each of my companions did the same, dropping back to the side of the man they selected. I ordered my man to surrender; his response, which was an immediate one, was the discharge of his pistol, which he must have had drawn and under his overcoat cape, wounding me severely through the body. I fired almost simultaneously, killing my adversary dead. One of the other men surrendered without resistance, the other sprang from his horse and, under cover of the woods on the right of the road, escaped. I succeeded in avoiding capture with a great deal of difficulty, owing to my wounded condition and the proximity of the enemy.

We had ridden a mile or two before I ascertained whom it was I had shot; I was told by the prisoner whom we captured that it was Lieutenant Meigs, of General Sheridan's staff.

My wound was so severe that I could not be moved from the first place of safety taken for six weeks, and did not return to the service for three or four months—the course of the ball having been diverted by a bone, I was told by my surgeon, alone saved my life.

G. W. Martin.

October 6th, 1865.

Personally appeared before me, a justice of the peace, for the county of Fauquier, and State of Virginia, G. W. Martin, whose name is signed above, and made oath that the above statement is true.

J. G. Beckham, J.P.

STATEMENT OF CAPTAIN A. D. PAYNE

G. W. Martin was an enlisted man in my company during the whole period of the war. The high character he always bore, and for which my knowledge of him enables me to vouch, together with the corroborating account of the two men who were with him, and which I have heard from them, assures me of the truth of his statement.

A. D. Payne,

Captain Company "H" (or Black-Horse Troop), Fourth Virginia Cavalry, Wickham's Brigade, Fitz Lee's Division, A.N.V.

Martin's affidavit and the statements by Beckham and Payne are from "The Killing of Lieutenant Meigs, of General Sheridan's Staff—Proof That It Was Done in Fair Combat," Southern Historical Society Papers 9, no. 2 (February 1881): 77–79.

In 1871, Montgomery received a letter from James A. Garrington, the orderly captured at the time of John Rodgers Meigs's death. In his letter, Garrington states that he believed that he shot Martin and that Martin had to have died of his wounds as when he last saw Martin, he looked like "a man who was about to give up the Gost. His brest was bare at the time and I observed that every time he breathed. blood and wind came out of the wound. So I think he Died."[1]

Montgomery wrote in return to Garrington asking him "to give me the whole truth, the whole story." Meigs wanted to hear "of the death of one of the honest truest & noblest youths who died in our countrys cause."[2] Garrington replied on April 3, 1871.

1. James A. Garrington to MCM, March 20, 1871, MCMC, MCMP, Ms. Div., LC. Garrington was born in Colburg, Ontario, Canada. He enlisted at Pittsburgh, Pennsylvania, in December 1861 for three years.
2. MCM to James A. Garrington, March 25, 1871, MCMC, MCMP, Ms. Div., LC.

James A. Garrington to Montgomery C. Meigs

Dear Sir, Warapun, Wisconsin, April 3, 1871

Yours of the 25th is at hand and I am verey glad to heare from you. And I am Sorey that I have never wrote to you all about this sad affare. For it appears by youre letter that you never have heard one word of trouth conserning the case. And you surprise me by saying that you thot I was killed. For I wrote a letter to Gen. Sheriden from camp Parrole. Annapolis M.d. two days before I left there for home. In 1865 I told him that I was going home to Canada and I asked him to be so kind as to answer me. and to direct his letter to me at Cobring. Canada. West. but he never answered my letter. And that caused me to suppose as much as eneything to have remained silent so long.—but I see now that I have acted verey rong in not writing to youre sellf all about it. In fact I wonder at my sellf now why I did not write to you about it.—but I will do all I can now towards repairing the rong that I have done by telling you all I know about the whole affare. I have nothing to give you. General. in proof of what I say but my word and you are welcome to that. And I hope that you will follow this thing up by finding out this Robert White [Wright] where we left the wounded man. We all took breakfast at the old man's house next morning. And I sat by the sid of the wounded man as he lay on a loung in the seting room. And fand him and keep the flys from bothering him for over an hour. And the old man was present and heird all the conversation that past between us. So the old man knows of course who shot the wounded man. And also that he the woundid man was the one who shot your son for he acknolleged to me there that it was he who shot the Lieut in over the Eye alltho they all climed to have done it as soon as they found out who it was that was killed. I was sorey that I told them who he was for they appeared to gloat over it so much.—And you will find that the Richmond sentinel lied in regard to the skin shot the man was shot as I said.—Why? Sir they had to stop at a mans house that night before we got inside there lines to get some wine or whiskey for him for he had fainted.—that dont look much like a skin shot does it sir? And they also lie in regard to the leader of the party ordering your son to surrender and that he answered by a shot. They never done enything of the sort. For they were disguised in Union uniforms. And we never knew but what they were Union Soldiers on till after the lieutenant was shot. In fact it was a willfoul murder. There was nothing soldierly about it.

I will now tell you all I know about it from we left Sheriden's Head Quarters at Harrisonburg.—on Sunday morning the 3rd day of October 1864. Your son sent for me and told me to tell his man to sadel up his mare and bring her round to his tent. And to also sadel up my horse. And that I had beter go to the sargent of the messengers and get another man to go with us for our Cavarly and the rebel cavarly were fighting out near bridge Water and that we would very likely have some fun before we got back to Head Quarters. So we started for the front. When we arived there we found the cavarly to work at each other. We met a squadron of the 9th N.Y. Cavarly I think.—flying in every direction who had ben defeated at the ford. The lieutenant rallyed them and drove the rebels back to there own side of the river. he then spent the rest of the day going from point to point where ever there was fighting going on and taking notes of what he see. And also hunting for fords and folowing up old roads and bridel paths to see where they run to. And making a map of the country. Catterson was not with us much during the day. The Lieut sent him several times with messages to officers of the cavarly. Catterson youst to be gone longer than there was eney use of when he was sent eneywhere with a message and I noticed that he liked to keep away from us as much as possible but I thot nothing of it. For I knew him to be sick in a privat way and I thot it hurt him to ride. And the Lieut noticed him and spoke to me about him and said he did not like the way he acted and sed he thot he acted cowardly and asked me if I had ever seen him tried. I sed no. that I never had for he had been away from us the most of his time on sick list at Washington. but I told the Lieut what was the mater with him. And he sed very likely thats what makes him act so. It was now late in the day and we started for Head Quarters. We rode fast for it was near dark. there was Five cavarly men riding a head of us about a 100 [?] rode in advance for several miles We met and past a regiment of our cavarly with their bundles of hay. going out on picquet. They also had past the Five men ahead of us. there was also a party of our cavarly probably 15 in number with an officer riding at the head in sight of us on the manourveur. We finaly catched up to the Five men in our front. they had at this time brought ther horses down to a walk and so had we. For the purpose of puting on our ponchos for it had begun to rain at this time. And the Lieutenant spoke to them something about the wether that we were going to have a heavey night's rain of it or something to that effect. I noticed them particularly as I did Every one that I saw at that time for. Mosbeys men were very troubelsome to us in the valley. And my inlistment was written less than too months of being out. And also I knew that there was no parrole or Exchange at that time. So that caused me to be verey sharp on the look out. so as not to be surprised by the rebels.—But in as sharp as I thot my self I never mistrusted the Five men we were riding with for they were drest in our Cavarly Clothing. They rode U.S. horses and had McClellen sadels And Eaven our regulation spurs and U.S. belts.—We rode a long a few minouts. and came to a road that led off to our right from the main pike. The five men tourned down it. The Lieutenant tournd down also. And sed to me this is the road we came in yesterday isent it? I beleave I sed I was not shoure. We advanced down it a short disttance. And we went to pass them on there left. Catterson at this moment was 3 or 4 rods in our rear. the men gave way to the right apearntly to let us pass. at that moment I sopose thay had there revolvers drown and consealed under there ponchoes. And I sopose they gave the wink to each other and fired for there revolvers all went off at the same time. I was not looking at them when they fired. I knew nothing of it ontill my horse gave a dodg and there pistols went off my cap was shot off my head and my horse wounded in the neck And the Lieutenant killed instantly. he never spoak or knew in this world who killed him. I did not see which of them shot him. I drew my revolver as quick as I could. as I done so I saw a man point his revolver at me. I fired at him and hit him at the same time. there was a nother man who caught hold of my bridal rane. I cocked my pistol. and pulled triger at him but unfortunately mist fire. And I shall allways think that its missing fire was the cause of my being captured. for I had a good horse. and if I had got started. It would have been good horses that would have caught me. At about the time that my revolver mist fire. I was hit on the head by one of them. I suppose with a revolver. and my revolver wrencht out off my hand.—the man who had hold of my bridal leveld his revolver at me. I threw up my hand and told him not to shoot. that I would surrender but he did shoot wounding me slightly in the neck. he cocked his revolver Again and was in the act of shooting me when the man I had shot cried outs. Dont kill him. I want him to be taken to Head Quarters for the other man has escaped meaning Catterson. They then took my bridal out of my hand. And leed my horse ordering me to put in spurs or they would blow my brains out.—All this took less time than it takes to tell it General. but of course you know all that.—they took me back up the road to the maine pike where we had just left. and Crost the maine pike And threw down the fence and took me in through a feeild to the woods. I saw the party of cavarly which was marching in our rear before we left the pike coming down the pike.—as we crost it.—but of course they did not know that the men that were with me were rebels. And they were not close enough to see what was up. I never saw Catterson after the shooting commenced. The rebels toold me that he never drew his revolver or fired a shot but roun his horse up to the fence and threw himself over the fence. . . . We arived at old Robert Whites [Wrights] about too o clock that morning. after breakfast they took me out to the stabil and robed me of 45 dollars my watch pocket knife and boots. they all belonged to Wickem's brigade and were detailed as Scouts from Gen. Wickem or Gen. Erley. I don't know which. they all lived rite a round there. and were well acquainted with the country. they also made there brags to me that they would take Gen. Sheriden out of his bed some night. And hang the dam son of a b—ch for burning there houses barns and wheat stacks. And I thot very likely they would. For I knew that Sheriden was not on the lookout for Eney such thing. For he never had more than one soldier on guard at his tent at night.—They took me to headquarters and a General by the name of Rosa—or Roser. Cross Questioned me in regard to

Sheriden. his move ments. strenth of army. when he was going to advance from Harrisonburg, &c.—They sent me from there up to Stanton and put me in a log penn. . . . They took me from there to richmond and confined me in Liby prison.—And what our soldiers soufered there. words faill to tell—I was a prisoner three months and 12 days.—I was exchanged on the 15th of Feb. 1865 and sent to Annapolis M.d. and there Discharged. I was sick all that year but began to pick up a little in 1866 but I never will injoy the same good health. I had before I boarded at Richmond's Friend ——— Hotel. for the intertainment of Union Soldiers.—You spoke of youre sons revolver being found with him. And that there was one shot fired out off it.—I don't know how that could be. For he was killed so quick. altho one of the rebels toold me that he did fire a shot as he was falling from his horse—bout I did not believe it.—altho he might—but it dont seem possible.—I know that he allways caried his revolver while out riding out of the holster stuck down in his belt in frount so as to have it handy in case of kneed but I dont see how he could have got it out of there even. For I think he was shot in the braine.—I think it more likely however that he fired that shot in the skirmish at the foord near bridg water alltho I dont remember of his shooting eney there. Alltho there was a good Chance for shooting. but we charged with drawn sabers.—please inquire of those who found the Lieut. how they found the revolver in what position, &c.—You speak of your son being a brave nobel youth.—He was indeed and it was well known by everey man in Sheridan's army.—And as for his treatment And behavior to me and All others with whom he came in contact with. As an officer and a gentleman there was none beter.—And General I feeal in my heart that I am not in eney way to blaim in this sad affare. I feeal that I have done my duty as a friend and A soldier to youre son and the flag of this country. And I have nothing to say farther than this that as to my behavior as a soldier why I refere you to allmost eney of the staff officers of Gen. Pleasanton or Gen. Sheriden for I belonged at Cavarly Corps Head Quarters the moast of my inlistment And they will very likely remember me for I have had maney a loondely ride. And in many a hot place with them. . . . Now I hoope General that I have not bored you with this long way of teeling you the history of the will foul murder of your son. And hope you will excuse my bad spelling and writing.—should you have eney fourther questions to ask. ask them. and I shall be happy to answer them to the best of my Knoledg. I remain yours. very resspectfully yours.

James A. Garrington. late of Co. M. 6. U.S. Cav.

P.S. If you should write to me a gaine be so kind as to let me know if I am entiteld to a land warrant the same as U.S. soldiers. before the war if so. how I am to get it. and oblidg me direct as before—

Autograph letter signed, MCMC, MCMP, Ms. Div., LC.

Additional information came some four years later. On October 5, 1875, Robert K. Carter, a schoolmate of Meigs who served as a Confederate surgeon near Harrisonburg, Virginia, at the time, wrote an account of the shooting.

On the night of October 2, 1864, the two had "a delightful social evening" along with Meigs's associates at Sheridan's headquarters. Based on his friendship for Meigs, "having been so bonded to him in early life for three years," Carter gathered information to ascertain the facts from the individuals "actually engaged in the affair." His letter records that Meigs was shot by members of the "12th Va. Reg. Cav." who were on a "mission . . . not to engage the enemy or in any way to harrass them, but simply to find out their positions." According to Carter, Meigs fired first, after "a gust of wind blew up the Rubber cloth of one & thus revealed the Grey Suit to Meiggs. . . . No sooner did he see it than he ordered them to halt & with the word fired one shot which took effect in shoulders of one of the men. The sound of the pistol had scarely died away when one of the others reached forward and shot him, killing him instantly."[1]

Benjamin Franklin Shaver claimed that it was he who shot Meigs in the head after Meigs shot Martin. To further suppport his statement, Shaver said that following the fight, he examined Martin's pistol and found the "caps burst, but the loads still in it."[2] He also claimed that after he shot Meigs in the head, Campbell shot Meigs in the body.

There is no evidence that Montgomery C. Meigs ever determined the exact circumstances of his son's death. But perhaps he had gathered enough evidence to satisfy himself and others that his son died valiantly on a rainy, autumn evening on the Swift Run Gap Road, near the small town of Dayton, Virginia.

---

1. R. K. Carter to Friend Hampton, October 5, 1875, and Walter McFarland to Joseph H. Taylor, April 17, 1876, Montgomery C. Meigs Papers, Correspondence, Military Orders and Related Matters, 1876–82; W. W. Parmer to Geo. W. Cullum, February 26, 1878, John Rodgers Meigs File, Cullum Correspondence, Special Collections Division, USMA.

2. P. C. Kaylor, "The Killing of Lieutenant Meigs, 1864," in *The Virginia Valley Records* (Baltimore: Genealogical Publishing Company, 1930), ed. John Wayland, p. 192.

APPENDIX:

GLOSSARY OF NAMES, PLACES, AND PHRASES

Adams, Julius W. (ca. 1840–65): Graduated from the USMA in 1861 and was assigned to the infantry. He served as assistant instructor of infantry tactics at the academy from July 1861 to 1862.

Agathos: The protagonist in *Agathos, and Other Sunday Stories,* a moral tale by the British Anglican clergyman Samuel Wilberforce (1805–73), published in England in 1840 and subsequently in the United States.

Alexander Allen and Company: Lumber merchants and wood products fabricators, 56 W. Falls Avenue, Baltimore, Maryland.

Alexandria and Washington Railroad: Began operations in 1858. It ran along the Washington and Alexandria turnpike from the Long Bridge to the city of Alexandria.

Allen, Edward Jay (1830–1915): A businessman who was in the construction business, Allen was taken into custody by Confederate authorities in Virginia. After being exchanged, he recruited and commanded the One Hundred Fifty-fifth Infantry Regiment, Pennsylvania Volunteers, until mustered out on July 21, 1863, on account of illness. He was an assistant engineer and agent for JRM during 1864.

Anderson, Robert (1805–71): Graduated from the USMA in 1825. He was in command at Fort Sumter, South Carolina, when it was attacked on April 12, 1861.

annual reports: Montgomery C. Meigs wrote reports for the year ending September 30, 1859, that centered on the Washington Aqueduct and Fort Madison construction projects.

Anthony's: Edward Anthony (1818–88) was an early U.S. daguerreotyper, with a business located at 308 Broadway, New York City. He was one of the principal dealers in daguerreotype and photographic equipment and supplies.

appointment of instructor: JRM served as acting assistant professor of mathematics at the USMA during the 1861–62 academic year.

Araby: A house and farmstead built about 1780 on the eastern side of the Monocacy River in Frederick County, Maryland, near the intersection of present-day I-270 and Baker Valley Road. *See also* Thomas, C. Keefer.

Astor House: A hotel located at 228 Broadway, New York City.

Atlanta: See *Fingal.*

Atlantic: A steamer purchased by the United States government in 1863 and renamed the U.S.S. *Commodore Read*

Aunt: Ann Elizabeth Hodge Rodgers. *See also* Rodgers, John.

Aunt Annie: *See* Meigs, Ann Wilcocks Ingersoll.

Aunt Emily: *See* Meigs, Emily Skinner.

Aunt Jerusha: Probably Caroline Rodgers (1819–83), sister of the elder Louisa Rodgers Meigs, but possibly Jerusha Elizabeth Turner Meigs, wife of William Montgomery Meigs, a younger brother of MCM.

Aunt Mollie: *See* Meigs, Mary Craythorne.

Aunt Sarah (in Hattie Coyle letter of December 1859): *See* Edmond, Sarah H.

Aunt Sarah (in JRM letter of August 8, 1864): *See* Rodgers, Sarah Perry.

Aunty Trowbridge: This could refer to either of two women. *See* Trowbridge, Catherine Whipple Sibley; Rodgers, Catherine (or Katherine) Trowbridge.

Averell, William Woods (1832–1900): Graduated from the USMA in 1855 and was assigned to the cavalry. He became brigadier general, U.S. Volunteers, on September 26, 1862. He was commander of the Second Cavalry Division, engaged in operations in West Virginia from April 26, 1864, to May 18, 1865.

Andrew B.: Presumably Andrew Coyle Bradley (ca. 1844–1902), an associate justice of the Supreme Court of the District of Columbia.

Back Creek: Waterway located in Bath County, Virginia.

Bailey, Fannie: *See* Bailey, Fanny.

Bailey, Fanny: Sister of Frederick William Bailey.

Bailey, Frank (b. ca. 1847): A son of Mrs. Margaret L. Bailey of Washington, D.C.

Bailey, Frederick William (ca. 1842–78): A son of Margaret L. Bailey of Washington, D.C. He graduated from the USMA in 1865 and was assigned to the infantry, but he was discharged at his own request in 1870. He died in Turkey in 1878.

Bailey, Lewis (1795–1870): The farmer and innkeeper of Bailey's Inn.

Bailey, Ray Tompkins (b. 1843): The fifth son of the proprietor of the inn at which Meigses, Macombs, and Pynes stayed.

Bailey's Inn: An inn and farm operated by Lewis and Mariah Bailey located near the intersection of Columbia Turnpike and the Leesburg-Alexandria road (now Bailey's Crossroads), Fairfax County, Virginia.

Baker, Edward Dickinson (1811–61): Served in both the House of Representatives and the U.S. Senate before being commissioned a colonel in the Seventy-first Pennsylvania Volunteer Infantry Regiment in June 1861. He was in direct command of the Union forces engaged at the Battle of Ball's Bluff.

Baker, Lafayette Charles (1826–68): A mechanic and entrepreneur, he was employed as an intelligence agent by the War Department in 1861 and became head of the bureau of secret service in February 1862. He was commissioned colonel, First District of Columbia Cavalry, on May 5, 1863, and became brigadier general, U.S. Volunteers, on April 26, 1865.

Ball's Bluff: Located in Loudoun County, Virginia. On October 21, 1861, an engagement ensued there when Federal forces crossed the Potomac River and were moving toward Leesburg, Virginia, but were repulsed by Confederate forces.

Baltimore Street: An east-west street running across the center of southern Baltimore city and intersecting Frederick Avenue.

Barbette: An artillery-piece carriage with a picot for traversing on emplacements behind the parapets of fortresses.

Barnum's American Museum: Established by Phineas Taylor Barnum in 1841 at Broadway and Ann Street in New York City.

Bartlett, William Holmes Chambers (1804–93): Graduated from the USMA in 1826. He was assigned to the Corps of Engineers. He served as professor of natural and experimental philosophy at the academy from 1836 to 1871.

Bayard, George Dashiell (1835–62): Graduated from the USMA in 1856. He was a brigadier general, U.S. Volunteers, when mortally wounded at the Battle of Fredericksburg.

Beauregard, Pierre Gustave Toutant (1818–93): Graduated from the USMA in 1838. He resigned from the U.S. Army on February 20, 1861, and was appointed brigadier general in the Confederate army. A synopsis of his official report of the Battle of Bull Run, dated October 14, 1861, was published in the *Richmond Dispatch,* October 23, 1861; the full report was sent to the Confederate Congress as part of the report of the secretary of war in January 1862 and was then published by order of the Congress.

Benham, Henry W. (1813–84): Graduated from the USMA in 1837. He was assigned to the Corps of Engineers. During the Civil War he was in command of the engineer brigade of the Army of the Potomac and received the brevet rank of major general in both the regular and volunteer establishments.

Benjamin, Judah Philip (1811–84): Served in the U.S. Senate from 1853 until February 4, 1861. Subsequently he served as attorney general, secretary of war, and secretary of state, in succession, in the Confederate government. In her January 12, 1861, letter to JRM, Mary Craythorne Meigs refers to Benjamin's speech of December 31, 1860, concerning proposed amendments to the Constitution, and to Senator Doolittle's and Senator Baker's replies on January 2 and 3, 1861.

Betts, John Edgar: A student from Northumberland County, Virginia, in the select course at Columbian College in 1859.

Beverly: A town in Randolph County, West Virginia.

Bien, Julius (1826–1909): Born in Germany but immigrated to New York City, where he established himself as a lithographer and map publisher.

Bier, Philip George (1841–64): A member of the Twelfth West Virginia Infantry Regiment. He was appointed aide-de-camp to Jeremiah C. Sullivan on December 8, 1863. In March 1864 he was promoted to captain and assistant adjutant general, Department of West Virginia. He was killed at the Battle of Cedar Creek, Virginia, on October 19, 1864.

Big Sandy River: A tributary of the Ohio River that flows through adjoining parts of Kentucky and West Virginia, forming part of the border between the two.

Binney, Joseph Getchell (1807–77): An ordained minister, Binney was the fourth president of Columbian College, serving from 1854 to 1858.

Blagden, Mary: The daughter of Thomas Blagden (1803–70), a lumber merchant and landholder. In the 1850s the Federal government acquired title to the Blagden farm south of the Eastern Branch (now the Anacostia River) as the site for the Asylum for the Insane (St. Elizabeths Hospital), and Blagden relocated to a farmstead beyond the city limits.

Bocock, Representative Thomas Stanhope (1815–91): A Democrat from Virginia, Bocock was the principal speaker at the February 1860 dedication of Clark Mills's statue of George Washington. He subsequently served as a member and speaker of the Confederate Congress.

Bolivar Heights: Located in Jefferson County, West Virginia.

Bonham, Milledge Luke (1813–90): Served as a U.S. Representative from South Carolina, 1857–60. He was a brigadier general in the Confederate army and was in command of a brigade of South Carolina infantry at the Battle of Bull Run.

Boonsborough: Located in Washington County, Maryland.

Bowen, Edgar Campbell (1842–1927): As a USMA cadet, he was tried by a general court-martial, found guilty, and sentenced to be dismissed from the academy; however, the punishment was mitigated to confinement to camp during the summer encampment. He graduated from the USMA in 1861.

Bowen, Nicholas (ca. 1836–71): Graduated from the USMA in 1860. He was assigned to the topographical engineer corps. He served as an assistant topographical engineer, making reconnaissances and fortifying positions in the Western Virginia campaign from July 15, 1861, to March 5, 1862. For information on the photo processed map of Jefferson County and adjacent counties "received from Lt. Nicholas Bowen in November 1862," and for other maps prepared by Bowen, see *A Guide to Civil War Maps in the National Archives* (1986).

Bowman, Alexander Hamilton (1803–65): Graduated from the USMA in 1825. He was assigned to the Corps of Engineers. He served as superintendent of the United States Military Academy from March 1861 to July 1864.

Bowman, Eulalie: The daughter of Colonel and Mrs. Alexander H. Bowman, she may have been engaged to George D. Bayard (1835–62), who was mortally wounded at the Battle of Fredericksburg.

Bowman, Mary Louise Collins (b. ca. 1821): Wife of Alexander H. Bowman.

Bradford, Augustus Williamson (1806–81): Lawyer who was elected Unionist governor of Maryland in 1861 and reelected in 1863.

Bradley, Charlie: Probably Charles S. Bradley (ca. 1842–1916), a son of Charles Bradley, secretary of the Franklin Insurance Company, and Catherine Coyle Bradley, who resided at 309 C Street, North, Washington, D.C.

Branham, J. Richardson: Resided at 454 E Street, North, across from the Leonidas Coyle family, who lived at 449 E Street, between Sixth and Seventh Streets, Washington, D.C.

Breckinridge, John Cabal (1821–75): Lawyer and politician. He served as a U.S. senator from Kentucky until expelled for support of the Confederate cause. He was appointed a Confederate brigadier general in 1861 and subsequently promoted to major general. For a time in 1864 he was the commander of the Department of West Virginia, until called to join Lee's forces around Richmond.

Brevoort House: A hotel at 11 Fifth Avenue, New York City.

brick walls: Mentioned in the September 22, 1859, letter from MCM to JRM, these were structures that with a channel conveyed towboat traffic from the Chesapeake and Ohio Canal in the District of Columbia to the Alexandria and Washington Canal in Alexandria (now Arlington) County, Virginia, located slightly upstream from the present-day Francis Scott Key Bridge.

bridge over Rock Creek: This bridge was replaced by the Pennsylvania Avenue Bridge, NW, now called the Montgomery C. Meigs Bridge. The original bridge had tubular ribs through which ran the public water supply.

Bridgewater: Located in Rockingham County, Virginia.

Brisbin, James Sanks (1837–92): Editor, newspaper publisher, and supporter of abolitionist policies. Commanded a detachment of twenty recruits for the Second U.S. Artillery Regiment. MCM, who had traveled to northern Virginia to view the Battle of Bull Run, encountered Brisbin and his men. Impressed by Brisbin's conduct and spirit, Meigs invited him to call at the quartermaster's office after his return to Washington. Attempting to obtain a pass to his home to recuperate from his wounds, Brisbin did call upon Meigs, who sent the young officer to his home and had him attended by his own physician.

Brooke, Catherine L.: Mistress of a boarding and day school at 138 Pennsylvania Ave., Washington, D.C.

Brown, Albert Gallatin (1813–80): U.S. congressman and senator from Mississippi. He was a captain in the Confederate army and served in the Confederate legislature.

Brown, Harvey (ca. 1796–1874): Graduated from the USMA in 1818. He was in command of U.S. troops in Washington, D.C., and at Fort McHenry, when he was ordered to take command of the expedition to hold Fort Pickens for the Union.

Brown's pike.: A two-edged blade attached to a six-foot-long ash handle. John Brown intended to arm slaves with these weapons. After the Civil War began, Governor Joseph E. Brown of Georgia had a similar weapon designed for use by Confederate troops not otherwise armed.

Brown[s] Mountain: Located in Pocahontas County, West Virginia.

Bruton, Thomas I. (b. 1842): Born in Florida but appointed to the USMA from Georgia and admitted on July 1, 1859. He resigned one week after Georgia seceded from the Union.

Buchanan, James (1791–1868): Fifteenth president of the United States.

Buckhannon River: A tributary of the Tygart, rising in Randolph County, West Virginia, and flowing generally north through Upshur and Barbour Counties.

bugbear: Any source, real or imaginary, of needless fright or fear; a goblin that eats up naughty children.

Bulltown: Scene of a battle between Confederate and Union forces on October 13, 1863. The town is now part of the Bulltown Historic District near Burnsville Lake, Braxton County, West Virginia.

Burnside, Ambrose Everett (1824–81): Graduated from the USMA in 1847. He was assigned to the artillery. He resigned from the army in 1853 to manufacture firearms. He was a major general, U.S. Volunteers, in command of the Army of the Potomac from November 1862 to January 1863. He afterward served as a departmental or corps commander.

Butler's doings: Refers to certain activities of Benjamin Franklin Butler (1818–93), a Massachusetts lawyer and state legislator in Massachusetts, 1840–60. Butler was a delegate to the Democratic convention in 1860 and to the succeeding convention in Baltimore that nominated John C. Breckinridge. He was brigadier, later a major, and then a general, U.S. Volunteers, 1861–65, and served as commander, Army of the James, 1863–64. JRM was referring specifically to the operations of the Army of the James under Butler following a landing at Bermuda Hundred,

Chesterfield County, Virginia, on May 5, 1864, which was countered and stopped by a smaller Confederate force under the command of P. G. T. Beauregard.

Cabin John Bridge: The bridge and conduit for the Washington Aqueduct under construction in 1859 across Cabin John Creek in Montgomery County, Maryland, approximately three miles upstream from the boundary line between Maryland and the District of Columbia.

Cacapon bridges: Constructed for the Baltimore and Ohio Railroad across the Cacapon River near its confluence with the Potomac near the village of Great Cacapon, Morgan County, West Virginia, and across the Little Cacapon River near its confluence with the Potomac and the village of Little Cacapon, Hampshire County, West Virginia.

Calbraith: *See* Rodgers, Calbraith Perry.

Cameron, Simon (1799–1889): U.S. senator from Pennsylvania. Cameron was an anti-Buchanan Democrat from 1845 to 1849 and a Republican from 1857 to 1861. He served as secretary of war from March 1861 until removed by Lincoln for corruption in January 1862, when he was appointed minister to Russia.

Campbell, F. M.: Enlisted at Richmond, Virginia, on April 15, 1862, for three years or for the duration of the war.

Campbell, Jacob Miller (1821–88): Mustered into Federal service as colonel in the Fifty-fourth Pennsylvania Volunteer Infantry Regiment, in February 1862. He was assigned to the Eighth Army Corps in April 1863 and was a commander of a brigade guarding the Baltimore and Ohio Railroad. He was subsequently awarded the brevet rank of brigadier general, U.S. Volunteers, for gallant and meritorious conduct at the Battle of Piedmont.

Campbell, John (1821–1905): Graduated from the Albany (New York) Medical College in 1843. He was appointed a contract surgeon in 1847 and was assistant surgeon at the USMA from June 2, 1857, to July 16, 1860.

Camp Northwest: Colonel William L. Jackson's base in August 1863, somewhere along or adjacent to the road between Huntersville, West Virginia, and Warm Springs, Virginia.

canal: The 185-mile Chesapeake and Ohio Canal from Washington, D.C., along the Potomac River to Cumberland, Maryland.

Captain Franklin: *See* Franklin, William Buel.

carriages, siege or field: Carriages for siege and garrison artillery were much heavier and less mobile than those for field artillery pieces. For a 12-pounder gun, for example, the field carriage weighed about 1,200 pounds and the siege carriage, nearly 1,000 pounds more.

carte de visite: Small visiting-card photographic portrait, usually 4 ½" x 2 ½" in size. Cartes de visite were inexpensive, lightweight, and easy to collect. They were very popular during the 1860s in part because famous people, such as English royalty, permitted their portraits to be taken and distributed.

Carter, Robert K.: A native of Frederick County, Virginia, Carter was a junior at Columbian College for the 1857–58 school year, at the time JRM was a sophomore.

Carusi, Louis [or Lewis] (b. ca. 1804): Listed in the 1860 census for the District of Columbia as a professor of dancing. He and his brother Nathaniel were proprietors of Carusi's Washington City Assembly Rooms, located at Eleventh and C Streets.

Cass, Lewis (1782–1866): Soldier, governor of Michigan, U.S. senator, and diplomat. He was the grandfather of Harry and Sue Ledyard, friends of JRM and Mary Montgomery Meigs.

Catlin, Robert (1841–1903): Born in Illinois but appointed from Washington Territory to the USMA in 1863. He was assigned to the artillery, Army of the Potomac, June 1863–August 1864. He was severely wounded and lost his left leg in military operations for the occupation of the Weldon railroad, near Petersburg, Virginia. His post–Civil War service included academic and administrative posts at the USMA.

Chambers, David Abbott: Resident of Washington City, Chambers was a member of the junior class of Columbian College, 1858–59.

Chambliss, William Parham (1827–87): Promoted to captain with the Fifth Cavalry Regiment on April 6, 1861, and brevetted lieutenant colonel on June 28, 1862, for gallant and meritorious service at the Battle of Gaines' Mill, Virginia. He was assistant instructor of cavalry at the USMA in 1863.

Chandler, Zachariah (1813–79): A Republican U.S. senator from Michigan, Chandler charged during a debate on December 27, 1861, that West Point had turned out nearly as many traitors as loyal men.

Chapman, William B.: A native of Massachusetts, Chapman was admitted to the USMA in 1858 but resigned on June 16, 1859. He was readmitted but again resigned on November 1, 1860.

Charlestown tragedies: The attack with the accompanying deaths, trial, and execution of John Brown as a result of his raid on the arsenal at Harpers Ferry. John Brown was executed December 2, 1859, at Charles Town, West Virginia.

Charlier, E.: *See* Ledyard, Henry Brockholst "Harry."

Chase, Salmon P. (1808–73): Secretary of the treasury from 1861 to 1864 and chief justice of the United States from 1864 to 1873.

Chateau: Bailey's Inn, Fairfax County, Virginia.

Church, Albert E. (1807–78): Graduated from the USMA in 1828. He served as assistant and later full professor of mathematics there from 1828 to 1831, and from 1833 to 1878.

Clitz, Henry B. (1824–88): Graduated from the USMA in 1845. He was wounded and captured at the Battle of Gaines' Mill, Virginia, in 1862. After being paroled he served at the academy as commandant of cadets and instructor of artillery, infantry, and cavalry tactics from October 1862 to July 1864.

Coal Knob: Located fifteen miles from Lewisburg, West Virginia, on the Kanawha Pike, in Tucker County.

Cobb, Howell (1815–68): Democratic U.S. congressman from Georgia. He served as secretary of the treasury, 1857–60. He chaired the Montgomery, Alabama, convention that established the Confederacy, and he served in the Confederate army from 1862 to 1865.

Col Baker: *See* Baker, Edward Dickinson.

collodion: A yellowish, viscous, highly flammable solution of paroxylin in ether and alcohol; used in the manufacture of photographic film, in engraving and lithography, and in medicines.

Columbiad: A large, cast-metal, smoothbore U.S. cannon developed by Colonel

George Bomford and first used in the War of 1812. The 8-inch Columbiad was 124 inches long and weighed 9,240 pounds.

Columbian College: Predecessor of George Washington University, Washington, D.C.

Columbus: Located in Muscogee County, Georgia, the town was the home of Henry Meigs, uncle of JRM. *See* Meigs, Henry Vincent.

Comstock, Cyrus Ballou (ca. 1831–1910): Graduated from the USMA in 1855 and was assigned to the Corps of Engineers. He served as aide-de-camp to Ulysses S. Grant from March 29, 1864, to July 15, 1866.

Conrad's Ferry: Crossed the Potomac River between Montgomery County, Maryland, and Loudoun County, Virginia.

Conway, Joseph E.: JRM wrote to General Benjamin F. Kelley on February 26, 1864, requesting that Lieutenant Joseph E. Conway, First New York Veteran Cavalry Regiment, be appointed acting assistant engineer and detailed to duty at the headquarters of the First Division of the Department of West Virginia. His request was approved. Conway was killed in action at Martinsburg, West Virginia, on July 25, 1864.

Coolidge, Richard H. (ca. 1820–66): Born in New York and appointed assistant surgeon in the U.S. Army in August 1841. He served as attending surgeon to army personnel in the Washington, D.C., area from May 1854 to December 1856 and was ordered to return to duty in Washington in June 1858.

Corlies, Benjamin F.: The husband of Susan Meigs, a distant cousin of MCM, Corlies was arrested on September 12, 1861, by the Federal marshal in New York City because he had lithographed and printed bills for the Bank of Pittsylvania, Chatham, Virginia, on the order of a suspected Confederate agent. Corlies's business partners testified to his loyalty, and Secretary of State Seward ordered his release on condition of his taking the oath of allegiance to the Union, which he did on September 18, 1861.

Cousin Eliza: Wife of Christopher Raymond Perry Rodgers.

Cousin Fanny: Frances Henley Higbee.

Cousin Hen: *See* Smith, Henrietta Elizabeth Henley.

Cousin Raymond: *See* Rodgers, Christopher Raymond Perry.

Cox, Samuel Sullivan "Sunset" (1824–89): A lawyer and newspaper editor, Cox was a Democratic member of Congress from Ohio from 1857 to 1865 (and later a Democratic member from New York). He defended the conduct of the war by General McClellan in a speech on the floor of the House of Representatives on January 31, 1862.

Coyle, Harriet (b. ca. 1820): Wife of Leonidas Coyle. Mr. Coyle was a partner in the hardware firm of Campbell and Coyle. The Coyles lived at 449 E Street, between Sixth and Seventh Streets, Washington, D.C.

Coyle, Harriet L. "Hattie" (b. ca. 1841): Oldest child of Leonidas and Harriet Coyle. In her letter of October 18, 1859, to JRM (MCMP, Ms. Div., LC), the elder Louisa Rodgers Meigs described Hattie Coyle as looking "quite pretty, more genteel than I ever saw her."

Cozzens Hotel: Lincoln registered at this hotel in West Point, New York, at 3 A.M., June 24, 1862.

Craige, Francis Burton (1811–75): An editor, lawyer, and legislator, Craige was elected from North Carolina as a Democrat to the U.S. House of Representatives and served from 1853 to 1861. He was a delegate to the North Carolina secession convention and to the Confederate Provisional Congress.

Craige, James Alexander (1841–1914): Entered the USMA in June 1860 but resigned on April 22, 1861. He subsequently served as a commissioned officer in the Confederate army and was badly wounded at the Third Battle of Winchester, September 19, 1864.

Crawford, Thomas (1813–57): Commissioned by MCM to design the marble pediment and bronze doors of the Capitol and a statue of Armed Freedom for the dome.

Crook, George (1818–90): Graduated from the USMA in 1852. He was in command of the Kanawha District, West Virginia, from February to June 1864 and was subsequently in command of the troops of the Department of West Virginia, July–August 1864.

crossed lines: Some individuals wrote across a previously written page, perpendicular to the other writing, thus making it difficult to read the page in either direction.

Crozet, Claudius (1790–1864): A professor at the USMA in the 1820s. He worked as an assistant to MCM on the work of the Washington Aqueduct. He was at West Point in January 1860 as a member of a board appointed to revise the instruction program.

Cullum, George W.: Author of *Systems of Military Bridges* . . . . which was formally published as no. 4 of the professional papers of the U.S. Corps of Engineers, with the title *Description of a System of Military Bridges* (1863).

Cumberland: A Union sloop sunk by the C.S.S. *Virginia* on March 8, 1862.

Custer, George Armstrong (1839–76): Graduated from the USMA in 1861. He was assigned to the cavalry. He was in command of the Third Division, cavalry corps, under Sheridan, in the Shenandoah campaign, from October 2, 1864, to March 26, 1865.

Cuyler, James Wayne (1840–83): Graduated from the USMA in 1864.

Davies, Thomas Alfred (1809–99): Graduated from the USMA in 1829. He served on the Wisconson frontier but resigned in 1831 to accept a civil engineer position. After the Battle of Bull Run, he served on duty in the defenses of Washington, was made brigadier general, U.S. Volunteers, in 1862, and served in Kentucky, Missouri, Kansas, and Wisconsin. He was brevetted a major general in August 1865.

Davis, Jefferson (1808–89): U.S. senator from Mississippi prior to serving as president of the Confederate States of America, 1861–65.

Delafield, Richard (1798–1873): Graduated from the USMA in 1818. He served as superintendent of the academy from September 8, 1856 to March 1, 1861, for all but six days. He was commandant of the engineer corps, 1864–66.

Denison, Jerusha Butler: Married Gideon Denison in 1780; she made her home with Commodore and Mrs. John Rodgers after the death of her husband.

De Russy, Rene Edward (ca. 1790–1865): Graduated from the USMA in 1812. He was a member of the Corps of Engineers. He served as superintendent of the academy from 1833 to 1838 and was in command of the Corps of Engineers at headquarters in Washington, D.C., from December 1858 to January 1861.

document numbers 20.29.22: Mentioned in a letter from MCM to JRM, April 7, 1860, this apparently refers to the following documents published in March 1860 by the U.S. Senate, 36th Congress, 1st Session: Senate Executive Document No. 20, Serial 1031, Message of the President . . . communicating . . . information in relation to the heating and ventilating of the Capitol Extension . . . , Jan. 25, 1860; Senate Miscellaneous Document No. 29, Serial 1038, Letter from the Superintendent of the Capitol Extension . . . in relation to the Dome and porticos of the Capitol, February 29, 1860; and Senate Executive Document No. 22, Serial 1031, Message of the President . . . communicating . . . information in relation to the marble columns for the Capitol Extension March 16, 1860.

"Don Juan": A musical arrangement for piano of an aria or arias from the opera *Don Giovanni* by Wolfgang Amadeus Mozart; mentioned by Hattie L. Coyle in a December 1859 letter to JRM.

Doolittle, James Rood (1815–97): Elected as a Republican from Wisconsin to the U.S. Senate in 1857 and reelected in 1863.

"Do they think of me at home?": A song mentioned by JRM in a letter to his mother, September 8, 1864, and popular with both Union and Confederate forces. The words were written by Caroline Atherton Briggs Mason (1823–90) and set to music by Sidney Martin Grannis (1827–post-1884); the song was published by Oliver Ditson and Company, Boston, in 1852.

Draco or Dracon: A late seventh-century B.C. Athenian statesman noted for a severe code of laws.

Dr. Johnson: Samuel Johnson (1709–84), the English author.

Droop Mountain: The battle of Droop Mountain (Pocahontas County, West Virginia) was fought November 6, 1863, during Averell's second raid against Lewisburg, West Virginia.

Dry Tortugas: A group of islets or keys in the Gulf of Mexico, about seventy miles west of Key West, Florida.

Dunaway, Wayland Fuller: A freshman at Columbian College in 1859 from Lancaster Court House, Virginia.

Dupont, Rear Admiral Samuel Francis (1803–65): Commander of the South Atlantic Blockading Squadron, September 1861 to July 1863, when he asked to be relieved following an unsuccessful attack upon Charleston.

Early, Jubal Anderson (1816–94): Graduated from the USMA in 1837 and assigned to the artillery. He resigned in 1838 but served again in the Mexican War, returning to civilian life in 1848. In 1861, he entered the Confederate army as colonel of the Twenty-fourth Virginia Infantry regiment; he was subsequently promoted brigadier, then major, and later lieutenant general. In 1864 he commanded the Second Corps of Lee's army.

Eastman, Chris: Daughter of Seth Eastman (1808–75), who graduated from the USMA in 1829.

Eastman, Robert Langdon (ca. 1836–65): Graduated from the USMA in 1861 and was assigned to the infantry. He was brevetted captain for gallant and meritorious service at the Battle of Malvern Hill, Virginia, and served as assistant professor of drawing at the academy from September 1862 to March 1865.

Edmond, Sarah H. (b. ca. 1817): Like Mrs. Leonidas Coyle, Sarah Edmond was born in Vermont. She was living at the Coyle residence in Washington, D.C., at the time of the 1860 census. "Aunt Sarah" of JRM's friend Harriet "Hattie" L. Coyle.

Elfers, Henry (ca. 1843–93): A member of Company D, Twenty-eighth Ohio Infantry Regiment. He was wounded at the battle of Princeton, West Virginia, on May 17, 1862, but returned to his regiment after hospitalization and furlough. He was mustered out of service with his regiment at the conclusion of his three-year enlistment, on July 23, 1864.

Emlen: *See* Meigs, Samuel Emlen.

enemy refugees and deserters: JRM interviewed at least three refugees and two deserters around April 23, 1864. They provided general information on Confederate troops and presumed knowledge about Union troops.

Ericsson, John (1803–89): A Swedish-born engineer and inventor who became a U.S. citizen. He designed and built the U.S.S. *Monitor.*

evacuation of Manassas: President Jefferson Davis had directed Joseph E. Johnston on February 20, 1862, to prepare to withdraw the Confederate forces from northeastern Virginia. Johnston began to reposition the troops from Fairfax and Prince William counties to new sites south of the Rappahannock River on March 9, 1862.

Eyre, Ellen: Daughter of the Philadelphia banker Manuel Eyre. She was the wife of Charles Bell Gibson.

Falling Waters: A village on the Potomac River in Berkeley County, West Virginia, and the area across the river in Washington County, Maryland, where there was a bridge over the Chesapeake and Ohio Canal to a ferry landing for the crossing of the Potomac. Union forces had a bridge there in 1863 but destroyed it at the time of the Gettysburg campaign; however, the Confederates succeeded in recrossing there on their retreat in July 1863. General Franz Sigel directed that a pontoon bridge be laid at Falling Waters prior to his ill-fated campaign up the Shenandoah Valley in May 1864.

falls: Great Falls on the Potomac River, Montgomery County, Maryland, about ten miles upstream from the boundary line between Maryland and the District of Columbia.

Fanny: Probably a reference to Frances Henley Higbee.

"Fantasie from Charles 5th": Refers to *Fantaisie Brillante,* arranged by Henri Herz (1803–88), based on the opera *Charles V* by Jacques Francois Fromental Halevy.

Farnsworth, Charles (ca. 1836–67): Severely wounded near Moorefield, [West] Virginia, in April 1862, Farnsworth rejoined his company after convalescent leave and was promoted to major on March 30, 1863. Captured in the skirmish at Harpers Ferry, he was held in Libby Prison, Richmond, Virginia, until exchanged and paroled in March 1864.

F.B.: *See* Schaeffer, Francis B.

Field, Josiah Horace Vincent (ca. 1842–64): Assigned as chief of ordnance, Department of West Virginia, May 10, 1864, to July 16, 1864, the date of his death from disease at Cumberland, Maryland.

Fingal: A British-built, schooner-rigged, iron-screw steamer that sailed from Greenock, Scotland, on October 10, 1861, with a cargo of arms, powder, and other ordnance stores. It successfully evaded the Union blockading force and arrived at Savannah, Georgia, on November 13, 1861, but was blockaded there. In September 1862 the vessel was converted to an ironclad gunboat and ram and renamed the C.S.S. *Atlanta.* Engaged by the U.S. monitors *Weehawken* (Captain John Rodgers) and *Nahant* (Commander John Downes) in Wassaw Sound, Georgia, on June 17, 1863, the *Atlanta* surrendered after being struck by four of the five shots fired by the *Weehawken.* It was subsequently towed to Port Royal, South Carolina, and repaired. As the U.S.S. *Atlanta,* the steamer participated in the James River campaign in 1864.

Flag: Formerly the *Phineas Sprague,* the U.S.S. *Flag* was purchased in Philadelphia on April 26, 1861, and commissioned on May 28, 1861. It was 195 feet long, with a beam of 30 feet, 10 inches. In 1865, it was decommissioned and sold.

Flirtation: Refers to Flirtation Walk, a wooded promenade on the academy campus.

Flora Temple: A celebrated race horse.

Floyd, John Buchanan (1806–63): A lawyer and legislator, Floyd served as governor of Virginia from 1849 to 1852, and as secretary of war from 1857 to December 29, 1860, when he resigned over President Buchanan's refusal to order the evacuation of Fort Sumter. He raised and commanded a brigade for the Confederate army and was commissioned a major general by the Virginia Assembly.

flying artillery: In January 1861, while serving as instructor at the academy, Charles Griffin (1826–67), who graduated from the USMA in 1847, was placed in command of an artillery company. The unit consisted of 4 guns, 69 horses, 70 enlisted men, and 3 additional officers. It was ordered to Washington by the War Department.

Foote, Andrew Hull (1806–63): Commander of naval operations on the Upper Mississippi from 1861 to 1862; relieved because of ill health.

foreign visitors: Mentioned by MCM in a June 1, 1860, letter to JRM, these were eighteen Japanese officials and fifty-three servants who visited Washington, D.C., from May 14 to June 7, 1860, to exchange ratifications of the Treaty of Amity and Commerce signed in Japan by Townsend Harris, as U.S. envoy, on July 29, 1858.

Forsyth, George A. (1837–1915): Mustered into Federal service in September 1861, as a first lieutenant, in the Eighth Illinois Cavalry Regiment. He was promoted to captain, Company G, in February 1862, and to major in September 1863. He was detached from his regiment for duty on General Sheridan's staff in July 1865, which position he retained until May 1865.

fort at the Tortugas: As part of the U.S. coastal defenses, army engineers in 1846 began the construction of Fort Jefferson on Garden Key in the Dry Tortugas coral keys to control navigation in the Gulf of Mexico. The fort was never completed.

Fort Donelson: Union Army forces under Ulysses S. Grant with the assistance of naval forces under Andrew H. Foote captured Fort Donelson on the Cumberland River in Tennessee, February 12–16, 1862.

Fort Hatteras: A North Carolina fort that surrendered to Union forces following shelling by naval vessels, August 28, 1861.

Fort Lafayette: A coastal defense fort two hundred yards off shore adjacent to Brooklyn, New York, on Hendricks Reef; it was started as Fort Diamond in 1812 and renamed Fort Lafayette in 1823. It was used as a Federal prison during the Civil War.

Fort Pickens: Located at Santa Rosa Island, Florida, the fort was built from 1829 to 1834 to control the entrance to Pensacola Bay. Its occupation by Union forces during the spring of 1861 denied the Confederates on the mainland the use of Pensacola harbor.

Fort Pike: Located in Louisiana, thirty miles east of New Orleans, this fort was constructed between 1819 and 1827 to guard the nine-mile-long pass connecting Lakes Pontchartrain and Borgne. It was seized by Louisiana troops on January 10, 1861.

Fort Put: Refers to Fort Putnam, on Crown Hill overlooking West Point and the Hudson River. It was built during the Revolutionary War to protect the defenses along the river from an enemy assault from the south and west.

Fort (Zachary) Taylor: Located at Key West, Florida, the fort was a double-casemated brick fortress, construction of which had begun in 1845. It was situated on a sandy shoal about a quarter of a mile offshore, commanding the water approaches to the town.

Fothergill: A process introduced by Thomas Fothergill, it involved coating a wet collodion photographic glass plate with albumen, which was then washed off, leaving just enough material in the collodion base to allow action by the developing agent.

found: Cadet slang for being found academically unproficient and hence rejected for admission or dismissed from the academy.

Fountain Hotel: Located on Light Street near Baltimore Street in Baltimore, Maryland.

Franklin: A town in Pendleton County, West Virginia.

Franklin, William Buel (1823–1903): Graduated from the USMA in 1843. He was assigned to the topographical engineers corps and promoted to captain July 1, 1857, for fourteen years' continuous service. He succeeded MCM as superintending engineer in charge of the Capitol extension and of the Post Office extension in November 1859.

Fred (letter of September 13, 1859): *See* Bailey, Frederick William.

Fred (letter of October 7, 1859): *See* Trowbridge, Frederick C.

Fred Rodgers: *See* Rodgers, Frederick.

Freeman, Charles A. (1837–1902): Mustered into service as first lieutenant, Company F, First [West] Virginia Infantry, at Camp Carlisle, Wheeling, West Virginia, on October 29, 1861. Freeman was captured by the Confederates in December 1861; he was exchanged in November 1862 and rejoined his regiment. General Kelley named him an aide-de-camp on the Department of West Virginia staff in February 1863. He remained on the staff as aide-de-camp or acting assistant adjutant general until he was honorably discharged in 1864.

Freeman, F. Major: A junior at Columbian College in 1858 from Frankfort, Kentucky.

Frémont, John Charles (1813–90): Appointed to the topographical engineers corps in 1838, he established a reputation as an explorer of the West and was a major participant in the conquest of California from Mexico. Appointed major general, U.S. Volunteers, in 1861, he commanded the Western Department but was subsequently relieved in November of that year because of differences with the administration over policy toward the abolition of slavery. Placed in command of the Mountain Department (of western Virginia) in 1862, he was out-generaled by Thomas J. Jackson in the spring and again relieved of command and spent the rest of the war awaiting orders.

French, Clara: A daughter of John W. French, chaplain of the USMA.

French, John W. (d. 1871): Graduated from USMA in 1856. He served there as chaplain and as professor of geography, history, and ethics.

French, Mary: A daughter of John W. French, chaplain of the USMA.

French, William Henry (1815–81): Graduated from the USMA in 1837 and was assigned to the artillery. He was promoted to major general, U.S. Volunteers, in November 1862. He was in command of a division in the Middle Department, Eighth Army Corps, and was commander of the Harpers Ferry District, June 27–30, 1863. French was guarding lines of communication along the Potomac River early in July 1863. On July 7 his troops were transferred to the Third Corps, Army of the Potomac, and he was placed in its command.

French Zouaves: Troops of Napoleon III who wore a mixture of African and European dress. They fought in the Crimean War in every battle except Balaclava. They were considered fierce, disciplined, and elite.

Fristoe, Edward T. (1830–92): Graduated from the Virginia Military Institute in 1849 and received a graduate degree from the University of Virginia in 1855. He was professor of mathematics at Columbian College from 1855 to 1860. At the beginning of the Civil War he was professor of mathematics and head librarian at the University of Missouri, when he left the university to join the Confederate Army, rising to the rank of colonel. He returned to Washington, D.C., as professor of chemistry at Columbian University in 1865.

Frostburg: A town in Allegany County, Maryland.

Fulton, Robert (1765–1815): An engineer and inventor, Fulton had worked on the development of submarine mines and torpedoes before building the successful steamboat *Clermont* in New York.

Galena: An ironclad screw steamer, the U.S.S. *Galena* was launched at Mystic, Connecticut, on February 14, 1862, and commissioned at the New York Navy Yard on April 21, 1862. Captain John Rodgers, the brother of Louisa Rodgers Meigs, was given command of the *Galena* on April 24 and joined the naval squadron in Hampton Roads in support of General McClellan's Peninsula campaign.

Gardner, Robert Skiles (ca. 1839–98): A captain and assistant quartermaster, U.S. Volunteers, at Clarksburg, West Virginia. He enlisted as a sergeant in Company F, Twenty-third Ohio Infantry Regiment, in June 1861, and promoted to regimental quartermaster. His promotion to captain and assistant quartermaster, U.S. Volunteers, was dated June 18, 1862. He was promoted brevet major in March 1865 and mustered out of service in March 1866.

Garrard, Kenner (1827–79): Graduated from the USMA in 1851. He was assigned to the artillery and later served in the cavalry. He held various positions at the academy; he was subsequently a major general, U.S. Volunteers, and a major general in the U.S. Army.

Garrington, James A.: Born in Colburg, Ontario, Canada, Garrington enlisted at Pittsburgh, Pennsylvania, in December 1861 for three years.

Gaskell, William Penn (b. ca. 1837): Mustered into service on August 9, 1862, at Philadelphia as a second lieutenant in Captain Wrigley's Independent Company, Pennsylvania Infantry (Acting Engineers). He was promoted to first lieutenant on December 16, 1862, and to captain and commanding officer of the company on March 30, 1863. He entered the U.S. General Hospital at Annapolis, Maryland, because of illness in June 1864 and was honorably discharged because of physical disability on July 5, 1864.

Gell, Sir William (1777–1836): A British traveler and archaeologist who published *Pompeiana: The Topography, Edifices, &c* (1817–19) and the two-volume *Pompeiana: The Topography, Ornaments, &c* (1832), on the results of his and others' excavations at Pompeii.

General Scott: *See* Scott, Winfield.

Gibson, Charles Bell: The son of William Gibson, Charles Gibson was a graduate of the University of Pennsylvania and became a surgeon and professor of surgery at Hampden-Sydney College. He served briefly as surgeon general for Virginia at the beginning of the Civil War and later was head of the Officer's Hospital in Richmond.

Gibson, Charles Henry (1840–1911): Graduated from the USMA in 1861. He served in the cavalry during the Civil War.

Gibson, William (1788–1868): A surgeon and professor of surgery at the Medical School of the University of Pennsylvania.

Gibson, William Eyre (1841–?): Admitted to the USMA in 1859 but resigned on April 22, 1861, to fight for the Confederacy.

Gillespie, George L. (1841–1913): Graduated from the USMA in 1862 and was assigned to the Corps of Engineers. He was in command of an engineer company in the Army of the Potomac from 1862 to 1863.

Gilmore Street: A north-south street running between Carroll and Druid Hill Parks, east of and parallel to Fulton Avenue (modern U.S. Route 1), Baltimore, Maryland.

Goddard, Elizabeth Cass Ledyard (1840–1918): Eldest daughter of Henry Ledyard and Matilda Cass Ledyard, she married Francis Wayland Goddard of Providence, R.I., on April 9, 1862.

Graff, Frederic (1817–90): An engineer, Graff succeeded his father as chief engineer of the Philadelphia water department.

Graff, Frederick (1774–1847): Civil engineer and superintendent of the Philadelphia water system, 1805–47.

Grafflin: The firm J. C. Grafflin and Company (Baltimore Bag Factory) was located at 75–77 South Street, Baltimore, Maryland.

grand order: Refers to Lincoln's General War Orders No. 1, January 27, 1862, and No. 2 and No. 3, March 8, 1862, in particular.

Grandin, J. Marsdin: Chaplain, Thirty-third Virginia Infantry; captured near Bolivar Heights.

Grandmother (letter of September 13, 1859): *See* Rodgers, Minerva Denison.

Greeley, Horace (1811–72): Editor of the radical antislavery *New York Tribune.*

Greene, Oliver Duff (ca. 1833–1904): Graduated from the USMA in 1854 and was assigned to the artillery. He served as first lieutenant in command of a battery of the Second Artillery that supported Richardson and Davies along Bull Run. He was appointed assistant adjutant general in 1861.

Green Spring Run: Located in Mineral County, West Virginia.

guns of the 3rd of June: Battle of Cold Harbor, Hanover County, Virginia, June 1–3, 1864. The major assault was made in the early morning of June 3, 1864.

Gurley, John Addison (1813–63): A Universalist Church pastor, a newspaper owner, and editor. He served as Republican member of Congress from Ohio from 1859 to 1863. His speech attacking the leadership of General McClellan in the U.S. House of Representatives, on January 29, 1862, evoked the reply from Congressman Samuel S. Cox.

Gwin, Willie: Possibly a son of Senator William Gwin (1805–85) of California.

Gynn Falls: A stream flowing through northwestern and southwestern Baltimore, emptying into the Middle Branch of the Patapsco River, Baltimore Harbor.

Hale, John Parker (1806–73): Elected to the U.S. Senate from New Hampshire as an antislavery candidate in 1847; he served until 1853. He was reelected to the Senate in 1855 and 1859.

Hall, Ellen (b. ca. 1841): Daughter of David Aiken Hall (b. 1795), a Washington lawyer. The Halls lived at the northwest corner of Third and C Streets, Washington, D.C.

Hall, Mary Louise Anderson Alexander: Wife of James Hall (1793–1868), author, attorney, judge, and banker. Their son, James Harrison Hall (ca. 1845–93), graduated from the USMA in 1865.

Halleck, Henry Wager (1815–72): Graduated from the USMA in 1839. He was a member of the Corps of Engineers from 1839 to 1854. As a second lieutenant, he served as assistant professor of engineering at the academy from July 6, 1839 to June 28, 1840. When he resigned in 1854, he began to practice law. President Lincoln appointed Halleck a major general in the U.S. Army on August 19, 1861. He was in command of the Department of the Missouri, the Department of the Mississippi, and subsequently general in chief of the Armies of the United States. In March 1864, he was succeeded in that post by Ulysses S. Grant. Thereafter Halleck served as chief of staff, U.S. Army.

Hamanassett: A rural holding of about thirty-seven acres in Delaware County, Pennsylvania, about eighteen miles from Philadelphia, that Dr. Charles D. Meigs acquired in 1856. On it he built a house, barn and stable, tenant house, spring house, ice house, and workshop.

Hardee, William Joseph (1815–73): Graduated from the USMA in 1838. He was commandant of cadets at the academy from July 1856 to September 1860 and instructor of infantry tactics from July 1856 to September 1860. He resigned from the U.S. Army and served first as brigadier, then as major, and finally as lieutenant general in the Confederate army.

Harman, Asher Waterman: Colonel in the Twelfth Virginia Cavalry.

Harper's Weekly: This publication carried a picture of "Old John Brown" on November 5, 1859, p. 712. For the woodcut copy of the Washington statue, see the issue dated Feb. 25, 1860, pp. 113–14.

Harry: *See* Ledyard, Henry Brockholst "Harry."

Hascall, Herbert Austin (ca. 1835–90): Graduated from the USMA in 1856. He was assistant professor of mathematics there from January 1860 to 1861.

Hassler, Charles William: Originally from Washington City, Hassler was listed as member of the senior class in the Columbian College catalog for the years 1858–59.

Hatcher, Thomas Clinton Lovett (ca. 1839–61): A resident of Loudoun County, Virginia, Hatcher was a member of the junior class at Columbian College, 1858–59. The *Democratic Mirror* (Leesburg, Virginia) reported in March 1861 that Hatcher attended a reception following Lincoln's inauguration where Lincoln said to him: "Stranger, I never allow anyone taller than myself to pass me unchallenged. I acknowledge myself beaten." Hatcher died in the Battle of Ball's Bluff.

Haupt, Herman (1817–1905): Graduated from the USMA in 1835. Haupt was in charge of transportation and construction of U.S. military railroads.

Hayes, Rutherford B. (1822–93): lawyer, statesman, and president of the United States.

H.C.: *See* Coyle, Harriet L.

Henderson, Thomas, M.D. (1789–1854): An assistant surgeon in the U.S. army from 1833 to 1854 and sometime professor of medicine at the Columbian College, Washington, D.C. A new edition of his book, *Hints on the Medical Examination of Recruits of the Army . . . as revised by himself and Dr. Richard H. Coolidge, Assistant Surgeon, U.S. Army,* was published in 1856. The pagination of the 1856 revised edition of Henderson's book agrees with the mention by MCM of "Marshall," p. 113.

Henley: *See* Smith, J. Henley.

Henry, Joseph (1797–1878): A scientist who conducted research in the field of electromagnetism. He served as the first secretary of the Smithsonian Institution, 1846–78. He also served as president of the National Academy of Sciences.

Higbee, Frances Henley: The wife of the Reverend Edward Higbee. Her mother was Eliza Denison Henley (1788?–1856), the sister of Minerva Denison Rodgers, JRM's grandmother.

Hilgard, Julius Erasmus (1825–91): Born in Bavaria, Hilgard studied civil engineering in Philadelphia and became an assistant to Alexander E. Bache, the organizer and head of the U.S. Coast Survey. During the Civil War, Hilgard was responsible for supplying the army and navy with topographical, geographical, geological, and tidal information.

Hodge, Mrs. J. Ledyard: Sister-in-law of Ann Rodgers.

Holabird, Samuel Beckley (ca. 1826–1907): Graduated from the USMA in 1849 and was assigned to the infantry. He was adjutant to Richard Delafield at the academy from September 1859 to May 1861.

Holgate, Asa H. (1838–80): Graduated from the USMA in 1863. He was assigned to the Corps of Engineers.

Holt, Joseph (1807–94): Served briefly as secretary of war for President Buchanan after the resignation of John B. Floyd on December 29, 1860, and as judge advocate general from 1862 to 1875.

Hooker, Joseph (1814–79): Graduated from the USMA in 1837. He was in command of the Army of the Potomac from January 1863 to June 1863.

Hookstown Road: This may refer to the modern Reisterstown Road (modern Maryland Route 140).

house at Arlington: An estate and house on the Virginia shore of the Potomac River opposite Washington, D.C. The house was built by George Washington Parke Custis, the adopted son of George Washington. Mary Randolph Custis Lee, the wife of Robert E. Lee, inherited a life interest in the estate upon the death of her father. The Lees left the estate in May 1861. Most of George and Martha Washington's furniture and china were transferred to the U.S. Patent Office for safekeeping in January 1862. The federal government acquired title to the Arlington estate, and at the instigation of MCM, the War Department began interring Union soldiers there in May 1864, thus establishing Arlington National Cemetery.

Howell, Charles W. (1842–82): Graduated from the USMA in 1863. He was assigned to the Corps of Engineers.

Hunt, Henry Jackson (1819–89): Graduated from the USMA in 1839. He commanded part of the troops that reinforced Fort Pickens. He commanded the artillery battery that supported Colonel Israel Richardson's brigade near Blackburn's Ford on Bull Run, July 21, 1861. He was actively engaged in many battles, including the Battle of Gettysburg.

Hunt, Mary B. (Craig): The second wife of Henry Jackson Hunt and sister of Presley O. Craig (ca. 1834–61), who was killed at Blackburn's Ford, at the Battle of Bull Run.

Hunter, David (1802–66): Graduated from the USMA in 1822. He succeeded Franz Sigel as commander of the Department of West Virginia in May 1864 and served until August 8, 1864.

Huntington, Adoniram Judson (1818–1903): A tutor in Latin and Greek, 1843–46; professor of Latin and Greek, 1846–49, 1852–59, 1866–82; and professor of Greek, 1882–1900, all at Columbian College.

Hyer, George: Hyer and his wife were from Madison, Wisconsin. They were the parents of third-classman Joseph Keyes Hyer (ca. 1842–82), who graduated from the USMA in 1865.

Imboden, Francis (Frank) Marion (1841–1929): Served as a volunteer aide-de-camp in Henry A. Wise's legion of western Virginia troops and was captured by Union Forces at the battle of Roanoke Island, North Carolina, in February 1862. He was exchanged and subsequently served as captain, Company H, First Virginia Partisan Rangers (later the Eighteenth Virginia Cavalry Regiment), commanded by his brother, John Daniel Imboden. Captured near Piedmont, Virginia, by Union forces on June 5, 1864, Francis Imboden remained in a prisoner-of-war camp until June 14, 1865, when he was released after taking an oath of allegiance.

Imboden, John Daniel (1823–95): Lawyer, local and state government official. He organized and commanded the Staunton Artillery Battery during the first year of the war. Commissioned colonel on May 7, 1862, he recruited and commanded the

First Virginia Partisan Rangers, the nucleus for what became the Northwestern Brigade with Imboden as brigadier general. Imboden's command at times operated independently in the Shenandoah Valley and eastern West Virginia, but at other times formed part of larger forces commanded by Generals Breckinridge and Early.

Indpt Co PV: Independent Company Pennsylvania Volunteers (Acting Engineers), mustered into Federal service at Philadelphia on August 9, 1862, and mustered out at Harpers Ferry on June 20, 1865. The company was assigned to various commands in the Department of West Virginia during much of its service.

insane asylum: The Hospital of the Insane of the Army, Navy, and the District of Columbia, later called St. Elizabeths Hospital. It opened its doors to patients in 1855.

iron bridge: This bridge was replaced by the Pennsylvania Avenue Bridge, NW, subsequently called Montgomery C. Meigs Bridge.

Irving, Washington (1783–1859): American author.

Jackson statue: The equestrian statue of Andrew Jackson in Lafayette Park, Washington, D.C., by Clark Mills.

Jackson, William Lowther "Mudwall" (1825–90): The second cousin of General Thomas J. "Stonewall" Jackson, he was a lawyer and a public official in western Virginia. He enlisted as a private in the Confederate army in 1861 but soon rose to the rank of colonel of the Thirty-first Virginia Infantry Regiment and later served on T. J. Jackson's staff. He subsequently recruited the Nineteenth Virginia Cavalry within Federal lines in West Virginia and was elected its colonel. Part of the brigade of General A. G. Jenkins, this unit under Jackson was subsequently active in the Shenandoah Valley campaign of 1864.

Jennie: *See* Rodgers, Jane Perry.

Jillson, William E.: Professor of rhetoric, belles lettres, and modern languages in 1859 at Columbian College.

Johnson, Reverdy (1796–1876): Whig senator from Maryland, 1845–49; attorney general of the United States, 1849–50; state senator in Maryland, 1860–61. Johnson was counsel for the defense in the Dred Scott case.

Johnston, Joseph Eggleston (1807–91): Graduated from the USMA in 1829 and was appointed to the artillery. He resigned from the U.S. Army and was appointed a major general in Virginia state forces and a brigadier general in the Confederate army. He commanded the Confederate forces at the Battle of Bull Run after joining Beauregard and was in command of Confederate forces in northern Virginia after the battle.

J.O.K.: James Oscar Kirk, a junior at USMA in 1859 from Lancaster County, Virginia.

Jones, William Edmonson "Grumble" (1824–64): Graduated from the USMA in 1848. He was assigned to the mounted rifles, with service primarily on the frontier. He resigned in 1857. In May 1861 he was commissioned as a major in the Confederate forces; he was promoted to colonel in September 1861 and to brigadier general in September 1862. He served under J. E. B. Stuart at Gettysburg in 1863 and under Longstreet in eastern Tennessee in early 1864, returning to the Shenandoah Valley in the spring of 1864.

Kanawha River: The principal river in southwestern West Virginia, flowing past Charleston to the Ohio River.

Kate: *See* Rodgers, Catherine (or Katherine) Trowbridge.

Kelley, Benjamin Franklin (1807–91): A merchant and a freight agent for the Baltimore and Ohio Railroad in Wheeling, West Virginia. He raised a volunteer regiment for the Union (the first south of the Mason and Dixon line), was wounded at the Battle of Philippi, and was subsequently commissioned a brigadier general, U.S. Volunteers. After recovering from his wound, Kelley commanded troops responsible for protecting the Baltimore and Ohio Railroad in West Virginia. After the Army of Northern Virginia invaded Maryland in June 1863, the War Department redesignated Kelley as departmental commander of the newly created Department of West Virginia, with headquarters at Clarksburg, West Virginia. Major General Franz Sigel relieved General Kelley as departmental commander in March 1864, but retained Kelley on active duty in command of troops guarding the railroad.

Kelley, William B. (ca. 1839–1926): Served on detached duty as aide-de-camp to his father, Benjamin F. Kelley, from September 1861 until he was mustered out of service at the expiration of his three-year enlistment, on November 30, 1864.

Kendrick, Henry Lane (ca. 1811–91): Graduated from the USMA in 1835. He was assigned to the infantry and, later, the artillery. He was assistant professor of chemistry, mineralogy, and geology at the academy from 1835 to 1847; and professor of chemistry, mineralogy, and geology from 1857 until shortly before his death.

Kenly, John Reese (1822–91): A Maryland lawyer who had served during the Mexican War, Kenly was commissioned as a colonel with the First Maryland Infantry Regiment in June 1861 and later promoted to brigadier general. He was in command of the Maryland brigade of French's division in the Middle Department, Eighth Army Corps. His troops held Harpers Ferry in June 1863.

Kershaw, Joseph B. (1822–94): A lawyer and legislator from South Carolina, Kershaw served in the Mexican War. He joined the Confederate army and rose to the rank of major general after service at Gettysburg and in battles in Virginia and Tennessee. After surrender, he was imprisoned at Boston before returning to South Carolina, where he served as a legislator and jurist.

Key West: An island and town, Key West is located at the southwestern-most point of the chain of keys that extend nearly one hundred fifty miles, curving around the tip of mainland Florida into the Gulf of Mexico.

King, William R. (1840–98): Graduated from the USMA in 1863. He was assigned to the Corps of Engineers.

Kirby's Lane: A short street running perpendicularly from Franklin Avenue, parallel to and east of Monroe and west of Fulton Streets (modern U.S. Route 1) in western Baltimore.

Knoxville: A town in Frederick County, Maryland.

Kurtz, John Daniel (ca. 1819–77): Graduated from the USMA in 1842. He was assigned to the Corps of Engineers. He served as assistant to the chief engineer, in Washington, D.C., from August 1861 to August 1866.

Lalie: *See* Bowman, Eulalie.

Larned, Lieutenant Colonel Sylvester: Member of the Second Michigan Volunteer Infantry Regiment. He was counsel for the accuser in the Dixon S. Miles court of inquiry, 1861.

Ledyard, Henry Brockholst "Harry" (1844–1921): Son of Henry Ledyard and Matilda Cass Ledyard. He graduated from the USMA in 1865 but resigned from the army in 1870 and was subsequently superintendent and then president of the Michigan Central Railroad. He was a student at a French preparatory school, the Charlier Institute, in New York City, prior to his appointment to USMA in 1861.

Lee, A. Nisbet (1843–79): A son of Captain Thomas J. Lee of the topographical engineers, he graduated from the USMA in 1865. He was tried by a general court-martial, found guilty, and sentenced to be dismissed from the academy; however, the punishment was mitigated to confinement to camp during the summer encampment.

Lee, Fitzhugh (1835–1905): Graduated from the USMA in 1856. At the beginning of the Civil War, Lee was appointed first lieutenant in the Confederate army. He rose quickly to the rank of brigadier general while fighting in major battles in Maryland and Virgiinia. He was the nephew of Robert E. Lee.

Lee, John Fitzgerald (ca. 1813–84): Graduated from the USMA in 1834. He was assigned to the artillery. He was judge advocate of the army from March 1849 to September 1862.

Lee, Robert E. (1807–70): Commander of the Army of Northern Virginia. Possibly the most revered general in American history though he served the Confederate cause in the Civil War.

Leo: Leonidas B. Coyle (ca. 1842–1906), younger brother of Harriet "Hattie" Coyle and later pastor of the Presbyterian Church, Bridgeton, New Jersey.

Leonard, Hiram (ca. 1805–83): Entered the U.S. army as a paymaster, U.S. Volunteers, in 1846 and was promoted to major paymaster on March 2, 1849, and lieutenant colonel and deputy paymaster general on September 6, 1862. In 1865 he was awarded brevet ranks of colonel and brigadier general. He retired from active duty in 1872.

letter to the *Tribune:* Secretary of War Edwin M. Stanton wrote to Charles Dana, managing editor of the *New York Tribune,* in February 1862, with his views about McClellan's credit for victory.

Lewis, Charles Edward (b. ca. 1844): Lewis was at the USMA in June 1861 but resigned on September 17, 1861, effective November 15, on account of "poor eyesight" strained by studying.

Lewisburg: A town in Greenbrier County, West Virginia.

Liberty Heights Avenue: An avenue (modern Maryland Route 26) running northwest from the intersection of Pennsylvania Avenue and Reisterstown Road near Druid Hill Park in northwestern Baltimore city.

life among the New Englanders: The elder Louisa Rodgers Meigs, her mother, one of her sisters, and her daughter Loulie had left Washington on July 9, 1864, and were spending the summer at Rye Beach, New York.

Lightburn, Joseph Andrew Jackson (1824–1901): Farmer, Union brigadier general, and Baptist clergyman, Lightburn served mostly in West Virginia during the war.

Lindenkohl, Adolph (1833–1904): Born and educated in Germany, Lindenkohl immigrated to the United States in 1852 and became a cartographer for the U.S. Coast and Geodetic Survey.

Little Falls Bridge: This bridge over the Potomac River was constructed during the early 1850s of heavy arch-reinforced timbers on tall stone piers. It connected the Chesapeake and Ohio Canal towpath in the District of Columbia to the Falls Bridge Turnpike in Alexandria (now Arlington) and Fairfax Counties, Virginia, at the location of the present-day "Chain Bridge."

Littell's Living Age: A weekly newspaper published in Boston beginning in 1844, it reprinted articles from British periodicals.

Lizzie: Elizabeth Cass Ledyard.

Lloyd's map: At least two maps of Virginia were prepared by James T. Lloyd and Henry H. Lloyd, respectively. One was entitled "Lloyd's official map of the state of Virginia. From actual surveys by order of the executive 1828 & 1829. Corrected and revised by J. T. Lloyd to 1861," colored, 34" x 47" (New York: J. T. Lloyd, 1861). The other was entitled "Lloyd's new war map of Virginia," colored, 16" x 27" (New York: H. H. Lloyd & Co., 1862). The Lloyds were located in New York City, J. T. Lloyd at 164 Broadway and H. H. Lloyd at 25 Howard Street.

Long Bridge: A bridge and causeways constructed across the Potomac River, connecting Maryland Avenue in the District of Columbia to the Columbia and Washington and Alexandria Turnpikes in Alexandria (now Arlington) County, Virginia, in close proximity to the present-day Fourteenth Street bridge.

Longstreet, James (1821–1904): Graduated from the USMA in 1842. With the onset of the Civil War, Longstreet was appointed brigadier general in the Confederate army. He served at the First Battle of Bull Run and subsequently served in several major battles, including Gettysburg. He was severely criticized for his failure of command at Gettysburg.

Loudoun Heights: Located east of Harpers Ferry and the Shenandoah River, Loudoun County, Virginia, and Jefferson County, West Virginia.

Loulie: *See* Meigs, Louisa Rodgers (1854–1925).

Louly: *See* Meigs, Louisa Rodgers (1854–1925).

Lynchburg: A town in Campbell County, Virginia.

Mackenzie, Ranald Slidell (1840–89): A nephew of Senator John Slidell of Louisiana, Mackenzie graduated from the USMA in 1862. He was assigned to the Corps of Engineers. He was brevet brigadier general, U.S. Army, and brevet major general, U.S. Volunteers, for Civil War service.

Macomb, Ann Minerva "Nannie" Rodgers (1824–1916): Younger sister of Louisa Rodgers Meigs. She married John Navarre Macomb (1811–89) in 1850.

Macomb, Christine Livingston: Newborn daughter of Ann Minerva "Nannie" Rodgers and John Navarre Macomb.

Macomb, John: Possibly John Navarre Macomb Jr.

Macomb, John N., Jr.: Applied for an appointment to the USMA in 1861 but did not receive one.

Macomb, Montgomery Meigs (1852–1924): Son of John Navarre Macomb and Ann Minerva "Nannie" Rodgers Macomb. He graduated from the USMA in 1874.

Mahan, Dennis Hart (1802–71): Graduated from the USMA in 1824. He was assigned to the Corps of Engineers and served from 1824 to 1832. He was at the academy (except for four years of detached service to study in France) as professor of engineering, from January 1832 to September 1871. He was the author of numerous textbooks on civil and military engineering.

Mama: *See* Coyle, Harriet.

map of the Upper Potomac: A manuscript map and a published map with insert entitled "Upper Potomac from McCoy's Ferry to Conrad's . . ." were prepared by the engineer department in 1863.

Markbreit, Leopold (ca. 1840–1909): A member of General Averell's staff.

Marriottsville: Located in Howard County, Maryland.

Marshall, Henry: Deputy Inspector of Hospitals, British Army, Marshall had written two volumes on this subject, *Hints to Young Medical Officers in the Examination of Recruits,* published in 1828, and *On the Enlisting, Discharging and Pensioning of Soldiers with the Official Documents on these Branches of Military Duty,* published in 1832.

Martin, George W. "Josh": Enlisted at Bealton, Virginia, on March 15, 1862, for three years or for the duration of the war. He was a prisoner of war and was paroled on May 6, 1865. At the time of his parole he was twenty-one years of age, stood 5 feet 11 inches tall, with dark hair and grey eyes, and had a florid complexion.

Martinsburg: Located in Berkeley County, West Virginia.

Marye, William A. (1840–1903): Graduated from the USMA in 1862. He was assigned to the ordnance corps. He retired from the U.S. Army as a colonel in the ordnance corps in 1902.

Maryland Heights: Located across the Potomac River from Harpers Ferry, West Virginia, in Washington County, Maryland.

Mason, James Louis (ca. 1817–53): Graduated from the USMA in 1836. He served in the engineers corps. He was a member of the board of engineers for fortifications on the Pacific Coast at the time of his death in San Francisco.

Massy: *See* Mrs. Granbury.

McCallum, Daniel Craig (1815–78): An architect, builder, and railroad engineer, McCallum was appointed director of military railroads on February 11, 1862.

McClausland, John (1836–1927): Taught mathematics at the Virginia Military Institute. He served in the Confederate army from 1861 to 1865.

McClellan, George B. (1826–85): Graduated from the USMA in 1846. He was assigned to the Corps of Engineers. He was appointed commanding general of the Army of the Potomac in August 1861, and general in chief of the Armies of the United States in November 1861. In November 1862, he was ordered to turn over the command of the Army of the Potomac to Ambrose E. Burnside.

McCoy's Ferry: Crossed the Potomac River between Washington County, Maryland, and Berkeley County, West Virginia.

McCrae, Tully (1839–1918): Graduated from the USMA in 1862. He was assigned to the artillery. He retired as a brigadier general, U.S. Army, in 1903.

McDaniel, John H.: A second-year student in the scientific course at Columbian College in the 1857–58 school year. He was from Saline County, Missouri.

McDermott, Patrick J. (b. ca. 1836): Mustered in as second lieutenant, Company A ("Capt. McDermott's Company"), Twenty-third ("Irish Regiment") Illinois Volunteer Infantry, on June 15, 1861; appointed first lieutenant, June 18, 1861, and acting captain, July 8 to September 21, 1861; held as prisoner of war, but paroled and promoted to captain, Company A, October 31, 1861. The Twemty-third Illinois Infantry Volunteer Regiment guarded the Baltimore and Ohio Railroad between Oakland, Garrett County, Maryland, and the village of New Creek, Mineral County, West Virginia.

McDowell, Irvin (1818–85): Graduated from the USMA in 1838. In 1861 he was assigned command of troops assembled south of the Potomac that became the Army of the Potomac. Criticized and relieved of command for his actions at Second Battle of Bull Run, he was exonerated by a court of inquiry. He was never afterward employed in the field.

McGinnes, John Randolph (1840–1918): Graduated from the USMA in 1863.

McIlvain, John (ca. 1827–70): Born in Belfast, Ireland, McIlvain was assigned to Company M, Sixth U.S. Cavalry, the regiment that provided the escort for General Sheridan's headquarters during the 1864 Valley campaign.

McKee, George Wilson (1841–91): Graduated from the USMA in 1863. He was assigned to the ordnance corps.

McKenzie. *See* Mackenzie, Ranald Slidell.

McMillan, James (ca. 1832–1906): Graduated from the USMA in 1856. He was assigned to the infantry. He served as Inspector General of Colonel Dixon S. Miles' division on July 21, 1861.

McNeill, John H. (1815–64): A Confederate partisan, he served in the Missouri campaigns of 1861–62. He was captured and imprisoned, but he later escaped to Virginia and organized a partisan band known as the McNeill Partisan Rangers. This independent group engaged Union forces in successful raids during the remainder of the war.

McVeigh, Townsend Jesse: A sophomore at Columbian College in 1859. He was from Loudoun County, Virginia.

Meade, George Gordon (1815–72): Graduated from the USMA in 1835. He was assigned to the artillery. He succeeded Joseph Hooker in command of the Army of the Potomac on June 28, 1863, three days before the Battle of Gettysburg.

Meigs, Ann Wilcocks Ingersoll: Wife of Dr. John Forsyth Meigs of Philadelphia; JRM's "Aunt Annie."

Meigs, Charles Delucena (1792–1869): The son of Josiah Meigs (1757–1822) and Clara Benjamin (d. 1849). Josiah was the brother of Return Jonathan Meigs (1740–1823) of Revolutionary War fame. Charles D. Meigs married Mary Montgomery (ca. 1792–1865) of Philadelphia in March 1815.

Meigs, Emily Skinner (1824–1905): Sixth child born to Dr. and Mrs. Charles D. Meigs; married Jonathan Williams Biddle in 1846. JRM's "Aunt Emily."

Meigs, Frank B. (1829–81): A younger brother of MCM.

Meigs, Harry Ingersoll (1847–1917): Son of Ann Wilcocks Ingersoll and Dr. John Forsyth Meigs of Philadelphia.

Meigs, Henry Vincent (1821–97): Fifth son of Charles Delucena and Mary Montgomery Meigs. He married Henrietta Hargreaves Stewart of Georgia, heiress to a cotton factory, and became manager of the factory. He subsequently served as a captain in the Confederate army. Henry and MCM did not reconcile their differences after the war.

Meigs, John Forsyth (1818–82): A younger brother of MCM. He was a physician in private practice in Philadelphia and a member of the college of physicians of Philadelphia Hospital.

Meigs, Louisa Rodgers (1817–79): The daughter of Commodore John Rogers and Minerva Denison Rodgers. Louisa married MCM on May 2, 1841. Though not particularly attractive physically, she had a strong personality and was known for her moral rectitude. For Louisa, hard work and self-improvement, belief in God, and the avoidance of idleness were the foundations of life. She and Montgomery were a loving couple who enjoyed horseback riding and attending social events. Their letters reveal their strong emotional ties with their families and close friends, and their devotion to their children.

Meigs, Louisa Rodgers (1854–1925): Youngest child of MCM and Louisa R. Meigs; known to her family as Loulie, or Louly.

Meigs, Mary Craythorne (b. 1838): Youngest child of Dr. and Mrs. Charles D. Meigs; JRM's "Aunt Mollie."

Meigs, Mary Montgomery (1843–1930): Oldest daughter of MCM and Louisa Rodgers Meigs, she was a frequent correspondent and a confident of her brother, JRM. At times coquettish in manner, she enjoyed going to the gymnasium and horseback riding. She married Joseph Hancock Taylor (1836–85) on March 30, 1864.

Meigs, Montgomery Cunningham (1816–92): A soldier and an engineer. His parents were Charles Delucena Meigs and Mary Montgomery Meigs. When Meigs was a child, his family moved from Georgia to Philadelphia. He entered the USMA in 1831. He graduated fifth in his class in 1836. At first assigned to the artillery, he transferred to the engineer corps. In addition to supervisory work on the General Post Office in Washington, D.C., he was responsible for the construction of the Washington Aqueduct, which included the construction of the Cabin John Bridge, and the building of the wings and the enlargement of the dome of the United States Capitol in the 1850s. For his prewar services and for his services at the beginning of the Civil War, he was promoted to the rank of brigadier general, in May 1861. He served as quartermaster general of the U.S. Army from June 10, 1861, to February 6, 1882. He was known for his integrity and honesty.

Meigs, Montgomery C., Jr. (1847–1931): Fourth child of MCM and Louisa R. Meigs, he studied engineering at the Lawrence Scientific School at Harvard University. Among other projects, he worked on improving the channel of the Mississippi River at Rock Island.

Meigs, Samuel Emlen (1828–1917): A younger brother of MCM, he applied for and received a commission as captain and assistant quartermaster, U.S. Volunteers, effective on November 14, 1861. He resigned and was honorably discharged on March 31, 1863, to return to his business affairs in Philadelphia. He married Cornelia, the daughter of William E. Rogers of Philadelphia.

Meigs, William Montgomery (1826–97): A younger brother of MCM.

Melvin, Thayer: A lawyer in civilian life, Melvin was mustered into Federal service as first sergeant, Company F, First [West] Virginia Infantry Regiment, at Wheeling on September 8, 1861. He was promoted to first lieutenant and transferred to General Kelley's staff on December 11, 1861. On July 14, 1862, he was promoted to captain and assistant adjutant general.

Merrimac: A wooden screw steamer, frigate rigged, built in 1855–56, the U.S.S. *Merrimac* had been scuttled and burned by U.S. forces at the Norfolk Navy Yard on April 20, 1861, but was raised and converted by the Confederacy to an ironclad ram, the C.S.S. *Virginia.*

Merritt's division: Wesley Merritt (1834–1910) graduated from the USMA in 1860. He was assigned to the cavalry. He was in command of the First Cavalry Division, of the Army of the Shenandoah, from August 1864 to March 1865.

Michie, Peter A. (1839–1901): Graduated second to JRM in the West Point class of 1863. He was assigned to the Corps of Engineers. Brevetted brigadier general for Civil War service, he served as instructor in service at West Point from 1867 to 1901. He authored textbooks on mechanics, physics, and astronomy.

Michler, Nathaniel (ca. 1827–81): Graduated from the USMA in 1849. He was assigned to the topographical engineers corps. He served as topographical engineer of the Army of the Cumberland from November 1, 1862, to June 1, 1863. He was captured near Rockville, Maryland, on June 28, 1863, while en route to join the Army of the Potomac. After being paroled, he was engaged in surveying Harpers Ferry and vicinity, from July 24 to September 18, 1863.

Middleway: Located in Jefferson County, West Virginia.

Miles, Dixon Stansbury (ca. 1804–62): Graduated from the USMA in 1824. He was assigned to the infantry. He was in command of the Fifth Division of McDowell's Army at the First Battle of Bull Run. The court of inquiry record regarding his conduct in that battle is filed in the National Archives, RG 153: Records of the Office of the Judge Advocate General (Army), Court Martial Case Files, 1809–94, Case II 498, and was published as Executive Document No. 7, U.S. Senate, 37th Cong., 2d sess., December 28, 1861.

Military Suspension Bridge pamphlet: On April 6, 1864, the Engineer Department requested the quartermaster general to furnish JRM with these pamphlets. MCM gave the pamphlets to JRM on April 11.

Mills, Clark (1810–83): A sculptor and bronze founder. In additional to the statues of Washington and Andrew Jackson, he later cast the statue of Armed Liberty for the Capitol dome.

Minnesota: A wooden screw steam frigate, the U.S.S. *Minnesota* was launched in 1855, grounded by the C.S.S. *Virginia* on March 8, 1862, and subsequently refloated.

Miss Clara: Presumably the daughter of Professor French.

Miss Hardee: Presumably the daughter of William Joseph Hardee (1815–73), commandant of cadets, USMA.

Miss Hattie: *See* Coyle, Harriet L.

Miss Hot Cakes: *See* Coyle, Harriet L.

Mississippi Tigers: Company G, Third Regiment, Alcorn's Brigade, was enlisted at Grenada, Mississippi, December 12, 1861.

Mohawk: A wooden screw steamer armed with five guns, commanded by Lieutenant Tunis Augustus Macdonough Craven (1813–64), the U.S.S. *Mohawk* sailed for Fort Jefferson on November 16, 1860, via Havana and arrived November 18, 1860.

Monitor: The original single-turret, iron and wood screw steamer, the U.S.S. *Monitor* was built under contract by John Ericsson and launched January 20, 1862. It engaged the C.S.S. *Virginia* at Hampton Roads, Virginia, on March 9, 1862. The vessel foundered off Cape Hatteras, North Carolina, on December 31, 1862.

Monocacy Junction: Now Frederick Junction, located in Frederick County, Maryland.

Monterey: Located in Highland County, Virginia.

Monty: *See* Meigs, Montgomery C., Jr.

Moor, August (ca. 1813–83): Born in Germany, Moor was mustered into Federal service in America as a colonel in the Twenty-eighth Ohio Infantry, in July 1861. He commanded the first brigade of the First Infantry Division under Franz Sigel at the Battle of New Market and under David Hunter at the Battle of Piedmont. He was mustered out of service in July 1864 and was subsequently awarded the brevet rank of brigadier general, U.S. Volunteers, for gallantry at the battles of Droop Mountain and Piedmont.

Moore, Julian Godwin: A sophomore at Columbian College in 1859. He was from Hertford County, N.C.

Moore, Samuel Preston (1813–89): He received his M.D. from the Medical College of the State of South Carolina in 1834 and became an assistant surgeon in the U.S. Army in 1835. From June 2, 1855, to April 2, 1860, he was senior surgeon at the USMA.

Mordecai, Alfred (1840–1920): Graduated from the USMA in 1861. He was assigned to the topographical engineers. He served as assistant professor of mathematics at the academy from September 1861 to June 1862.

Morgan, Edwin Barber, or Morgan, Edwin Denison: The Morgans were cousins.

Mosby, John Singleton (1833–1916): Enlisted as a private in the First Virginia Cavalry Regiment in 1861 and was commissioned first lieutenant in February 1862. In January 1863 he organized a unit of partisan rangers known as the 43rd Battalion, Virginia Cavalry, that operated against Union forces in northern Virginia until April 1865.

Mr. B.: President James Buchanan.

Mr. Bailey: *See* Bailey, Lewis.

Mr. Craige: *See* Craige, James Alexander.

Mr. Peale: *See* Peale, Titian Ramsay.

Mr. Taylor: *See* Taylor, William Bower.

Mr. Williams: *See* Williams, Robert.

Mr. Wood: *See* Wood, John.

Mrs. Elderkin: Wife of William A. Elderkin (1838–99), who graduated from the USMA in 1861 and served there as assistant professor of mathematics and as assistant instructor of artillery tactics during the war.

Mrs. Granbury: A "Jeannie Granbury, age 22, female, married, born in Virginia, died of 'child bed,' after 9 days' sickness," is listed in the *Federal Mortality Census*

Schedules, 1850–1880 (formerly in the custody of the Daughters of the American Revolution), and Related Indexes. The U.S. Census for 1860 lists a "Jeannie Granberry, female, 8 months of age in June 1860, born in the District of Columbia," as a resident in the household of Sara A. Massie, forty-one years of age, born in Virginia. A Sarah Massie, widow of Thomas J. Massie, lived at 386 Ninth Street, West, Washington, D.C.

Mrs. Lear: Fanny Henley Lear, niece of Martha Washington.

Mrs. Rowland: *See* Rowland, Catherine Armistead Mason.

Mrs. Sherman: Possibly the wife of the Washington lawyer Charles E. Sherman, who resided at 413 Third Street, West, Washington, D.C.

Naglee, Henry M. (1815–86): Graduated from the USMA in 1835. He was engaged in the relief of Harpers Ferry, after the Battle of Gettysburg.

National Rifles: A company of the volunteer militia of the District of Columbia, commanded by Captain Francis B. Schaeffer. Although an infantry company, it was armed with two mountain howitzers, revolvers, and sabres, by order of Secretary of War John B. Floyd.

Nautical Almanac: First published in 1855 for the United States Navy for the use of navigators.

New Creek: Located in Mineral County, West Virginia.

New Hope: Located in Augusta County, Virginia.

nine sheet map: The "Map of the State of Virginia Constructed in conformity to Law from the late Surveys authorized by the Legislature and other original and authentic Documents by Herman Boye" was copyrighted in 1826 and printed from nine copperplate engravings, on a scale of about 1:300,000, with an overall size of 60" by 93". It was distributed first in 1827. A revision supervised by Col. Claudius Crozet, with Ludwig von Bucholtz doing the actual mapping, was published in 1859.

Noble, Mason (b. ca. 1842): Son of the Reverend Mason Noble, pastor of the Fourth Presbyterian Church of Washington (1832–39) and later chaplain of the U.S. Navy Yard in Washington. Noble resided at 242 F Street, North, Washington, D.C.

Nookie: *See* Meigs, Mary Montgomery.

O'Keefe, Joseph (ca. 1841–65): Captain and aide-de-camp, New York Volunteers, April 1862, and major, Second New York Volunteer Cavalry, November 1864. He was an aide to Sheridan. He died on April 6, 1865, of wounds received at the Battle of Five Forks, Virginia.

Old Del: *See* Delafield, Richard.

Ogden, Martin: Son of the Washington lawyer Martin Ogden, the younger Ogden was a law student in 1860.

Olivier, Leonce N.: A resident of Louisiana, Olivier was admitted to the USMA in 1859 but he resigned in March 1861 and enlisted in the First Louisiana Infantry to fight for the Confederacy.

Opequon Creek: Located in Frederick County, Virginia.

order from Dept Headquarters: General Order No. 18, announcing assignments by Brigadier General Kelley to staff duty in the Department of West Virginia. JRM is listed as chief engineer.

order of the Secs: Refers to Engineer Order No. 4, October 26, 1860.

ordnance gun: A wrought-iron, muzzle-loading, rifled artillery piece with a 3-inch caliber. With one pound of powder and an elevation of 10°, it was capable of firing a 9-pound projectile to a range of nearly 2,800 yards.

other John: John Navarre Macomb (1811–89) or John Navarre Macomb Jr.

Overman, Mary: Wife of William V. Overman of Philadelphia, she was the mother of Lewis Cooper Overman (ca. 1843–99).

Owen, Augustus Allen: A sophomore at Columbian College in 1859 from Middlesex County, Virginia.

paces in a mile: When surveying on horseback, the rider measured the stride of a horse then calculated distances based on the number of paces multiplied by the length of the horse's stride.

Page, Charles Grafton (1812–68): Physician, inventor, and scientist, Page served as a patent examiner and as a professor of chemistry at Columbia College. He was a pioneer in the field of daguerreotyping and photography.

Pancoast, Joseph (1805–82): Graduated from the medical department of the University of Pennsylvania. He was a professor at the Jefferson Medical College and a surgeon at the Philadelphia Hospital.

Paper Mill Bridge: A wooden covered bridge located near the intersection of Florida Avenue ("Boundary Street") and P Street, NW, Washington, D.C., on the road between Georgetown and Bladensburg, near a paper mill built about 1800.

Parrott, Robert Parker (1804–77): Graduated from the USMA in 1824. He was assigned to the artillery and later the ordnance corps. He resigned from the army in 1836 and became superintendent of Gouverneur Kemble's West Point Iron and Cannon Foundry at Cold Spring, N.Y., three miles from West Point. He was the inventor of the Parrott gun and projectile.

Parrott gun: A cast-iron, rifled, muzzle-loading artillery piece reinforced by a heavy wrought-iron jacket at the breech, invented by Robert Parker Parrott, superintendent of the West Point Foundry at Cold Spring, New York. Parrott rifles, in 10-, 20-, 30-, 60-, 100-, 200-, and 300-pounder calibers, siege and garrison and field artillery carriages, were extensively used during the war. The 20-pounder, with a 2-pound charge of powder, fired at a 15° elevation, had a range of 4,400 yards; the 30-pounder, with a 3 ¼-pound charge, a range of 4,800 yards at 15° and 6,700 yards at 25°.

Parsons, Charles Carroll (1838–78): Graduated from the USMA in 1861. He was assigned to the artillery. He was principal assistant professor of geography, history, and ethics at the academy from March 1863 to September 1864.

Pavilion: Refers to the dancing pavilion built by George Washington Parke Custis in the 1820s or 30s on his property near the intersection of the Columbia Turnpike and the Georgetown-Alexandria Canal, on land now occupied by the Pentagon complex, Arlington County, Virginia.

Peace Convention: During February 4–27, 1861, southern and northern delegates met in the Willard Hotel in Washington, D.C., in an to attempt to establish a compromise to avoid war.

Peale, Titian Ramsay (1799–1885): Portrait painter and natural scientist, Peale was appointed an examiner at the U.S. Patent Office in Washington in 1849, where he served until 1872.

Petersburg: Located in Grant County, West Virginia.

Phipps, Frank Huntington (1843–1925): Graduated from the USMA in 1863. He was assigned to the ordnance corps.

Piatt, Donn (1819–91): A lawyer and journalist from Ohio, Piatt served as secretary of the United States legation in Paris from 1853 to 1855. In 1863, he served as inspector general and chief of staff to his fellow Ohioan Robert C. Schenck, at headquarters in Baltimore, Maryland.

Pleasonton, Alfred (1824–97): Graduated from the USMA in 1844. He served in the defenses of Washington, D.C., early in the war and served in the cavalry corps in the Shenandoah Valley campaign of 1864.

pledge: To save a cadet from dismissal for drinking, officials of the USMA created a plan to have an entire class pledge not to drink at West Point while on duty.

Pocahontas County: Located in West Virginia.

Pollard, John, Jr.: A member of the junior class of the USMA for the academic year 1858–59, Pollard was from King and Queen County, Virginia.

Pollock, Henry: A photographer and owner of a portrait gallery at 44 Lexington Street, Baltimore, Maryland.

pontoon bridge: A bridge constructed of flat-bottomed boats made of wood or framed canvas connected with balks (cross-ties) and covered with chess (flooring); once secured, the bridge was covered with dirt to reduce noise of troop movements and to minimize wear and tear. The wooden pontoon was used more frequently because of its durability, while the canvas pontoon had the advantage of being lighter and more easily handled and transported.

Pope, John (1822–92): Graduated from the USMA in 1842. He was assigned to the topographical engineers. He was a brigadier general in the U.S. Army, in command of the Army of Virginia at time of the Second Battle of Bull Run, August 29–30, 1862. Pope's dispatch claiming victory after the first day of battle appears in *The War of the Rebellion: Official Records.*

Post: The *New York Post,* edited by William Cullen Bryant (1794–1878), strongly supported abolition of slavery.

President's fast day: On March 30, 1863, President Lincoln proclaimed April 30, 1863, "as a day of national humiliation, fasting, and prayer."

Price, Sterling (1809–67): A lawyer who was elected as a Democrat to the U.S. House of Representatives from Missouri in 1845. He resigned in 1846 to serve in the Mexican War. Although he had opposed secession of his state, after the Civil War began he was appointed major general in command of the Missouri militia and subsequently opposed Union forces. After early successes in Missouri, he retreated into Arkansas.

prizefight: In a championship fight on April 17, 1860, the British boxer Tom Sayers fought a draw with the American boxer John Camel Heenan. The bout lasted thirty-seven rounds, fought over two hours and six minutes.

proclamation about blankets: MCM wrote to Secretary of War Simon Cameron on September 25, 1861, recommending that a policy of furnishing recruits with blankets be adopted.

Professor Church: *See* Church, Albert E.

Professor Henry: *See* Henry, Joseph.

Professor Page's: *See* Page, Charles Grafton.

proper channel: MCM wrote to Lieutenant Colonel Rene Edward De Russy on September 26, 1859, requesting authorization to send a package of shirts to JRM. He was given permission to send the shirts through the quartermaster at West Point.

pulling out plebes, or pulling plebes: This refers to the trick of grabbing a cadet by his ankles and pulling him out of his bed and onto the ground. Another term used was "yanked out."

Punch: A popular English magazine that contained illustrations, articles, and reports on various subjects, including politics. It was noted for its satire.

Pyne, Emma: The wife of the Reverend Dr. Smith Pyne.

Pyne, Reverend Smith, D.D. (1803–75): Rector of St. John's Episcopal Church, Lafayette Square, Washington, D.C., from 1845 to 1864. His residence was at the corner of H Street and Vermont Avenue, NW.

Pythagoras (b. ca. 570): A Greek philosopher.

Quincy, Walter C.: Assistant master of road of the Baltimore and Ohio Railroad in 1867. He lived at the Three Tuns Hotel in Baltimore.

qui vive: French for "on the alert"; to be quick and sharp; sentinel-like.

Rabb, James David (1840–63): Graduated from the USMA in 1863. He was assigned to the Corps of Engineers and was stationed in Louisiana, constructing defensive and fieldworks. He died in New Orleans two months after graduating from West Point.

railroad meeting: In early 1862, Secretary of War Edwin M. Stanton requested a meeting of northern railroad executives to discuss the rate schedule for military transportation.

Ray: *See* Bailey, Ray Tompkins.

Reid, James Riley (1837–1917): Graduated from the USMA in 1863. He served in the infantry during the Civil War but resigned from the army on March 14, 1866.

Relay House: A junction point between the Baltimore and Ohio Railroad and the Washington Branch Railroad. It was located nine miles west of Baltimore.

report of the committee: U.S. House of Representatives, Committee on Expenditures in the War Department, Report No. 566 (May 24, 1860), 36th Congress, 1st Session, 1860.

reservoir: The Washington Aqueduct system being constructed under MCM's supervision included a "Receiving Reservoir" (now called the Dalecarlia Reservoir) covering forty-six acres, partly in the District of Columbia and partly in Maryland, approximately nine miles downstream from the dam at Great Falls and the "Distributing Reservoir" of forty-two acres in Georgetown, District of Columbia, about two miles from the Receiving Reservoir.

retrenchment: Henry Wilson (1812–75), Republican senator from Massachusetts, served as chairman of the Committee on Military Affairs and the Militia. In that capacity, on December 11, 1861, he introduced Senate Bill 101 to increase the number of cadets at the military academy. The committee reported the bill favorably on December 17, 1861, and debate began on December 27, at which time several radical Republican senators spoke in opposition.

Richardson, Israel B. (ca. 1815–62): Graduated from the USMA in 1841. He was assigned to the infantry. After resigning from the army in 1855, he reentered U.S. service as colonel, Second Michigan Volunteer Infantry Regiment, on May 25, 1861. At the First Battle of Bull Run, he commanded the fourth brigade of the first (Tyler's) division, which was ordered to defend Blackburn's Ford between Centreville and Manassas, Virginia. Richardson's brigade was part of the fifth division of McDowell's Army, commanded by Colonel Dixon S. Miles. Richardson was mortally wounded at the Battle of Antietam.

Ricketts, James Brewington (1817–87): Graduated from the USMA in 1839. He was assigned to the artillery. He was wounded at the First Battle of Bull Run. Later he became commander of the third division of the Sixth Army Corps, in March 1864, in the campaign against Petersburg, Virginia. The division was ordered north in June 1864 to aid in the defense of Washington against Jubal Early.

Riggs, George W. (1813–81): Partner in the banking firm of Corcoran and Riggs, Washington, D.C., 1840–48. He purchased Corcoran's interest in 1854 and continued the firm as Riggs and Company.

Rockfish Gap: Located in Augusta and Nelson Counties, Virginia.

Rodgers, Augustus Frederick (1829–1908): Youngest son of Commodore John and Minerva Denison Rodgers, he was an assistant in the coast and geodetic survey.

Rodgers, Calbraith Perry (ca. 1845–78): Second-oldest son of Robert Smith and Sarah Perry Rodgers.

Rodgers, Catherine (or Katherine) Trowbridge (b. 1829): Daughter of Mr. and Mrs. Charles C. Trowbridge of Detroit and widow of Louisa Rodgers Meigs's younger brother, Henry "Hal" Rodgers (1822–54).

Rodgers, Christopher Raymond Perry (1819–92): Son of George Washington Rodgers and Anna Marie Perry Rodgers, he was the nephew of Commodore John Rodgers. He served as commandant of midshipmen at the United States Naval Academy from 1860 to 1861 and served as superintendent from 1874 to 1878.

Rodgers, Frederick (1842–1917): Eldest son of Robert Smith and Sarah Perry Rodgers, he graduated from the U.S. Naval Academy in 1861. His Civil War service included duty with both the West Gulf and North Atlantic Blockading Squadrons.

Rodgers, George Washington (1822–63): Son of George Washington Rodgers (1787–1832), younger brother of Commodore John Rodgers (1773–1838). He married Anna Marie Perry (1797–1856), the younger sister of Oliver Hazard Perry (1785–1819) and Matthew Calbraith Perry (1794–1858). The younger G. W. Rodgers was first cousin of the younger John Rodgers (1812–82) and of Louisa Rodgers Meigs (1817–79), and brother of Christopher Raymond Perry Rodgers (1819–92).

Rodgers, Jane Perry (1852–1920): Fourth child and only daughter of Robert Smith and Sarah Perry Rodgers.

Rodgers, John (1812–82): Son of Commodore John Rodgers and older brother of Louisa Rodgers Meigs, he was appointed a midshipman in the U.S. Navy in 1828. In 1861, he was in charge of procuring and operating ironclad vessels for service on inland rivers and was later ordered to duty with the South Atlantic Blockading Squadron. He was commissioned captain in 1862 and commodore on June 17, 1863. In February 1863, he was in command of the U.S.S. *Weehawken,* an ironclad monitor. His wife was Ann Elizabeth Hodge Rodgers (b. 1823).

Rodgers, Minerva Denison (1784–1877): Daughter of Gideon and Butler Denison of Norwich, Connecticut, who established their family at Sion Hill, near Havre de Grace, Harford County, Maryland in 1795. She married Commodore John Rodgers (1773–1838) in 1806. The couple had eleven children.

Rodgers, Robert Smith (1809–92): Brother of Louisa Rodgers Meigs. In December 15, 1841, he married Sarah Perry. He was commissioned as major and mustered into Federal service on December 24, 1861, with the Second Eastern Shore Maryland Regiment. After the regiment was ordered to western Maryland in July 1863, Rodgers served for several months, beginning in October, as the commander of the troops along the Baltimore and Ohio Railroad near Martinsburg, West Virginia, and for a while as commander of the third brigade of the first division in the Department of West Virginia. He later reverted to regimental commander of the Second Eastern Shore Maryland Regiment and participated with it during the summer of 1864 in campaigns under Generals David Hunter and Philip H. Sheridan. He was mustered out of service at the expiration of his three-year term of enlistment in October 1864. "Uncle Robert" to JRM, who visited with him as circumstances permitted during the fighting in West Virginia and Virginia in 1864.

Rodgers, Sarah Perry (b. 1818): Wife of Robert Smith Rodgers.

Rodgers, William Ledyard (1860–1944): Eldest child and only son of Captain John Rodgers and Ann Elizabeth Hodge Rodgers.

Roe's: A hotel located on the grounds of the USMA, it was less expensive than the newer Cozzens Hotel located in Buttermilk Falls, present-day Highland Falls, New York, approximately two miles south of the academy.

roommate (of JRM): *See* Catlin, Robert.

Rosecrans, William Starke (1819–98): Graduated from the USMA in 1842. He was assigned to the Corps of Engineers from 1842 to 1854, when he resigned his commission for a civilian career as an engineer and entrepreneur. He served as volunteer aide-de-camp to George B. McClellan and was commander of a brigade in McClellan's West Virginia campaign. He was in command of the Department of Western Virginia from September 21, 1861, to April 7, 1862.

Rosser, Thomas Lafayette (1836–98): Left the USMA in April 1861, just three weeks prior to graduation, to join the Confederate army. In October 1864, he was in command of Jubal Early's cavalry in the Shenandoah Valley.

Rowland, Catherine Armistead Mason: Wife of Isaac S. Rowland, who had raised and commanded volunteer forces under Winfield Scott in the Black Hawk and Mexican Wars. She was the daughter of Stevens Thomson Mason, first governor of Michigan.

Rowland, Thomas (1842–74): Born in Detroit, Michigan, he was appointed to the USMA at large. He was at the head of the class when he resigned on April 26, 1861. He subsequently joined the Virginia forces to serve in the Confederate army.

Rulison, William H. (ca. 1821–64): A physician. At the time of his death at Smithfield, Jefferson County, West Virginia, on August 29, 1864, he was medical director for the chief of cavalry, Alfred T. A. Torbert, in Sheridan's Middle Military Division.

Russell, William Howard (1820–1907): Correspondent in the United States for the *Times* of London, from March 1861 until April 1862. His dispatches, reprinted in U.S. newspapers, gave offense to both North and South.

St. John's Episcopal Church: Located on Lafayette Square at Sixteenth and H Streets, NW, Washington, D.C.

Samson, George Whitefield (1819–96): Pastor of the E Street Baptist Church, in Washington, D.C., from 1850 to 1852, and again in the late 1850s, when he was appointed president of Columbian College and professor of moral and intellectual philosophy. He resigned in 1871.

Sanibel Island: An island in the Gulf of Mexico, opposite the Caloosahatchee River, Lee County, Florida.

Satterlee, Richard Sherwood (1798–1880): After practicing as a medical doctor in Seneca, New York, and Detroit, Michigan, Satterlee entered the U.S. Army as an assistant surgeon in 1822 and became a major surgeon in 1832. He was the senior officer of the medical board that examined JRM at West Point in September 1859.

Saturday Club: The club met in Washington area homes mainly on Saturday afternoons and was made up of individuals with an interest in science and public affairs. This informal group was the forerunner of the Philosophical Society of Washington, D.C.

Saxton, Rufus (1824–1908): Graduated from the USMA in 1849. He served as assistant instructor of artillery tactics at the academy from 1859 to 1860. He was quartermaster on several staffs from 1861 to 1862 and was in command of defenses at Harpers Ferry from May to June 1862.

Scammon, Eliakim Parker (1816–94): Graduated from the USMA in 1837. He was assigned to the artillery. He later became a topographical engineer. He was commissioned colonel in the Twenty-third Ohio Volunteers in June 1861 and was promoted to brigadier general, U.S. Volunteers, on October 15, 1862. Scammon was commander of the District of the Kanawha, West Virginia, from November 1862 to February 3, 1864, when he was captured by Confederate guerrillas.

Schaeffer, Francis B.: Commanded the National Rifles.

Schaeffer, George C. (1814 or 1815–73): A chemical researcher and teacher, Schaeffer was an examiner in the U.S. Patent Office and also served as professor of materia medica at Georgetown College, 1854–58. He was fired from the Patent Office in 1858 for rejecting too many applications, but he was rehired in 1865 as librarian and served until his death.

Schenck, Robert Cumming (1809–90): A Whig congressman from Ohio from 1842 to 1851, and minister to Brazil from 1851 to 1853, he was a brigadier general, U.S. Volunteers, from 1861 to 1862. As a major general, he was commander of the Middle Department and the Eighth Army Corps, with headquarters in Baltimore from 1862 to 1863, until he resigned in December and returned to Congress as a Republican. The Middle Department consisted of New Jersey, Pennsylvania, Delaware, the eastern shores of Maryland and Virginia, West Virginia, and the counties of Cecil, Harford, Baltimore and Anne Arundel in Maryland.

Schmalcalder compass: A diopter magnetic compass, equipped with a prismatic aiming device fastened to the magnetic needle, this was invented by Charles Augustus Schmalcalder in 1812. Compasses of this type were requested by JRM in September 1863 for surveys in West Virginia.

Scott, Winfield (1786–1866): General in Chief of the Army, with the rank of major general, from July 1841 to November 1861.

Sedgwick, John (1813–64): Graduated from the USMA in 1837. He was assigned to the artillery. He was division and corps commander of the Army of the Potomac, 1861–64, up to the time he was killed at the Battle of Spottsylvania, in May 1864.

"See the *Conquering* hero comes": The first line of the chorus in part 3 of the oratorio *Judas Maccabaeus,* composed by George Frideric Handel (1685–1759), with a libretto by the Reverend Thomas Morell (1703–1784). The text of the chorus reads as follows: "See the conqu'ring hero comes! Sound the trumpets, beat the drums. Sports prepare, the laurel bring. Songs of triumph to him sing. See the godlike youth advance. Breathe the flutes, and lead the dance. Myrtle-wreaths and roses twine, To deck the hero's brow divine."

sending them off in January: Cadets were dismissed for academic deficiency.

severe fight: The Battle of Piedmont (Augusta County, Virginia), which occurred on June 5, 1864. Union forces under David Hunter defeated a smaller Confederate force commanded by William E. "Grumble" Jones, who was killed in the fighting.

Seward, William H. (1801–72): Secretary of state, 1861 to 1869. Senator Seward's speech on January 12, 1861, appears in the *Congressional Globe* (1861), pp. 341–44.

Shaver, Benjamin Franklin: Enlisted in the Union army on May 22, 1861, at Harrisonburg, Virginia, for a period of twelve months. He reenlisted in April 1862 for the duration of the war. He was a prisoner of war at the Confederate States of America Army Hospital in Farmville, Virginia, in April 1865, when he was paroled.

Shaw, Sidney F. (ca. 1832–1909): Mustered into U.S. service at Wheeling, West Virginia, in Company G., Fifteenth [West] Virginia Infantry, in August 1862. He was promoted to captain and company commander in October 1862. He was detached from duty with his company and placed in charge of constructing blockhouses along the Baltimore and Ohio Eailroad in April 1864, a position in which he remained until appointed chief engineer of the Department of West Virginia in the late fall 1864.

Shepherdstown: Located in Jefferson County, West Virginia.

Sheridan, Philip Henry (1831–88): Graduated from the USMA in 1853. In addition to other service, he was commander of the Army of the Shenandoah, August 4–7, 1864, and of the Middle Military Division from August 7, 1864, to March 25, 1865.

Sherman, William Tecumseh (1820–91): Graduated from the USMA in 1840. He succeeded Ulysses S. Grant in March 1864 as commander of the Military Division of the Mississippi, the western theater of operations. Union forces under Sherman's command had initiated the campaign against Confederate forces under Joseph E. Johnston that were defending northern Georgia and the city of Atlanta on May 7, 1864. The Army of the Tennessee reached the vicinity of Resaca, Georgia, but withdrew to Snake Creek Gap and entrenched on May 9 without dislodging Johnston's forces.

Shute, Samuel Moore (1823–1902): Taught Greek and Latin languages and English language and literature at Columbian College from 1858 to 1902.

Sibley, Ebenezer Sprote (ca. 1805–84): Graduated from the USMA in 1827. He was assigned to the artillery. In 1856 he was assigned to duty in the Office of the Quartermaster General in Washington, D.C.

Sigel, Franz (1824–1902): A native of Baden, Germany, and a graduate of the Karlsruhe Military Academy. With the outbreak of the American Civil War, he organized a regiment of infantry. He was assigned to relieve Benjamin F. Kelley as commander of the Department of West Virginia in February 1864, but in turn was relieved by General David Hunter after the Battle of New Market, May 15, 1864.

Slidell, Matilde: Possibly the daughter of John Slidell, Confederate diplomat.

Smith, Cassie Selden (1837–1907): Married Edmund Kirby Smith (1824–93), an 1845 West Point graduate and Confederate general, in September 1861, in Lynchburg, Virginia, where the general was convalescing from serious wounds received at the First Battle of Bull Run.

Smith, Frederick Augustus (1812–52): Graduated from the USMA in 1833. He was assigned to the Corps of Engineers. He was supervising engineer of the preliminary surveys for the Potomac Aqueduct in 1852 at the time of his death and was succeeded by MCM.

Smith, Henrietta Elizabeth Henley: Louisa Rodgers Meigs's first cousin, she married a Washington, D.C., lawyer, Jonathan Bayard Harrison Smith (1810–89), in 1842. Jonathan Smith was the son of Samuel Harrison Smith, who had edited the *National Intelligencer,* and was later president of the Washington branch of the U.S. Bank.

Smith, J. Henley (ca. 1844–1907): Son of Louisa Rodgers Meigs's first cousin, Henrietta Elizabeth Henley Smith, and Jonathan Bayard Harrison Smith. The younger Smith was a Washington lawyer.

Smith, Samuel Crawford (1838–1916): A resident of Berks County, Pennsylvania, Smith was mustered into service at Philadelphia on August 9, 1862, as third sergeant of Captain H. E. Wrigley's Independent Company, Pennsylvania Infantry (Acting Engineers). He was promoted to first lieutenant on March 30, 1863, and to captain and company commander on July 6, 1864. He was mustered out of service at Harpers Ferry, West Virginia, on June 20, 1865.

Snikers Ferry: This refers to Snickers Ferry (Parker's Ford), located in Virginia.

Somerset: By this, JRM is referring to the Battle of Logan's Cross Roads (or Mill Springs), about ten miles from Somerset, Kentucky, where on January 19, 1862, Union forces under George H. Thomas defeated a Confederate army commanded by George B. Crittenden.

South Peak: Now called Sharp Top Mountain (3,875 ft.), this topographical feature is located in the Peaks of Otter Recreational Area, on the Blue Ridge Parkway.

"*Spingler*": Possibly a cottage on the resort at Bailey's Crossroads, Fairfax County, Virginia, or the Ledyard residence in Washington, D.C.

Spirit of the Times: This journal began publication in 1831 and billed itself as a gazette of the literary, fashionable, and sporting world. It reported on races and published engravings of horses and actresses.

Stanton, Edwin McMasters (1814–69): Served briefly as U.S. Attorney General in the waning months of the Buchanan administration and was appointed by Lincoln to succeed Simon Cameron as secretary of war on January 13, 1862.

Star: See *Washington Star.*

Starr, William C. (ca. 1823–97): Mustered into Federal service as lieutenant colonel,

Ninth [West] Virginia Infantry Regiment, at Point Pleasant, Mason County, [West] Virginia, on November 2, 1861. He was captured at the Battle of Port Republic on June 9, 1862, but was exchanged and rejoined his regiment in August. Franz Sigel designated Starr acting provost marshal, Department of West Virginia, on April 9, 1864, and judge advocate general, Army of the Shenandoah, on June 3, 1864. He was mustered out on October 31, 1864, when the Ninth West Virginia Infantry was consolidated with another regiment.

statue of Washington: The bronze equestrian statue of George Washington by Clark Mills, located in Washington Circle at the intersection of Pennsylvania Avenue and K and 23rd Streets, NW, Washington, D.C., was dedicated by President James Buchanan on February 22, 1860.

Steamer *Joseph Whitney:* On January 5, 1861, Winfield Scott ordered a company of the Second U.S. Artillery commanded by Brevet Major Lewis Golding Arnold transferred from Fort Independence, Boston, Massachusetts, to Fort Jefferson, Florida. The company arrived on the steamer *Joseph Whitney* at Fort Jefferson on January 18, 1861.

Steele, Reginald H.: Principal of Rugby Academy, 395 Thirteenth Street West, Washington, D.C.

Steilberg, William (1841–1927): Born in Hanover, Germany, Steilberg enlisted as a private in Company A, Twenty-eighth Infantry Regiment. He was mustered out with his regiment at the completion of his three-year enlistment on July 23, 1864.

stereographs: Single or double pictures for a stereoscope.

Stevens battery: In 1842 the Congress authorized the construction of an ironclad steam vessel by the brothers Robert Livingston Stevens (1787–1856) and Edwin Augustus Stevens (1795–1868), engineers and inventors from Hoboken, New Jersey. Despite appropriations of more than half a million dollars, the navy had not accepted the Stevens battery and the project had been abandoned at Robert Stevens's death. In 1861 Edwin Stevens offered to complete the vessel, but the navy was unwilling to accept his proposed terms.

Stewart, William D. (d. 1886): Physician, major, and surgeon, U.S. Volunteers, from 1861 to 1865. He was assigned to duty in the Department of West Virginia adjutant general's office, July 30, 1863, and served there in various medical assignments until August 1864, when he was reassigned to duties with the provost marshal general, Washington, D.C.

Stringham's victory: On August 27–28, 1861, a naval and army expedition, Commander Silas H. Stringham and General Benjamin F. Butler commanding, captured Forts Clark and Hatteras in North Carolina. Commodore Samuel Barron was in charge of Confederate naval defenses. Both Stringham and Barron had long careers in the United States Navy before Barron resigned in April, 1861.

Strother, David Hunter (1816–88): Author, illustrator, and soldier. During the Civil War he served on staff duty primarily in West Virginia and northern Virginia. As an illustrator he used the name "Porte Crayon."

Strother, Philip Williams (1839–1922): Son and grandson of congressmen from Virginia. He was educated at schools in that state and at Columbian College. Admitted to the bar in 1861, he was practicing his profession in Giles County, Virginia, prior to enlisting in the Confederate army. He served until badly wounded at the Battle of Spotsylvania Court House in 1864.

Stuart, James Ewell Brown "Jeb" (1833–64): Graduated from the USMA in 1854. He was assigned to the cavalry. He was aide-de-camp to Robert E. Lee on the expedition to suppress Brown's raid on Harpers Ferry, Virginia in 1859. He resigned from the U.S. Army in May 1861 to serve in the Confederate army.

"Suaviter in modo": From the phrase *suaviter in modo, fortiter in re,* meaning "Gentle in manner, firm in action."

Sue: *See* Tomkins, Susan Livingston Ledyard.

Sullivan, Jeremiah Cutler (1830–90): Commanded the First Division, Department of West Virginia, during 1863 and 1864. He was wounded at the Battle of Piedmont in June 1864 and was subsequently "awaiting orders" after Sheridan relieved David Hunter of command in the Shenandoah Valley.

Sumner, Charles (1811–74): Elected to the U.S. Senate from Massachusetts as a Republican in 1857, 1863, and 1869.

suppers: The food at West Point was based on army field rations and therefore was mostly beef. On top of being boring, the food and the cleanliness standards were of very poor quality and unhealthy. In order to get special foods and treats, the cadets risked getting demerits and possible expulsion to dine at Benny Havens, the eating establishment and watering hole well known in West Point history and lore, located in Highland Falls, New York. A cadet could purchase oysters, buckwheat cakes, and a mug of a brew called "flip," a mixture of liquor or wine, sugar, beaten eggs, and nutmeg, among other foods and drink. Another way to get better food was to prepare food in the fireplace in a cadet room.

Suter, Charles R. (1842–1920): Graduated from the USMA in 1862. He was assigned to the Corps of Engineers. He was assistant engineer, then chief engineer, in the Department of the South from 1863 to 1864. In 1864 he became a captain in the Corps of Engineers.

Taylor, John Rodgers Meigs: Joseph and Mary Taylor's first child, born January 13, 1865. Appointed to the USMA at large, he graduated in 1889. He retired as a major for disability in 1914, but served during World War I and retired again as a colonel in 1918. He died in Washington, D.C., on March 31, 1949, without issue.

Taylor, Joseph Hancock (1836–85): Graduated from the USMA in 1856. He was assigned to the cavalry. He was the son of Joseph Pannell Taylor (1796–1864), commissary general of the Union army, and nephew of President Zachary Taylor. He married JRM's sister Mary Montgomery Meigs on March 30, 1864.

Taylor, Joseph Pannell (1796–1864): Brother of President Zachary Taylor, he began military service as a second lieutenant in 1813. He had advanced to colonel, commissary general, in September 1861, and to brigadier general in February 1863. His son, Joseph Hancock Taylor, married JRM's sister Mary Montgomery Meigs on March 30, 1864.

Taylor, William Bower (1821–95): A physicist who studied civil engineering, Taylor was more interested in scientific investigations and editorial activities. He was appointed an examiner in the U.S. Patent Office in 1854, where he remained until 1877, when he became editor of publications for the Smithsonian Institution.

Thomas: A servant of the Meigses. He accompanied MCM on the secret mission to secure Fort Pickens.

Thomas, C. Keefer [or Kiefer?] (dates unknown): A native of Frederick but a Baltimore merchant who had acquired Araby, the eighteenth-century house and farmstead, in about 1860.

Thomas, Charles William (1833–82): Graduated from the USMA in 1855. He was assigned to the infantry. He served field duty as quartermaster in the Third and then the Seventh Army Corps, 1862 and 1863, respectively. He was in charge of Quartermaster Depot, Baltimore, Maryland, from October 1863 to September 1864.

Thomas, George Henry (1816–70): Graduated from the USMA in 1840. He was a brigadier general, U.S. Volunteers, in command of a division in the Army of the Ohio and was engaged in combat at the Battle of Mill Springs, Kentucky, January 19–20, 1862.

Thomas, Lorenzo (ca. 1805–75): Graduated from the USMA in 1823. He was assigned to the infantry. In addition to other service, he served as chief of staff to Winfield Scott from 1853 to 1861. He was the organizer of United States Colored Troops during 1863–65.

threatened dangers from England: On November 8, 1861, the captain of the Union warship *San Jacinto* stopped the British mail steamer *Trent* and removed two Confederate commissioners, James M. Mason and John Slidell, who were en route to England. The incident, known as the *Trent* Affair, caused unrest among the British people and government. Secretary of State William H. Seward ordered the two men released, thus averting a more serious crisis.

Tibbits, William Badger (1837–80): Captain, then major in the Second New York Infantry Regiment, from May 1861 to May 1863, when he was mustered out of service. He was a colonel in the Twenty-first New York Cavalry Regiment from February 1864 to October 1865.

Tomkins, Susan Livingston Ledyard (1844–77): Second daughter of Henry Ledyard and Matilda Cass Ledyard; twin sister of Henry Brockholst "Harry" Ledyard.

Torbet, Alfred Thomas Archimedes (1833–80): Graduated from the USMA in 1855. He was assigned to the infantry. He was chief of cavalry, Middle Military Division, from August 8, 1864, to January 10, 1865.

Tortugas: *See* Dry Tortugas.

Totten, Joseph Gilbert (1788–1864): Graduated from the USMA in 1805. He was assigned to the Corps of Engineers. He was in command of the Corps of Engineers and in charge of the engineer bureau from 1838 to 1864.

Townsend, Ebenezer Grant (1813–87): A clergyman who was wounded in action at Lookout Mountain, Tennessee, on November 24, 1863. No record was found of a recommendation by MCM in the summer of 1864, as requested by JRM. However, MCM recommended in February 1865 that Townsend be promoted to major and transferred to a mounted unit, and, in October 1866, that he be commissioned in the Veteran Reserve Corps. In both letters of recommendation he mentioned the friendship of JRM and Townsend's deceased son.

Townsend, Hurlbut G. (d. 1862): Appointed to the USMA from New York. He stood fourth in his class at the time of his death from typhoid fever, in December 1862.

traitors from the cabinet: Secretary of the Treasury Howell Cobb of Georgia, Secretary of War John B. Floyd of Virginia, and Secretary of the Interior Jacob Thompson of Mississippi resigned from President Buchanan's cabinet in December 1860 (Cobb and Floyd) and January 1861 (Thompson).

Tribune: The *New York Tribune,* edited by Horace Greeley (1811–72), strongly supported abolition of slavery.

Trinity: Trinity Protestant Episcopal Church, established in 1828, erected a new building in 1851 at the northeast corner of Third and C Streets, NW, Washington, D.C.

Tripler, Charles Stuart (1806–1906): A surgeon in the U.S. army from 1830 to 1906, Tripler was promoted to brigadier general in March 1865, for faithful and meritorious service during the Civil War. He was the author of *Manual of the Medical Officer of the Army of the United States: Part I, Recruiting and Inspection of Recruits* (1858).

Trowbridge, C. C.: *See* Trowbridge, Charles C.

Trowbridge, Catherine Whipple Sibley (d. 1880): Wife of Charles C. Trowbridge and close friend of Louisa Rodgers Meigs.

Trowbridge, Charles C. (1800–83): He was a Detroit banker, and the members of the Trowbridge family were good friends of the Meigses while MCM was stationed in Detroit, Michigan.

Trowbridge, Frederick C. (b. ca. 1837): Son of Mr. and Mrs. Charles C. Trowbridge.

Trowbridge, William Petit (1828–92): Graduated from the USMA in 1848. He was assigned to the Corps of Engineers. He resigned on December 1, 1856, and was appointed professor of mathematics at the University of Michigan. He served as assistant to Alexander D. Bache, chief of the U.S. Coast Survey, from 1857 to 1861. During the Civil War, Trowbridge was in charge of the Engineer Agency, New York City, responsible for supplying engineering materials for the field armies.

Twining, William Johnson (1840–82): Graduated from the USMA in 1863. He was assigned to the Corps of Engineers. His service during the Civil War was with the Departments of Ohio and Cumberland and with the Twenty-third Army Corps.

Uncle: *See* Meigs, John Forsyth.

Uncle Frank: Franklin Bache Meigs (1829–81), a younger brother of MCM.

Uncle Gus: *See* Rodgers, Augustus Frederick.

Uncle Henry: *See* Meigs, Henry Vincent.

Uncle John: *See* Rodgers, John.

varicocele: A varicose condition of the spermatic veins of the scrotum.

Virginia Military Institute burning: General Hunter ordered the burning of VMI because of the participation of the cadets in battle against Union forces and because the buildings were being used as a Confederate arsenal.

Wabash: A 3,274-ton screw steamer frigate, the U.S.S. *Wabash* was built at the Philadelphia Navy Yard, between 1854 and 1856.

Wadsworth, Anne: Possibly the daughter of Alexander Wadsworth and Louisa Dension Wadsworth, Louisa Rodgers Meigs's aunt.

Warm Springs: Located in Bath County, Virginia.

Washington Star: A newspaper published in Washington, D.C., in the nineteenth and twentieth centuries.

Waynesboro: Located in Augusta County, Virginia.

Weber, Max (1824–1901): Born in Germany, Weber graduated from a German military academy and fought with Franz Sigel in the 1848 revolution. He then immigrated to New York. On May 9, 1861, he was commissioned as colonel, Twentieth New York Infantry Regiment, and was promoted to brigadier general, U.S. Volunteers, on April 28, 1862. Severely wounded at the Battle of Antietam, he was later assigned to the Department of West Virginia under Sigel and continued there under Hunter and Sheridan.

Weehawken: An ironclad monitor. On her passage from New York to Hampton Roads, January 19–22, 1863, the U.S.S. *Weehawken* encountered a severe gale, but Captain John Rodgers reported to Secretary of the Navy Gideon Welles that the ship was strong and safe. During the attack at Charleston, with Rodgers in command, the U.S.S. *Weehawken* was struck fifty-three times by Confederate fire.

Weir, Robert Walter (1803–89): Professor of drawing at the USMA from 1834 to 1876.

Weitzel, Godfrey (1835–84): Graduated from the USMA in 1855. He was assistant professor of engineering at the academy from September to December 1859 and from August 1860 to June 1861.

Welcker, George Lewis (ca. 1812–48): Graduated from the USMA in 1836. He served in the Corps of Engineers. As a captain, he was in charge of repairs of the St. Augustine seawall, Florida, a project he was involved with at the time of his death in Savannah, Georgia.

Welles, Gideon (1802–72): Secretary of the navy from 1861 to 1869.

West, Edward Walter (1836–1916): A lawyer from New York. Among other service, he was aide-de-camp to Brigadier General Silas Casey, who commanded the second division in the Army of the Potomac. The division was severely mauled by a Confederate forces at Seven Pines, Virginia, on May 31, 1862. West's map of the Battle of Fair Oaks (Seven Pines), was published in the *New York Times* on June 13, 1862.

Wetmore, Captain Henry Shepard (1841–96): A cadet at the USMA from July 1, 1858 to August 15, 1861. He commanded the Ninth Ohio Battery, also known as Wetmore's Battery, at the Battle of Logan's Cross Roads (or Mill Springs), Kentucky, January 19, 1862.

Wharton. JRM may be referring to Henry Clifton Wharton, who graduated from the USMA in 1858. He was commissioned first lieutenant in March 1863. He was brevetted captain in December 1864 and later commissioned lieutenant colonel.

Whistler, Joseph N. G. (ca. 1822–98): Graduated from the USMA in 1846. He was assigned to the infantry. He served as an assistant instructor of infantry tactics at the academy from September 1861 to March 1863.

Wickham, Williams Carter (1820–88): A lawyer and planter, Wickham was commissioned lieutenant colonel of the Fourth Virginia Cavalry in September 1861. Later appointed brigadier general, he served with Jubal Early in the Shenandoah Valley Campaign of 1864.

Wide-awakes: Radical Republicans opposed to slavery who were famous for organizing night parades. They were early supporters of Lincoln.

Williams, John Benson (ca. 1836–1903): Graduated from the USMA in 1861. He served in the infantry, in the campaigns of the Army of the Potomac during 1861 and 1862. He was on sick leave from December 1862 to February 1863, when he was dismissed from the U.S. Army.

Williams, Robert (ca. 1825–1901): Graduated from the USMA in 1851. He served as assistant instructor of cavalry and infantry tactics at the academy, from late December 1857 to May 1861.

Williamsport: A town in Washington County, Maryland.

Willie: *See* Rodgers, William Ledyard.

Wilson, John Moulder (1837–1919): Graduated from the USMA in 1860. He was assigned to the artillery but transferred to the Corps of Engineers in July 1862. He served as a topographical engineer at the Battle of Antietam before serving as assistant professor of Spanish at the academy, from March to June 1863. He later had a distinguished army career, including service as the chief of the Corps of Engineers.

Wood, John (b. ca. 1827): A photographer in Washington, D.C., in the 1850s and 1860s. At some point he was employed at the Capitol and had a room fitted up for making albumen negatives and prints. He was apparently also a sometime employee of Matthew Brady in Washington during the Civil War and is credited with assisting in photographing some of the scenes of the Antietam battlefield in 1862.

Wood, Robert Crooke (ca. 1800–69): A graduate in 1821 of the Columbia Medical School in New York, he was appointed assistant surgeon in the U.S. Army in 1825. He was promoted to major surgeon on July 4, 1836, and in the late 1850s was serving in the surgeon general's office in Washington. He was subsequently promoted to colonel and served as the assistant surgeon general, U.S. Army, from June 1862 to October 1865.

Woodruff, Israel Carle (1815–78): Graduated from the USMA in 1836. He was assigned to the artillery. He soon after was transferred to the Corps of Topographical Engineers. He served as assistant to the chief of topographical engineers in Washington, D.C., from 1857 to 1863. After the consolidation of the topographical engineers with the Corps of Engineers, he served as assistant to the chief engineer in Washington, D.C., from 1863 to 1866.

Worden, John Lorimer (1818–97): Commanded the *Monitor* during its fight with the C.S.S. *Virginia.* He was subsequently given command of another monitor, the U.S.S. *Montauk.* Worden's reports of the cruise of the *Montauk* from New York to Hampton Roads and to Beaufort, South Carolina, on December 30, 1862, and January 5, 1863, are published in the *Official Records . . . Union and Confederate Navies,* Series 1, vol. 8, 360–62.

Wright, Horatio Gouverneur (1820–99): Graduated from the USMA in 1841. He was assigned to the Corps of Engineers. He was assistant to the chief engineer, Washington, D.C., 1856–61. Subsequently, as a brigadier and major general, U.S. Volunteers, he was a division and a corps commander in the Army of the Potomac. In July 1864, he commanded the Sixth Corps and was assigned to the defense of Washington and the pursuit of Jubal Early's forces.

SELECTED BIBLIOGRAPHY

Manuscript and Archival Sources

Primary manuscript collections include the Montgomery C. Meigs Papers and the Rodgers Family Papers, located in the Manuscript Division of the Library of Congress. In addition to the letters and other documents of John Rodgers Meigs, the Montgomery C. Meigs Journals, the Montgomery C. Meigs Diaries, and the correspondence of Montgomery C. Meigs with his father, Charles D. Meigs, are significant. Information on the early Meigs family life is contained in the Rodgers Family Papers, in the correspondence of Louisa Rodgers Meigs, especially with her sister, Ann Minerva "Nannie" Rogers Macomb, and their mother, Mary Cunningham Montgomery Meigs.

Other sources include: Letters Sent by the Baltimore and Ohio Railroad, January 1, 1859–February 9, 1867, at the Maryland Historical Society; the William Henry French Papers and the George W. Cullum Correspondence Papers, in the Special Collections and Archives in the United States Military Academy Library.

Many National Archives and Records Administration record groups were searched to locate documents, to provide historical context, and to identify individuals. The following record groups yielded significant documents and information:

RG 15: Records of the Veterans Administration. Pension Case Files, Pension Application Files for the Civil War.

RG 29: Records of the Bureau of the Census, United States census records for the years 1850 and 1860 for the District of Columbia, Maryland, and Virginia.

RG 77: Records of the Office of the Chief of Engineers. Letters Sent to Engineer Officers, 1812–69; Engineer Orders and Circulars, 1811–68; Letters Received, 1826–66; Letters Received, 1832–65; Letters Sent to Others than the Secretary of War, 1829–67; Letters Sent to Engineer Officers, 1812–69. For maps located in RG 77 see Civil Works Map File, Virginia: 1.140: General Information, the Shenandoah Valley and the Upper Potomac Region, maps US 224-1 and US 224-2; 1.141: Battlefields in the Shenandoah Valley; and 1.142: Operations and Defenses in the Shenandoah Valley and the Upper Potomac Region. Reference numbers are to *A Guide to Civil War Maps in the National Archives* (1986).

RG 92: Records of the Office of the Quartermaster General. Letters, Endorsements, and Reports Sent to the Secretary of War and Heads of Departments, December 1848–December 1863; Montgomery C. Meigs Letterbooks, Consolidated Correspondence File, 1794–1890.

RG 94: Records of the Adjutant General's Office. From Orders, 1797–1939: Orders and Circulars, 1797–1910. From Records Relating to the United States Military Academy, 1803–1917: Letters Sent Relating to the Military Academy, 1812–67; Correspondence

Relating to the Military Academy, 1819–66; Application Papers of Cadets, 1814–66; Military Academy Orders, 1814–67; Orders and Circulars, 1832–1910; Register of Cadets, 1803–65; Monthly Class Reports and Conduct Rolls, 1831–66; Papers Relating to the Examination of Cadets, 1816–65. From Records of Appointment, Commission, and Personal Branch, 1783–1917: Letters Received, 1863–94. From Records of the Civil War, 1861–65: Staff Papers; Generals' Reports of Service. From Records Concerning Medical Personal, 1839–1914: Personal Papers, Medical Officers and Physicians, prior to 1912. From Records of the Record and Pension Office: Carded Military Service Records, 1784–1904, Civil War, 1861–65.

RG 107: Records of the Secretary of War. Telegrams Received by Major General Halleck, September 21, 1863–October 30, 1863.

RG 109: War Department Collection of Confederate Records. Records of the Fourth Virginia Cavalry and First Virginia Cavalry. Compiled Service Records of the Confederate Soldiers Who Served in Organizations from the State of Louisiana, Microfilm Publication M 320, Roll 87; Compiled Service Records of the Confederate Soldiers Who Served in Organizations from the State of South Carolina, Microfilm Publication M 267, Rolls 89 and 93.

RG 112: Records of the Bureau of Naval Personnel. Logs of United States Naval Ships and Stations, 1801–1946.

RG 153: Records of the Office of the Judge Advocate General (Army), Court Martial Case Files, 1809–94. The Miles Court of Inquiry record was published as Executive Document No. 7, U.S. Senate, 37th Cong., 2d sess., December 28, 1861.

RG 393: Records of United States Army Continental Commands, 1821–1920. From the Records of Middle Department and Eighth Army Corps (P), 1861–66: Dispatches Sent, June–July, 1863; Press Copies of Telegrams Sent, July 1863–August 1866; Registers of Letters Received, June 1862–April 1864; Letters Received, 1863–66; Press Copies of Telegrams Sent, July 1863–August 1866; Press Copies of Telegrams Received, July 1863–January 1865; Special Orders, June 1862–May 1864; General Orders Issued, August 1864–July 1865. From the Records of Department of West Virginia (P), 1863–65: Letters Sent, June 1863–January 1866; Letters Received, 1863–65; Special Orders, July 1863–December 1864; Lists and Registers Relating to Furloughs, Discharges, and Leaves of Absence, 1863–65.

RG 404: Records of the United States Military Academy. Post Orders, 1838–1904; Proceedings of the Medical Board, June 1856–August 1862; Letters Received from the Engineer Department, 1838–66; Registers of Punishments, October 1837–October 1900; Orders of the Commandant of Cadets, 1861–1910. (The contents of RG 404 are located in the Archives of the United States Military Academy.)

Secondary Sources

For information on the life of Montgomery C. Meigs, see Sherrod E. East, "The Banishment of Captain Meigs," *Records of the Columbia Historical Society* 40–41 (1940): 97–143; David W. Miller, *Second Only to Grant: Quartermaster General Montgomery C. Meigs, A Biography* (Shippensburg, Pa.: White Mane Press, 2000); and Russell F. Weigley, *Quartermaster General of the United States Army: A Biography of M. C. Meigs* (New York: Columbia University Press, 1959). For Meigs's work on the Capitol see Wendy Wolff, ed., *Capitol Builder: The Shorthand Journals of Montgomery C. Meigs, 1853–1859, 1861* (Washington, D.C.: U.S. Government Printing Office, 2001).

For Meigs family genealogy see the following, located in the Library of Congress: Henry B. Meigs, *Meigs Chart of American Ancestry* . . . (Baltimore: by the author, Herald Building, 1902); Joe Vincent Meigs, *The Meigs Tribe, . . . Table Made by Joe Vincent Meigs in April, 1905* (Boston: by the author, 1905); "Record of the Descendants of Vincent Meigs Who Came From Dorcetshire, England to America About 1635," typescript prepared by Return Jonathan Meigs (1901), 9, 1935.

Information on Columbian College is contained in various school catalogues, especially *A Catalogue of the Officers and Students of Columbian College, D.C., 1858–59* (Washington, D.C.: Thos. McGill, Printer, 1859), George Washington University Archives, Washington, D.C.

For information on the death of John Rodgers Meigs see John L. Heatwole, *The Burning: Sheridan's Devastation of the Shenandoah Valley* (Charlottesville, Va.: Rockbridge Publishing Company, 1998); P. C. Kaylor, "The Killing of Lieutenant Meigs, 1864," in *The Virginia Valley Records* (Baltimore: Genealogical Publishing Company, 1930), edited by John Wayland, 187–96; and "The Killing of Lieutenant Meigs, of General Sheridan's Staff—Proof That It Was Done in Fair Combat," *Southern Historical Society Papers* 9, no. 2 (February 1881): 77–79.

The most significant printed sources of official Civil War documents are *The War of the Rebellion: A Compilation of the Official Records of the Union and Confederate Armies* (Washington, D.C.: U.S. Government Printing Office, 1880–1901) and *Official Records of the Union and Confederate Navies in the War of the Rebellion* (Washington, D.C.: U.S. Government Printing Office, 1894–1922)

General reference books used include Mark Mayo Boatner III, *The Civil War Dictionary* (New York: Vintage, 1991); Patricia L. Faust, ed., *Historical Times Illustrated Encyclopedia of the Civil War* (New York: HarperPerennial, 1986); E. B. Long, with Barbara Long, *The Civil War Day by Day: An Almanac 1861–1865* (New York: Da Capo, 1971); James L. Mooney, ed., *Dictionary of American Naval Fighting Ships* (Washington, D.C.: Naval Historical Center, 1991); and C. Percy Powell, *Lincoln Day by Day: A Chronology, 1809–1865* (Dayton, Ohio: Morningside, 1991).

In addition to service records located at the National Archives, the following works were of significant importance in identifying individuals who fought in the Civil War: George Washington Cullum, *Official Register of the Officers and Cadets of the U.S. Military Academy* (West Point, N.Y., 1863); George Washington Cullum, ed., *Biographical Register of the Officers and Graduates of the U.S. Military Academy at West Point, N.Y. from Its Establishment, in 1802, to 1890 with the Early History of the United States Military Academy*, 3d ed., revised and extended (Boston: Houghton, Mifflin, 1891) and *Supplement*, vol. VI-A, 1910–1920, by Colonel Wirt Robinson (Saginaw, Mich.: Seeman and Peters Printers, 1920); Lewis R. Hamersly, *Records of Living Officers of the U.S. Navy and Marine Corps*, revised edition (Philadelphia: J. B. Lippincott, 1870); Francis B. Heitman, *Historical Register and Dictionary of the United States Army, 1789–1903* . . . (Urbana: University of Illinois Press, 1965); *Register of Graduates and Former Cadets, 1802–1960, of the United States Military Academy* (New York: West Point Alumni Association Foundation, n.d.); Jon L. Wakelyn, ed., *Biographical Dictionary of the Confederacy* (Westport, Conn.: Greenwood, 1977); Ezra J. Warner, *Generals in Blue: Lives of the Union Commanders* (Baton Rouge: Louisiana University Press, 1964) and *Generals in Gray: Lives of the Confederate Commanders* (Baton Rouge: Louisiana University Press, 1959).

For information on West Point history, the following proved most helpful: Alan and

Barbara Aimone, "Much to Sadden and Little to Cheer: The Civil War Years at West Point," *Blue and Gray Magazine* 9, no. 2 (December 1991): 10–28, 48–64; Stephen E. Ambrose, *Duty, Honor, Country: A History of West Point* (Baltimore: Johns Hopkins University Press, 1966); Thomas J. Fleming, *West Point: The Men and Times of the United States Military Academy* (New York: Morrow, 1969). A compilation of some of Thomas Rowland's letters describing life at West Point from 1859 to 1861 was published in the *South Atlantic Quarterly* 14 (July 1915): 201–19, 330–47. His letters relating to his Confederate service were published in the *William and Mary Quarterly Historical Magazine,* 1st ser., vol. 24, nos. 3 and 4 (1916): 146–53 and 232–38; vol. 25, no. 2 (1916): 73–82; and vol. 26, no. 4 (1917): 225–35. See also T. Harry Williams, "The Attack upon West Point during the Civil War," *Mississippi Valley Historical Review* 25, no. 1 (June 1938): 491–504.

Several biographical compilations were used also to identify various individuals. These include *Appletons' Cyclopedia of American Biography* (New York: D. Appleton, 1932); *Biographical Directory of the United States Congress* (Washington, D.C.: United States Government Printing Office, 1989); *Dictionary of American Biography* (New York: Scribners, 1928–); Leslie and Sidney Lee, eds., *Dictionary of National Biography Founded in 1882 by George Smith* (London: Oxford University Press, 1917–); *National Cyclopaedia of American Biography* (New York: J. T. White, 1921); Stewart Sifakis, *Who Was Who in the Union* (New York: Facts on File, 1988); Frederick A. Virkus, ed., *The Abridged Compendium of American Genealogy* (Baltimore: Genealogical Publishing Company, 1987); and *Who Was Who in America* (Chicago: Marquis, 1942).

For the actions of Congress see the *Congressional Globe* (Washington, D.C.: Congressional Globe Office, 1834–73).

In addition to the Records of the Columbia Historical Society, which include many articles on Washington, D.C., for early area history the following proved helpful: William H. Boyd, ed., *Boyd's Washington and Georgetown Directory* (Washington, D.C.: Wm. H. Boyd, 1860); Constance McLaughlin Green, *The Church on Lafayette Square* (Washington, D.C.: Potomac Books, 1970) and *Washington: Village and Capital, 1800–1878* (Princeton: Princeton University Press, 1962); Thomas F. Hahn, *Towpath Guide to the Chesapeake and Ohio Canal* (York, Pa.: American Canal and Transportation Center, 1974); Virginia Jeans Laas, ed., *Wartime Washington: The Civil War Letters of Elizabeth Blair Lee* (Urbana: University of Illinois Press, 1991); Fremont Rider, *Rider's Washington* (New York: Macmillan, 1924); C. B. Rose Jr., *Arlington County Virginia: A History* (Arlington, Va.: Arlington Historical Society, 1976); Melvin Lee Steadman Jr., *Falls Church by Fence and Fireside* (Falls Church, Va.: Falls Church Public Library, 1964); Benjamin Ogle Tayloe, *Our Neighbors on La Fayette Square: Anecdotes and Reminiscences* (Washington, D.C.: Junior League of Washington, D.C., 1982); Jane Chapman Whitt, *Elephants and Quaker Guns* (Arlington, Va.: Vandamere Press, 1984); and WPA, *Washington City and Capital* (New York: Oxford University Press, 1937).

For Maryland and other states: Henry P. Johnson, *Record of Service of Connecticut Men in the Army and Navy of the United States during the War of the Rebellion* (Hartford, Conn.: Lockwood and Brainard, 1889); Frederick Phisterer, *New York in the War of the Rebellion* (Albany, N.Y.: J. B. Lyon, 1912); Grace L. Tracey and John P. Dern, *Pioneers of Old Monocacy . . .* (Baltimore: Genealogical Publishing Company, 1987); Joseph A. Waddell, *Annals of Augusta County Virginia, 1736 to 1871* (Bridgewater, Va.: C. J. Carrier, 1958); L. Allison Wilmer, J. H. Jarrett, and George W. F. Vernon, *History and Roster of Maryland Volunteers, War of 1861–5* (Baltimore: Press of Guggenheimer, Weil, 1898–99); H. Wilson, comp., *Trow's New York City Directory, 1862–63* (New York: John Trow, 1863); and J. W. Woods, *Woods Baltimore City Directory, 1867–68* (Baltimore: J. W. Woods, 1868).

For railroads, forts, bridges, maps, guns, munitions, and music, see *Atlas Accompanying the Official Records of the Union and Confederate Armies* (Washington, D.C.: U.S. Government Printing Office, 1891–95); David Bosse, *Civil War Newspaper Maps: A Historical Atlas* (Baltimore: Johns Hopkins University Press, 1993); Michael W. Caplinger, *Bridges over Time: A Technological Context for the Baltimore and Ohio Railroad Main Stem at Harpers Ferry, West Virginia* (Morgantown, W. Va.: Institute for the History of Technology and Industrial Archaeology, 1997); Charles Eugene Claghorn, ed., *Biographical Dictionary of American Music* (West Nyack, N.Y.: Parker Publishing Company, 1973); *A Guide to Civil War Maps in the National Archives* (Washington, D.C.: U.S. Government Printing Office, 1986); Edward Hungerford, *The Story of the Baltimore and Ohio Railroad* (New York: Putnam, 1928); Albert C. Manucy, *Artillery through the Ages: A Short Illustrated History of Cannon, Emphasizing Types Used in America* (Washington, D.C.: U.S. Government Printing Office, 1949); Donald Beekman Myer, *Bridges and the City of Washington* (Washington, D.C.: U.S. Commission on Fine Arts, 1974); Warren Ripley, *Artillery and Ammunition of the Civil War* (New York: Van Nostrand Rheinhold, 1970); Walter W. Ristow, *American Maps and Mapmakers* (Detroit: Wayne State University Press, 1985); Robert B. Roberts, *Encyclopedia of Historic Forts: The Military, Pioneer, and Trading Posts of the United States* (New York: Macmillan, 1988); E. M. Sanchez-Saavedra, *A Description of the Country: Virginia's Cartographers and Their Maps, 1607–1881* (Richmond, Va.: Virginia State Library, 1975); Irwin Silber, ed., *Songs of the Civil War* (New York: Columbia University Press, 1960); Festus P. Summers, *The Baltimore and Ohio in the Civil War* (New York: Putnam, 1939); and Ames W. Williams, *The Washington and Old Dominion Railroad* (Arlington, Va.: Arlington Historical Society, 1989).

Additional sources include Roy P. Basler, ed., *The Collected Works of Abraham Lincoln* (New Brunswick, N.J.: Rutgers University Press, 1953); Marshall M. Brice, *Conquest of the Valley* (Verona, Va.: McClure Printing Company, 1965); William C. Davis, *Battle at Bull Run* (Garden City, N.Y.: Doubleday, 1977) and *The Battle of New Market* (Garden City, N.Y.: Doubleday, 1975); Richard R. Duncan, *Lee's Endangered Left: The Civil War in Western Virginia, Spring of 1864* (Baton Rouge: Louisiana State University Press, 1998); Cecil B. Eby Jr, *"Porte Crayon": The Life of David Hunter Strother* (Chapel Hill: University of North Carolina Press, 1960); Cecil B. Eby Jr., ed., *A Virginia Yankee in the Civil War: The Diaries of David Strother* (Chapel Hill: University of North Carolina Press, 1961); Stephen D. Engle, *Yankee Dutchman: The Life of Franz Sigel* (Fayetteville: University of Arkansas Press, 1993); Gary W. Gallager, ed., *Struggle for the Shenandoah: Essays on the 1864 Valley Campaign* (Kent, Ohio: Kent State University Press, 1991); Chester G. Hearn, *Six Years of Hell: Harpers Ferry during the Civil War* (Baton Rouge: Louisiana State University Press, 1996); John L. Heatwole, *The Burning: Sheridan's Devastation of the Shenandoah Valley* (Charlottesville, Va.: Rockbridge Press, 1998); Jim Dan Hill, *Sea Dogs of the Sixties: Farragut and Seven Contemporaries* (Minneapolis: University of Minnesota Press, 1935); Robert Erwin Johnson, *Rear Admiral John Rodgers, 1812–1882* (Annapolis, Md.: U.S. Naval Institute, 1967); Robert Underwood Johnson and Clarence Clough Buel, eds., *Battles and Leaders of the Civil War* (New York: Thomas Yoseloff, 1956); Virgil Carrington Jones, *Gray Ghosts and Rebel Raiders* (McLean, Va.: EPM Publications, 1988); Richard B. Kleese, *Shenandoah County in the Civil War: The Turbulent Years* (Lynchburg, Va.: H. E. Howard, 1992); Joseph Judge, *Season of Fire: The Confederate Strike on Washington* (Berryville, Va.: Rockbridge

Publishing Company, 1999); Michael G. Mahon, *The Shenandoah Valley, 1861–1865* (Mechanicsburg, Pa.: Stackpole Books, 1999); Howard Rollins McManus, *The Battle of Cloyds Mountain: The Virginia and Tennessee Railroad Raid, April 29–May 19, 1864* (Lynchburg, Va.: H. E. Howard, 1989); Roy Morris Jr., *Sheridan: The Life and Wars of General Phil Sheridan* (New York: Crown, 1992); Ray A. Neff, *Valley of the Shadow* (Terre Haute, Ind.: Rana Publications, 1989); ["Northern Virginia Daily"], *Standing Ground* (Strasburg, Va.: Shenandoah Publishing House, 1996); Charles Oscar Paullin, *Commodore John Rodgers . . . 1773–1838: A Biography* (Cleveland: Arthur H. Clark, 1910); Edward H. Phillips, *The Lower Shenandoah Valley in the Civil War* (Lynchburg, Va.: H. E. Howard, 1993); George E. Pond, *The Shenandoah Valley in 1864* (New York: Scribner, 1883); Philip H. Sheridan, *Personal Memoirs of Philip H. Sheridan, General United States Army* (New York: Da Capo, 1992); Mark A. Snell, *From First to Last: The Life of Major General William B. Franklin* (New York: Fordham University Press, 2002); Edward J. Stackpole, *Sheridan in the Shenandoah* (Harrisburg, Pa.: Stackpole Books, 1992); Stephen Z. Starr, *The Union Cavalry in the Civil War* (Baton Rouge: Louisiana State University Press, 1981), vol. 2, *The War in the East from Gettysburg to Appomattox, 1863–1865;* Gary C. Walker, *Yankee Soldiers in Virginia's Valleys: Hunter's Fiery Raid* (Roanoke, Va.: A and W Enterprise, 1989); Jeffrey C. Weaver, *The Civil War in Buchanan and Wise Counties: Bushwhackers' Paradise* (Lynchburg, Va.: H. E. Howard, 1994); Jeffrey D. Wert, *From Winchester to Cedar Creek: The Shenandoah Campaign of 1864* (Carlisle, Pa.: South Mountain Press, 1987) and *Mosby's Rangers* (New York: Simon and Schuster, 1990); Richard S. West, *Mr. Lincoln's Navy* (Westport, Conn.: Greenwood, 1976); and Harold R. Woodward Jr., *Defender of the Valley: Brigadier General John Daniel Imboden, CSA* (Berryville, Va.: Rockbridge Publishing Company, 1996).

INDEX

MARY A. GIUNTA is a historian formerly with the National Historical Publications and Records Commission, where she served in various administrative positions, including assistant director and acting director of the Publications Program, and as project director and editor-in-chief of the three-volume work *The Emerging Nation: A Documentary History of the Foreign Relations of the United States under the Articles of Confederation, 1780–1789* (1996), which was awarded the 1997 Thomas Jefferson Prize by the Society for History in the Federal Government. She currently works as a historian and writer in Washington, D.C., and Columbia Furnace, Virginia.

The University of Illinois Press
is a founding member of the
Association of American University Presses.

Composed in 10.5/13 Adobe Minion
by Jim Proefrock
at the University of Illinois Press
Designed by Dennis Roberts
Manufactured by Thomson-Shore, Inc.

University of Illinois Press
1325 South Oak Street
Champaign, IL 61820-6903
www.press.uillinois.edu